# GIS for SCIENCE

## Volume 3

## MAPS FOR SAVING THE PLANET

DAWN J. WRIGHT AND CHRISTIAN HARDER, EDITORS

Esri Press | Redlands, California

Christian Harder and Dawn J. Wright, eds.; *GIS for Science: Maps for Saving the Planet, Volume 3*; DOI: *https://doi.org/10.17128/9781589486713*
Library of Congress Control Number: 2019936340
ISBN 978-1589486713

Developmental editing by Mark Henry; Technology Showcase editing by Keith Mann; Copyediting by Joe DeLillo; Digital edition production by Victoria Roberts. Cover design and map by John Nelson.

On the cover: This vibrant representation of our planet paints on its surface the World Ecological Land Units—a segmentation of distinct bioclimate, landform, lithology, and land cover that forms the basic components of terrestrial ecosystem structure. Hillshading was derived from GEBCO 2020 elevation and bathymetric data. It was created in ArcGIS Pro. The aesthetic was inherited from the Half-Earth globe, designed by John Nelson of Esri and Estefanía Casal of Vizzuality, with the goal of invoking the sense that our shared planet is something beautiful and charming and worthy of sound stewardship.

See this book come alive at
**GISforScience.com**

# CONTENTS

# GIS FOR SCIENCE: A FRAMEWORK AND A PROCESS

*By Jack Dangermond, founder and president, Esri*
*and Dawn J. Wright, chief scientist, Esri*

Science—that wonderful endeavor in which someone investigates a question or a problem using reliable, verifiable methods and shares the result—has always been about increasing our understanding of the world. Early on, we applied geographic information systems (GIS) to science—to biology, ecology, economics, or any of the other social sciences. It wasn't until about 1993, when Michael Goodchild coined the term GIScience, that the world began to realize that GIS is a science in its own right. Today, we call this The Science of Where®. GIS incorporates sciences such as geology, geography, health and human science, data science, computer science, statistics, geovisual analytics, decision science, and much more. It integrates these disciplines into a kind of metascience, providing a framework for applying science to almost everything, merging the rigor of the scientific method with the technologies of GIS. The study of where things happen, it turns out, has great relevance.

We live in a world that faces more and more challenges. Even as a global pandemic continues to ravage human lives, the natural world remains under siege, with a global rate of species extinction that, according to the Intergovernmental Science-Policy Platform on Biodiversity and Ecosystem Services (IPBES), is at least tens to hundreds of times higher than at any time during the past 10 million years, and accelerating. During the coronavirus disease 2019 (COVID-19) pandemic, most African countries have reported reduced monitoring of the illegal wildlife trade. The protection of endangered species, conservation education, and anti-poaching operations also have faltered. Conversely, so few people understand how deterioration in the natural world can serve to light the fire of a new pandemic. Biodiversity—the variety of life on Earth from microscopic genes to entire ecosystems—is currently not a focus of the global response to factors threatening the lives and livelihoods of all creatures and organisms. As the venerable naturalist Edward O. Wilson has warned: "The only hope for the species still living is a human effort commensurate with the magnitude of the problem."

Part of this human effort is a global mobilization to identify and map the species that are at risk and not already safeguarded. This effort involves action maps, as The Nature Conservancy calls them, to show us where things are, and what we should protect, build, and invest in. These action maps can help us decide what to do in the face of trade-offs between wild grasslands and mineral exploration, or between transportation infrastructure and wildlife, etc. Such action maps are now well within our grasp as we continue to undergo a massive digital transformation. This transformation enables a science that increasingly helps us measure and analyze things and predict what will happen next. Just as important, science helps us design, evaluate, and ultimately weave all these pieces together in a protective fabric across the planet.

GIS provides a language to help us understand and manage inside, between, and among the organizations working toward a sustainable future. GIS also provides a framework in which we can compile and organize maps, data, and applications. This evolving technology helps us visualize and analyze relationships and patterns among datasets, perform predictive analytics, design and plan with the data, and ultimately transform our thinking into action. GIS empowers people to easily use geospatial information. As Richard Saul Wurman says, "Understanding precedes action." Esri is driven by the idea that GIS as a technology is the best way to address the immense challenges of today and the future.

> "Part of this human effort is a global mobilization to identify and map the species that are at risk and not already safeguarded."

This third volume of *GIS for Science* is full of new examples showing how GIS advances rigorous scientific research. The science-based organizations featured in these pages use ArcGIS as a comprehensive geospatial platform to support spatial analysis and visualization, open data distribution, and communication. In many cases, we use this research to preserve and restore iconic pieces of nature—revered and sacred places worthy of being set aside for future generations. These places belong to nature, but they also belong to science.

A central organizing principle of our work as scientists is the discipline of the scientific process. But science is also driven by the organic human instinct to dream, discover, understand, create, and help each other in times of great need. The Science of Where is a concept that brings these impulses together as we seek to support and transform the world through maps and analytics, connecting everyone, everywhere, every day through science and the urgent need to preserve the planet.

# INTRODUCTION BY THE EDITORS

*By Christian Harder and Dawn J. Wright*

This book is about science and the scientists who use GIS technology in their work. This contributed volume is for professional scientists, the swelling ranks of citizen scientists, and anyone interested in science and geography. Our world, now in its third decade into the 21st century, seems to be entering a crucial time in history in which humanity still can create a sustainable future and a livable environment for all life on the planet. But if we look critically at the facts, no informed observer can refute the reality that the current downward trajectory does not bode well.

As COVID-19 quickly grew into a global crisis in early 2020, we saw the GIS community pivot in response. At Esri, we were encouraged and frankly humbled by the often heroic work of GIScientists and others to confront a crisis unlike any seen in our lifetime. As that work continues globally, our objective in assembling this third volume of *GIS for Science* was to present other relevant and interesting stories about the current state of the planet in a time of great challenge. As such, this volume is indeed a storybook. It is not an atlas or a research monograph. Even so, we still looked for a cross section of sciences and scientists studying a wide range of problems.

GIS has found its way into virtually all the sciences, but the reader will notice that the earth and marine sciences are especially well represented, as are stories about the work of young people entering the scientific disciplines. Web GIS patterns and a simultaneous explosion of earth-observation sensors fuel this growth. Between all the satellites, aircraft, drones, and myriad ground-based and tracking sensors, the science community is now awash in data. Well-integrated GIS solutions integrate all this big data into a common operating platform—a digital, high-resolution, multiscale, multispectral model of our world.

Despite all these advances, science remains under attack on many fronts. From fake news to political pressure, science is still too often used as a political tool at a time when level-headed, objective scientific thinking is needed. We are convinced that GIS offers a unique platform for scientists to elevate their work above the fray. We invite you to read these stories in any order; the common thread is that all this work happens at the intersection of GIS and science. As you read through these stories, you'll see that the use of GIS as a cross-cutting, enabling technology is limited only by our imaginations.

In some cases, such as the fascinating work of Applied Ocean Sciences to map the soundscape of the global ocean as derived from marine mammals, fish, ships, wind, and waves, it's fair to say that scientific insight could only happen in the context of advanced spatial analytics within GIS. In other cases, such as the work of NASA's Earth Science Division and its partners to publish an astonishing volume of imagery, GIS is more of a mission-critical tool in turning scientific data into information products for public use. Here, GIS is a vital storytelling platform that communicates scientific insight to stakeholders in their communities.

It's impossible to describe the full breadth and scope of what GIS means for science and scientists without showing digital examples. So we have created a companion and complement to this book online. You can access it here:

## GISforScience.com

This website, comprising collections of digital data, stories, interactive web maps, Python notebooks, journal articles, and even videos, brings real-world examples to life and demonstrates the storytelling power of the ArcGIS® platform. The website also includes links to learning pathways from the Learn ArcGIS site (Learn.ArcGIS.com) and blogs related to the practical use of ArcGIS in each of the case studies.

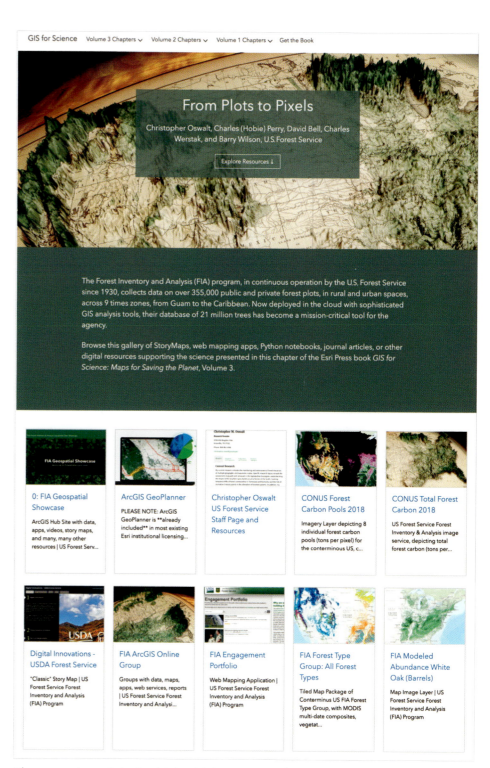

*The companion website for this book is in many ways the most important component of the project. Visit GISforScience.com to access more than 100 interactive web maps, apps, and stories created using ArcGIS StoryMaps, videos, and other digital resources described in the text broken down by chapter (e.g., "From Plots to Pixels").*

# FOREWORD

By Edward O. Wilson

In many respects, geography and mapping are not just basic to science and technology. They are the same thing. In molecular biology, for example, a process affecting the whole organism is first tracked to a particular organ or system of organs, and then to tissues and chemicals responsible for the activity. The impact of the process is then defined in the changes it induces in other parts of the whole organism.

When I step back and look from a bit of a distance at my own research, the social biology of ants, it is easy to translate most of what I have learned into geographic patterns—from which species inhabit which islands and continents, and, for reasons at the opposite extreme—to which microscopic organs and tissues orient a single ant through its life cycle.

If you put a dollop of honey on the ground a short distance from the edge of a fire ant nest, one or two ants patrolling the area are sure to find it, likely within a few minutes. After feeding a while, the lucky scout returns to the nest, dragging its extruded sting on the ground behind it. The scout is laying a faint but potent chemical trail that runs from the discovered food back to the nest. Depending on how hungry the colony is, a number of the workers run out along the trail in search of the newly discovered food.

Where in the body does the substance, in biology called a pheromone, originate? Which of the glands, from three at the base of the mandibles in the front to those at the base of the sting in the rear, produce the trail pheromone? By dissecting glands from freshly killed fire ants and drawing artificial trails, I discovered the source to be a tiny sliver near the sting called Dufour's gland. To that time, its role in ant biology had been unknown.

In another, more traditional exercise in geographical biology, I was able to shed light on the geological origin of the Greater Antilles, including Cuba, the Dominican Republic and Haiti, Puerto Rico, and Jamaica. Was this constellation a fragment of land that broke away millions of years ago from ancient Central America? Or did it arise from the sea from where it is located today? I obtained ant fossils preserved in amber approximately the same age as the origin of the Greater Antilles. The evidence I found favored the first hypothesis, a breakaway from Central America.

As biologists learn more and more about the biology and origins of Earth's present-day fauna and flora, they will also create an even stronger armamentarium of scientific and practical geography.

In this context, I am heartened to see today's scientists using GIS and other evolving technologies in their efforts to protect and restore biodiversity. Like my own work with the fire ants, the scientists featured in these pages are translating what they learn into geographic patterns for all of us to see, this time in the form of maps that are saving the world.

*Edward O. Wilson, research professor emeritus at Harvard University, is the guiding force that shapes the mission of the E.O. Wilson Biodiversity Foundation. In his long career, he has transformed his field of research—the behavior of ants—and applied his scientific perspective and experience to illuminate the human circumstance, including human origins, human nature, and human interactions.*

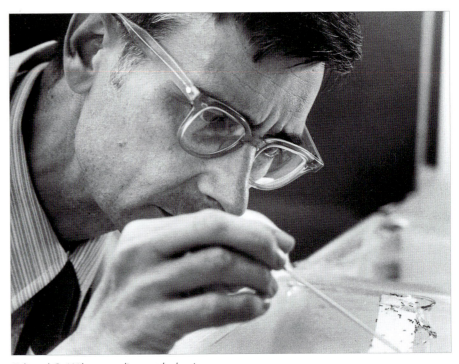

Edward O. Wilson studies ant behavior.

Wilson, age 92 at the time of printing, has never stopped working. Learn more about the work of his Half-Earth Project in the chapter titled "Mapping Half Earth."

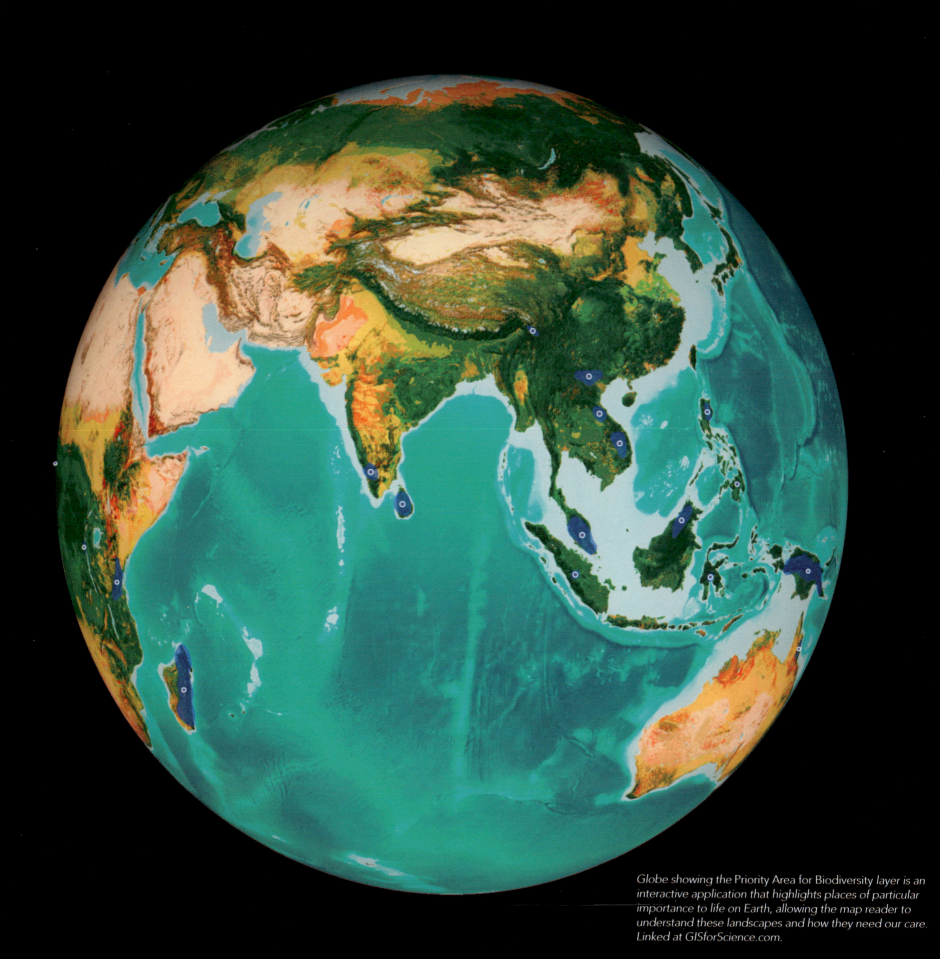

Globe showing the Priority Area for Biodiversity layer is an interactive application that highlights places of particular importance to life on Earth, allowing the map reader to understand these landscapes and how they need our care. Linked at GISforScience.com.

# GIS for SCIENCE

## Volume 3

MAPS FOR SAVING THE PLANET

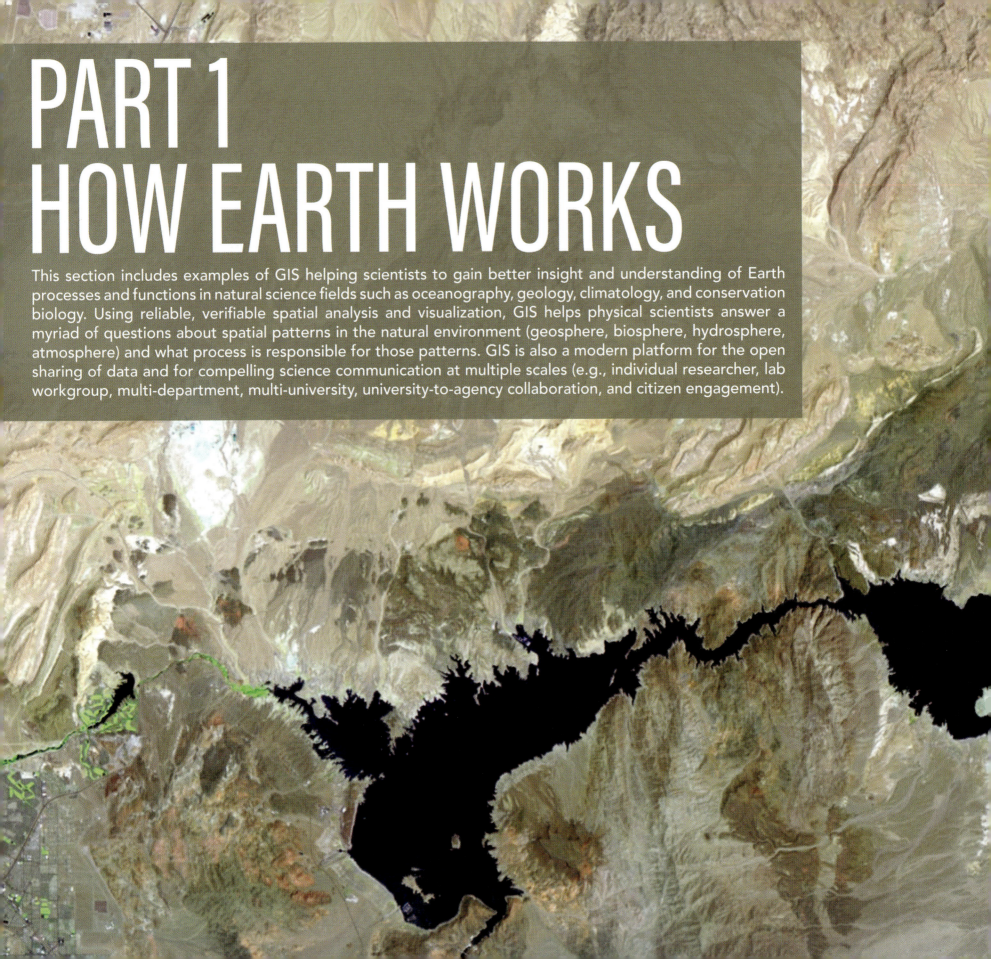

# PART 1
# HOW EARTH WORKS

This section includes examples of GIS helping scientists to gain better insight and understanding of Earth processes and functions in natural science fields such as oceanography, geology, climatology, and conservation biology. Using reliable, verifiable spatial analysis and visualization, GIS helps physical scientists answer a myriad of questions about spatial patterns in the natural environment (geosphere, biosphere, hydrosphere, atmosphere) and what process is responsible for those patterns. GIS is also a modern platform for the open sharing of data and for compelling science communication at multiple scales (e.g., individual researcher, lab workgroup, multi-department, multi-university, university-to-agency collaboration, and citizen engagement).

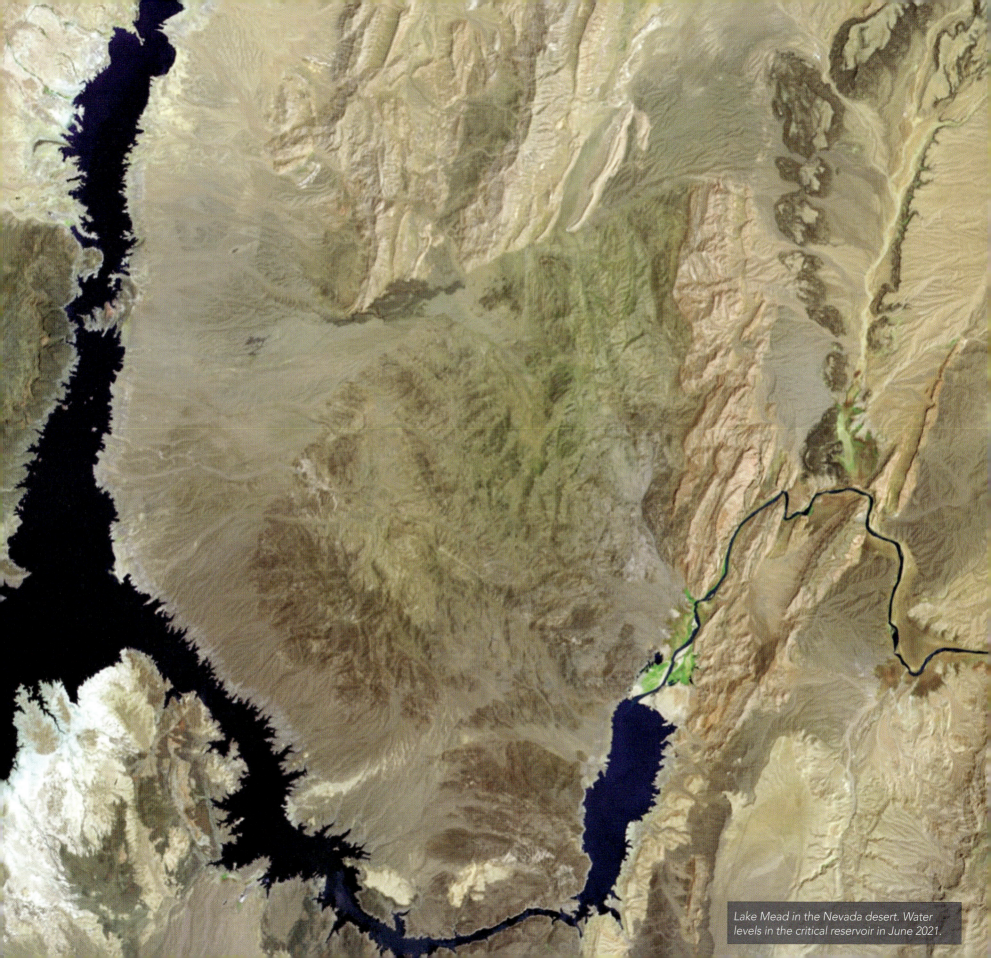

*Lake Mead in the Nevada desert. Water levels in the critical reservoir in June 2021.*

# EARTH'S COASTLINES

With approximately half the world's population living less than 65 miles from the ocean, coastal ecosystems are arguably Earth's most critical real estate. Yet coastlines are among the more difficult features to accurately map; until now, no comprehensive high-resolution geospatial dataset existed. This chapter presents a new map and ecological inventory of global coastlines developed by Esri, the U.S. Geological Survey, and other partners.

By Roger Sayre, Madeline Martin, Jill Cress, **U.S. Geological Survey;** Kevin Butler, Keith Van Graafeiland, Sean Breyer, Dawn Wright, Charlie Frye, Deniz Karagulle, **Esri;** Tom Allen, **Old Dominion University**; Rebecca J. Allee, Rost Parsons, **National Oceanic and Atmospheric Administration**; Bjorn Nyberg, **University of Bergen, Norway**; Mark J. Costello, **Nord University, Norway**; Frank Muller-Karger, **University of South Florida**; and Peter Harris, **GRID-Arendal, Norway**

Horseshoe Bay Beach, Bermuda.

# A COASTLINE JOURNEY

Imagine sailing every coastline of planet Earth. We are all familiar with the stories of intrepid seagoing explorers who set out from safe havens to cross the vast expanses of unknown oceans in hopes of discovering new lands. Their reasons were many—expanding empires, escaping persecution, seeking treasure, and satisfying basic human and scientific curiosity. They sailed with simple navigation tools and relied on maps adorned with drawings of fantastic sea creatures and the ominous warning "Here Be Lions" (contrary to popular opinion, only one surviving map, the 1510 *Hunt-Lenox Globe*, contains the phrase "Here Be Dragons"). These famous explorations were largely transoceanic—crossing the oceans to reach new lands.

*Cropped portion of the 1504 Hunt–Lenox Globe, showing the infamous "Here Be Dragons" text.*

The purpose of this journey is not to cross the ocean, but rather to sail all of its shores. This exploration will ply the waters along every coast, sailing parallel to them at a distance close enough to the coastlines to observe all of their magnificent characteristics. The voyagers are in for a treat because there is an astonishing diversity of coastal areas on Earth by any measure—rocky versus sandy, tropical versus polar, calm waters versus pounding waves, vegetated versus barren, long and straight versus twisty and tortuous, and yes, humans absent versus common. There are many ways of classifying coastal areas, and many criteria for distinguishing differences from one stretch of coastline to another.

Sailing all of the world's coastlines would take several years. By our own calculations (and yes, we mapped the coastlines of the world—more on that later), the planet has about 2.5 million kilometers (more than 1.5 million miles) of coastline. A ship traveling nonstop at speeds of 20 knots—a typical speed in the cargo container and cruise ship industries—would take almost eight years to complete the voyage. That doesn't include time spent at ports, island hopping, or slow-steaming in interesting or navigationally tricky areas.

The time and resources needed for such an endeavor make such a long voyage "running the coastline" impractical in ships or even aircraft. But orbiting satellites provide coverage of Earth in regular, repeating time cycles, meaning that satellite imagery exists for every stretch of coastline on the planet. So instead of a voyage by ship, the journey becomes one of looking at satellite images to capture every coastline of the Americas, Eurasia, Africa, Australia, and hundreds of thousands of big, small, and tiny islands in the global ocean.

Our team has already taken that journey through a collection of coastal satellite images, extracting new information about coastline positions to make a new high-resolution map called the *Global Shoreline Vector* (GSV). After publishing that coastline mapping work,[1] our team—as often happens on journeys—took an interesting detour. Our team had set out to characterize coastal environments, and the first step in doing that was to produce a new map of the global coastline. But having mapped the global shoreline in the finest detail to date, our team was ready to make the most definitive geographic characterization of global islands. We published the results[2] in Esri's *GIS for Science, Volume 2*, the predecessor to this volume. That work provided a definitive count and geolocation of global islands, and it happened as a fortuitous "by-product" of our global coastline inventory work. However, it also caused us to get temporarily sidetracked from our original intention of producing a definitive characterization of the world's coastlines.

Leaving aside the magical allure of islands, this chapter focuses on the task at hand, a characterization of global coastal environments. This work comes at a time when spacecraft have explored the heavens and submersibles have explored the seafloor. Our team was profoundly moved and inspired by the accomplishments of Dr. Kathryn Sullivan, the only person to walk in space (1984) and dive to the deepest part of the ocean (2020) in the Marianas Trench (see a related ArcGIS StoryMap linked within this book's companion website found at GISforScience.com). Although perhaps not as headline-grabbing as her outer-space and inner-space explorations, our team's virtual "surface-space" scientific exploration of the global shorelines is the first of its kind.

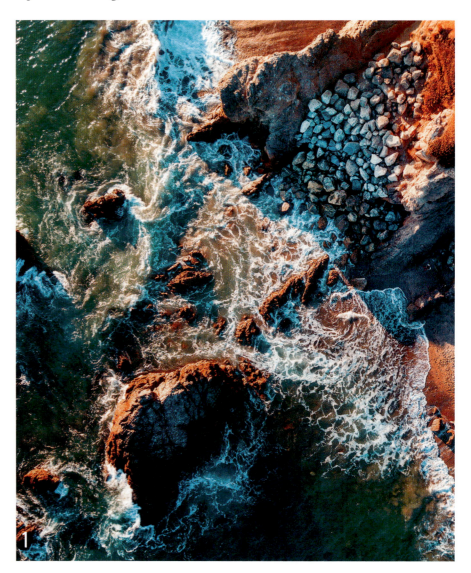

*In addition to the diversity of physical, ecological, and human factors used to classify coastal areas, important geographic and temporal dimensions help frame coastline classification efforts. These four images represent different spatial vantage points for coastal exploration that influence classification scale and granularity: 1) low altitude (Malibu, California), 2) on-shore (Portland Head, Maine), 3) high altitude (North Sentinal Island, Andaman Islands, India), and 4) off-shore (the Australian research vessel Sprightly).*

# COASTAL VARIATION

The coastline is universally understood as the intersection of Earth's land masses and oceans. Coastal areas are extremely important to humans, in that approximately half of the world's population lives within 100 km (62.1 mi) of the ocean.[3] The very survival of many human societies depends on the provision of ecosystem goods and services as benefits from coastal environments, and the supply of those benefits is a function of the health of the coastal systems. Highly degraded coastal ecosystems are compromised in their ability to satisfy human needs for food and other goods and services.[4, 5] To continue producing important benefits for humankind, coastal ecosystems must be well-managed, and that will require a comprehensive understanding of the types, distributions, ecology, and condition of Earth's coastal ecosystems. In other words, researchers will need well-documented maps that provide a digital canvas for ecosystems data.

Coastal settings exhibit spectacular geomorphological diversity (as shown in the coastal photographs on these two pages). Coastal classification aims to classify diversity and partition the coastline into distinct units with similar properties. It is expected that similar units will exhibit similar responses to environmental perturbations or management interventions. Classifying coastlines is an application-specific exercise, and coastal systems can be classified in numerous ways depending on the intended emphasis (e.g., geomorphological, biological, socioecological, or oceanographic). Because coastal areas are often densely populated and subject to natural and human-caused disasters, much coastal classification work has been focused on hazards, risk, vulnerability, and sensitivity. Many nations, for example, classify their coastlines using an environmental sensitivity index (ESI), which characterizes the sensitivity of coastal assets (biodiversity, centers of economic production, cultural features, etc.) to oil spills.[6] These classifications are intended as management tools for the protection of valuable coastal resources. Most coastline sensitivity classifications include, among other factors, aspects of the physical environmental settings in which coastlines occur.

Characterizing the geomorphological nature of coastal physical environments is key to understanding coastal variation,[7, 8] and there is a long history of coastal classification based on physical features. Either sea-level rise or land subsidence can cause submergence of a coastline, and a simple binary classification of submerged versus emerged coastlines was already in use at the start of the 20th century.[9] At coarse scales (e.g., thousands of kilometers), and from a geological perspective, tectonic activity is a primary determinant of coastal variation, as explained in the classic paper by Inman and Nordstrom.[10] The presence of coastal mountains, the width of continental shelves and continental plains, the existence of volcanic and barrier islands, the location of major river systems—all these features result from the movement of tectonic plates on the lithosphere. This type of genetic classification approach emphasizes the origin of coastal landforms. Classifications emphasizing geomorphological structure and processes have evolved from initial considerations of submergence and tectonic history to include emphases on dominant coastal processes,[11] coastal systems,[12] and the morphodynamic coevolution of form and process.[13]

Another fundamental class of approaches to understanding coastal variation deals with hydrodynamic forcing features. These approaches focus on how water movements shape coastlines. Many coastal areas are underlain by bedrock, which slowly erodes over time due to the actions of water and wind. This weathering process is erosional in nature, and these generally rocky coastlines are classified as erosional. In contrast, the sediments produced from weathering can be deposited in the coastal zone to form depositional environments. The sediments that are deposited along the shore create beaches, tidal flats, and dunes, and the sediments that are transported to the coast by rivers create deltas and estuaries. These coastal areas, built up over the long term from sediment deposition, are classified as depositional. These next images are examples of geomorphological diversity in coastal areas based on substrate type. Although the distinction between erosional and depositional coastlines has

*Chalky bluffs at the White Cliffs of Dover, England.*

long been understood and even mapped at very coarse scales,[10] the first higher-resolution global characterization and map of these two fundamental shoreline types was only recently produced.[14] Neilson and Costello[15] classified lengths of coastline on the mainland and islands of Ireland into cliff, rocky, stony, sandy, and muddy habitats as part of an ecological survey to design a national network of marine protected areas. Those erosional and depositional habitats were found to be primary determinants of species and ecological community occurrences.[16] Delving deeper into the classification of depositional environments, three hydrodynamic influences are of primary importance, as described in the classic paper by Boyd et al.[17] In this characterization of depositional environments, a coastal area is classified according to the relative influence of waves, tides, and rivers as hydrodynamic forces that move sediment. Coastal areas are then identified as wave-dominated, tide-dominated, or river-dominated. Using this framework, Harris et al.[18] classified Australian coastal depositional environments based on a quantitative analysis of wave power, tidal power, and river power. Nyberg and Howell[14] extended this previous work to classify the three hydrodynamic forces shaping depositional environments of global coastlines.

*Hexagonal basalt columns, Jeju Island, South Korea.*

*Sandy beaches abutting coastal mountains, National Park of American Samoa.*

*Rocky coasts and outcrops, Asilomar State Beach, California.*

# INTEGRATED AND MULTIDISCIPLINARY COASTLINE CLASSIFICATION

The classification approaches described earlier generally focus on a single classification variable, whether it is environmental sensitivity to a disturbance, tectonic setting, erosional versus depositional nature, or dominant hydrodynamic influence. In recognition of the strong physical, ecological, and human diversity associated with coastal environments, there have also been important attempts to classify coastal environments using multiple factors. Cooper and McLaughlin[19] presented a review of emerging coastal classification approaches and argued that advances in spatial analytical technologies (GIS and remote sensing) enabled increased focus on multifactor classification at that time. Evaluating and quantifying the classic "holistic" view of coastlines as integrated products of geomorphological and oceanographic processes[7] was more easily accomplished using geographic analysis and spatial technologies. These emerging technologies allowed for complex multidisciplinary analysis and mapping of the many physical, ecological, and human aspects of coastal environments. For example, Jelgersma et al.[20] classified low-lying deltaic areas using 18 variables, which were grouped into four main categories: Offshore water environment (six marine variables); Coastline properties (seven shoreline morphology variables); Deltaic system properties (four variables describing land and river conditions); and Human activity (one variable). This study was an early example of a complex delta classification approach aimed at including "everything that matters."

Globally comprehensive, standardized, integrated, high spatial resolution characterizations and maps of coastal systems are still relatively lacking. At the turn of the 21st century, a major World Resources Institute analysis concluded, "Information on the location and extent of coastal features and ecosystems types often provides the basis for subsequent analyses of condition of the ecosystem, relationships between different habitats, and overall trends. Yet, despite this fundamental importance, such information is incomplete and inconsistent at the global level."[4] In response to many similar calls for the detailed mapping of Earth's ecosystems, the Group on Earth Observations (GEO), a consortium of more than 100 nations and participating organizations seeking to advance the use of earth observation for environmental problem-solving, has commissioned a task (task T1 in the 2020-2022 GEO Ecosystems Implementation Plan published at www.earthobservations.org) to produce a standardized, robust, and practical classification and map of the planet's terrestrial, freshwater, and marine ecosystems. The work we have done to characterize coastal ecosystems, described next, was undertaken in response to that official commission from GEO and complements our earlier related GEO work to characterize global pelagic ecological marine units (EMUs).[21]

## Our approach—coastal segment units (CSUs) and ecological coastal units (ECUs)

Our team produced a standardized, consistent, high spatial resolution, and globally comprehensive characterization of the coastlines of the world to integrate coastal water properties, coastal land properties, and properties of the coastline itself. The team partitioned the coastlines of the world into 4 million 1 km or shorter segments and attributed those segments with data on 10 variables that describe the basic ecological settings in which the coastline segments occur. The partitioning and attribution have resulted in a new open data resource with which any 1 km coastline segment, anywhere in the world, can be queried, and the values for 10 ecologically meaningful characteristics of that segment returned.

Our primary objective in undertaking this work was to produce globally comprehensive data that describe the ecological settings of coastlines. A 1 km coastline segment may be an appropriate spatial resolution and analytical unit for coastal managers. As such, our team hopes that while the characterization is global, the resulting product may have management utility at place-based scales. Moreover, it appeared that 1 km is a large enough distance to aggregate summaries of other environmental data, yet small enough to show rich variety in those summaries. Integrating the data on features of the land, water, and coastline showed the potential of spatial data integration. The data product is intended to be useful to managers as an extensive inventory of the ecological settings of coastal areas. Every 1 km segment with a unique set of class labels for the 10 attributes is called a coastal segment unit (CSU). The data are also intended to be useful for coastal zone research, including coastal zone assessments related to the Sustainable Development Goals (SDGs) of the United Nations 2030 Agenda for Sustainable Development. (Readers can find a link to this document and other digital resources relating to the work at GISforScience.com.)

A secondary objective in undertaking this work was to identify groups of similar coastal areas based on their *aggregate* ecological setting, as determined from a global statistical clustering of all the segments data. The "all-in" clustering was performed with all 10 attributes included and with equal weighting of the attribute variables. The global clustering was exploratory in nature, and the preliminary results identified 16 distinct coastline environments. These 16 distinct coastline environments are called ecological coastal units (ECUs). The ECUs represent broad groups of globally similar coastal environments.

*Remote coastline, Eucla, Western Australia.*

# METHODOLOGY
## Initial plans and evolution of the segmentation approach

Our paper describing the development of the global shoreline vector[1] also presented a preliminary vision for how we would conceptualize and map ECUs, wherein we committed to develop global ECUs as an objective, quantitative segmentation of the global coastal zone into environmentally distinct and ecologically meaningful units. That initial vision included a notion to bound the global coastal zone as a set of nearshore coastal waters extending out to the 30 m bathymetric contour line, offshore coastal waters extending from the 30 m bathymetric contour out to the continental shelf edge, and coastal land areas extending from the coastline landward to the edge of the continental plain or to mountain systems. We then anticipated attributing the area inside this coastal zone with relevant ecological setting variables in a wall-to-wall (completely spatially tessellated) fashion. We held a workshop to elicit expert opinion on the specifics of bounding the coastal zone on land and in the water and to generate the list of variables that would adequately describe the coastal settings. (Workshop resources are linked at GISforScience.com.)

Initial attempts to delineate a global polygonal coastal zone area within which we would characterize coastal ecosystems proved challenging, mainly with respect to bounding the landside area. While the much narrower shore-zone concept is understood as that portion of the profile subject to wave action, the larger coastal zone is an onshore and offshore area influenced by proximity to the coast, with ill-defined limits on both sides.[22] Deciding how far out in the water and how far in on the land we should extend the coastal zone proved problematic due to the difficulty in knowing the nature and magnitude of coastal influence at progressively longer distances seaward or inland.

Ultimately, we abandoned the attempt to delineate a standardized polygonal global coastal zone within which we would map ecological settings in a wall-to-wall fashion as impractical and overly ambitious. We do see the value of such an effort and encourage future attempts to do so. For our purposes, however, we decided that adopting a coastline segmentation approach would provide a simpler and more practical opportunity to characterize coastal environments while still retaining the ability to include considerations of landside and waterside ecological settings. Coastal segmentation is focused on partitioning the coastline into a set of relatively homogenous reference units based on physical and ecological (and sometimes socioeconomic) properties and using these coastline units for resource management or research programs. Coastline segmentation approaches seek to identify similar coastal systems that may exhibit similar responses to disturbances or management interventions.[23]

Our work to identify the coastal segment units (CSUs) and ecological coastal units (ECUs) represents a global partitioning of the coastline into 1 km segments, an attribution of those segments with ecological setting data followed by a classification, and a spatial statistical analysis of the segment properties. The work proceeded in four steps:

### Step 1 - Development of the global shoreline vector (GSV)

To develop a strong global characterization of coastal ecosystems, we first needed to produce a new, high-resolution global shoreline vector (GSV) as the spatial and linework foundation for the effort. We produced that vector shoreline through a semiautomated interpretation of year 2014 annual composite Landsat images for every image on the planet that contained a coastline. That work was published in a special GEO Blue Planet (a GEO oceans initiative) issue of the *Journal of Operational Oceanography*.[1] It contains the details about how the GSV was extracted from

*Lava enters the sea on the shore of Hawaii Island, creating new low-erodibility coastline.*

imagery, cleaned, and rendered topological. Our team also presented our initial thinking about the development of global ECUs in that paper.

### Step 2 - Segmenting the GSV and identifying segment midpoints as the basic spatial analytical unit

The GSV was then segmented into approximately 4 million segments, most of which were 1 km in length, and none of which were greater than 1 km. A perfect segmentation of the GSV into only 1 km segments was not possible given the realities of coastline lengths. The segmentation was accomplished by starting at the origin vertex of every island or continental mainland polygon and inserting endpoints at every 1 km linear distance. The last segment before returning to the origin vertex was always shorter than 1 km. Islands whose shoreline lengths were less than 1 km also resulted in a number of segments that were less than 1 km. Approximately 4 million segments resulted from the segmentation process, and these features were considered the basic spatial analytical unit of the data development effort.

The midpoints of these segments were subsequently extracted into a global shoreline points dataset. The 1 km segments are always conceptually and numerically represented by their midpoints, and all of the attribution and spatial analysis was facilitated by using point features rather than line features. As such, the statistical clustering was in essence an analysis of the similarity of a set of global points, and any references we might make to "clustering the segments" in reality refers to clustering the data associated with the segment midpoints. A similar approach was taken in the development of the ecological marine units (EMUs),[21] where a 3D point mesh was developed and individual points spatially represented a cuboidal volume of water around the point.

## Step 3 – Variable selection, attribution of segments, and classification

The team selected 10 variables to represent the ecological settings of coastal environments—two land variables, five ocean variables, and three coastline variables (as seen in the map of coastal segment attributes). Each segment midpoint was then attributed with values for each of the 10 variables. The variables describing the ocean environment included an integrated measure of physical properties (temperature, salinity, and dissolved oxygen), and chlorophyll concentration, tidal range, wave height, and turbidity. The variables describing the land environment were climate setting, (as an integrated measure of long-term temperature and precipitation), and an index of erodibility. The variables describing the coastline

were its regional sinuosity, the slope profile of a 200 m perpendicular line segment extending 100 m landward and 100 -m seaward from every segment midpoint, and an index of river outflow. Following the attribution of the segments, they were classified by establishing class labels and associated ranges of class values for each of the 10 variables. Where possible, the class labels and class value ranges were consistent with the Coastal and Marine Ecological Classification Standard (CMECS)[24] nomenclature and breakpoints. The classification step resulted in the development of the coastal segment units (CSUs), defined as any segment with a unique set of class labels for the 10 attributes. The variables are more fully described later.

## Step 4 – Statistical clustering

Preliminary statistical clustering was conducted on all points, with all 10 attributes included, to group points with similar landside, waterside, and coastline properties. The clustering routine accommodated the mix of categorical and continuous variables in the 10 inputs and included an effort to reduce dimensionality. The statistical analysis is more fully described in a later section.

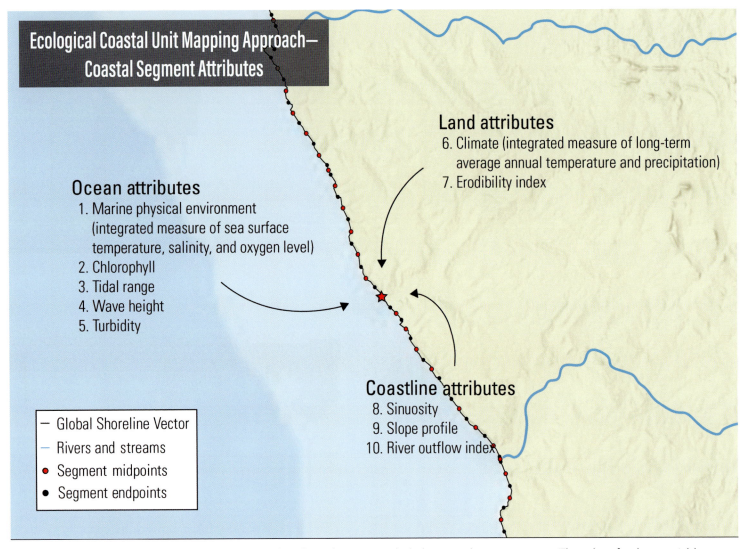

**Ecological Coastal Unit Mapping Approach— Coastal Segment Attributes**

**Ocean attributes**
1. Marine physical environment (integrated measure of sea surface temperature, salinity, and oxygen level)
2. Chlorophyll
3. Tidal range
4. Wave height
5. Turbidity

**Land attributes**
6. Climate (integrated measure of long-term average annual temperature and precipitation)
7. Erodibility index

**Coastline attributes**
8. Sinuosity
9. Slope profile
10. River outflow index

— Global Shoreline Vector
— Rivers and streams
• Segment midpoints
• Segment endpoints

*The set of 10 attributes for describing the aggregated ecological setting in which the coastal segments occur. The values for these variables, which describe characteristics of the adjacent ocean, the adjacent land, and the coastline itself, are drawn from a variety of data sources and attributed to the coastal segment midpoints.*

Rugged coastal geomorphic landforms in western Alaska.

# VARIABLE DESCRIPTIONS

The selection of variables (as shown in the tables) to represent the ecological settings in which coastlines occur was a subjective exercise supported by expert elicitation. The decision to include variables representing the landside, the waterside, and the coastline itself follows earlier integrative classification work[7,20] and acknowledges that coastal variation is multidimensional, with strong geomorphological and oceanographic drivers. This section describes each of the variables, their source data and how their values were attributed to the segments (segment midpoints):

## Ocean variables

### 1. Marine physical environment—*an integrated measure of sea surface temperature, oxygen, and salinity.*

*Temperature, primary productivity, and oxygen level vary spatially throughout the ocean and are recognized as key drivers of marine ecology.[25,26,27] Organisms have varying tolerances to salinity levels, and changing salinity is therefore another important driver of marine species distributions,[28] but salinity in the open ocean has long been recognized as relatively uniform and stable.[29] In the coastal zone, however, salinity levels are more variable due to the influences of tidal action, fluvial discharge, and evaporation; salinity is therefore recognized as a key driver of coastal zone ecology.[28]*

*In 2017, our team developed a comprehensive set of 3D global ecological marine units (EMUs)[21] from an analysis of NOAA World Atlas data on the marine physical and chemical environment. Data on the sea surface physical environment from the set of EMUs distributed along the coastlines were used to attribute the coastal segments. For every coastline segment, an integrated measure of temperature, salinity, and dissolved oxygen was obtained from the closest EMU point to the segment midpoint. The integrated measure is a categorical variable that expresses the combination of the three inputs: salinity, oxygen, and temperature. There were 23 classes of the integrated measurement, as presented in table 1.*

| Salinity Class | Dissolved Oxygen Class | Temperature Class |
|---|---|---|
| Euhaline | Highly Oxic | Superchilled |
| Euhaline | Oxic | Warm to Very Warm |
| Euhaline | Oxic | Moderate |
| Euhaline | Oxic | Moderate to Cool |
| Euhaline | Oxic | Cold |
| Euhaline | Oxic | Very Cold |
| Euhaline | Oxic | Superchilled |
| Euhaline | Hypoxic | Moderate to Cool |
| Euhaline | Severely Hypoxic | Cold |
| Euhaline | Severely Hypoxic | Very Cold |
| Polyhaline | Highly Oxic | Superchilled |
| Polyhaline | Oxic | Cold |
| Polyhaline | Oxic | Very Cold |
| Polyhaline | Severely Hypoxic | Cold |
| Polyhaline | Severely Hypoxic | Very Cold |
| Polyhaline | Anoxic | Cold |
| Polyhaline | Anoxic | Very Cold |
| Mesohaline | Highly Oxic | Very Cold |
| Mesohaline | Oxic | Moderate to Cool |
| Mesohaline | Oxic | Cold |
| Mesohaline | Oxic | Very Cold |
| Mesohaline | Hypoxic | Cold |
| Mesohaline | Severely Hypoxic | Very Cold |

Table 1

### 2. Chlorophyll a concentration

*Primary production is another key driver of marine ecology, and chlorophyll concentration as a measure of plankton abundance, the base of the ocean food web, is used as a proxy variable for representing primary production.[27,30] The data on chlorophyll a concentration were obtained from the European Space Agency Ocean Colour Climate Change Initiative (linked at GISforScience. com). This dataset contains merged chlorophyll measurements from 1997 to 2020 provided from SeaWiFS, MODIS, MERIS, and VIIRS sensors.[31] The chlorophyll variable is continuous and was used as such in our analyses. The chlorophyll data are in a 4 km*

| Chlorophyll Level | Concentration (µg/l) |
|---|---|
| Low | Less than 2.0 |
| Moderate | 2.0–5.0 |
| High | More than 5.0 |

Table 2

*raster format. We averaged the long-term (1997–2020) monthly mean chlorophyll values to identify a long-term annual average. The derived long-term annual average will likely have smoothed seasonally high and low values of chlorophyll, precluding straightforward classification into CMECS trophic productivity levels (oligotrophic, mesotrophic, and eutrophic). We therefore simply identified three chlorophyll classes (low, moderate, and high) according to natural breaks[32] in the chlorophyll data and did not assign a CMECS trophic productivity class label. The chlorophyll value attributed to each segment was the value from the raster cell whose center was closest to the segment midpoint. The three chlorophyll classes are presented in table 2.*

### 3. Tidal range

*Tidal activity is an important hydrodynamic driver of coastal zone ecology, mediating salinity levels,[33] nutrient availability,[34] temperature,[35] root zone aeration,[36] carbon flux,[37] species distributions,[38] and other environmental features of importance to organisms. The tidal range, the difference between low and high tide, establishes the vertical limits within which waves and tidal currents interact. Tides create a gradient of environmental conditions wherein species have adapted to various tidal regimes (e.g., see the paper on pulsing ecosystems[39] written by William Odum, his father Eugene Odum, and Eugene's brother Harold Odum. The two Odum brothers were seminal in establishing ecosystem ecology as a discipline and wrote the pioneering textbook Fundamentals of Ecology.[40]) Of course, tidal variation is also an extremely important and regular environmental phenomenon impacting human livelihoods and behavior. As Davis and Fitzgerald[8] noted, "the rise and fall of the tides is one of the major rhythms of planet Earth."*

| Tidal Range Category | Tidal Range (m) |
|---|---|
| Atidal | Less than 0.1 |
| Microtidal | 0.1–0.3 |
| Minimally Tidal | 0.3–1.0 |
| Moderately Tidal | 1.0–4.0 |
| Macrotidal | 4.0–8.0 |
| Megatidal | More than 8.0 |

Table 3

*Tidal range data were obtained from the French Space Agency (CNES) through the AVISO+ data dissemination program (website linked at GISforScience.com). The dataset used is the FES2014 finite elements solution. The data are derived from a compendium of global satellite altimetry measurements and are available in a 1/16th° (~6 km) raster grid. The tidal range data are continuous data and were used as such in the analysis. For subsequent classification grouping and labeling purposes, our team used six CMECS classes to describe tidal ranges (table 3). Each coastline segment midpoint was attributed with a tidal range value from the raster cell whose center was closest to the segment midpoint.*

### 4. Wave height

*In the coastal zone, wave energy is a key determinant of marine physical environmental structure through its influence on erosional and depositional processes.[17] Similarly, wave height and exposure influence the composition and distribution of biological assemblages.[41] Mean significant wave height data were obtained from the NOAA WaveWatch III® 30-year Hindcast Phase 2 resource. The data are available as a 30-minute (~55 km) global grid. The wave height data are continuous data and were used as such in the analysis. For the subsequent classification grouping and labeling purposes, the team used seven CMECS classes to describe wave height ranges (table 4). Each coastline segment midpoint was attributed with a wave height value from the raster cell whose center was closest to the segment midpoint.*

| Wave Height Category | Wave Height Range (m) |
|---|---|
| Quiescent | Less than 0.1 |
| Very Low Wave Energy | 0.1–0.25 |
| Low Wave Energy | 0.25–1.0 |
| Moderate Wave Energy | 1.0–2.0 |
| Moderately High Wave Energy | 2.0–4.0 |
| High Wave Energy | 4.0–8.0 |
| Very High Wave Energy | More than 8.0 |

Table 4

## 5. Turbidity

Turbidity in the coastal zone is a function of water-driven sediment movement in riverine discharge and shoaling waves. Turbidity influences the distribution of aquatic vegetation primarily through reduction of light.[42] Water turbidity is often used as a measure of water quality, and events such as cyclones, floods, and algal blooms can increase total suspended matter to levels detrimental to primary productivity and nutrient exchange.[43]

| Turbidity Level | Diffuse Attenuation Coefficient (m⁻¹) |
|---|---|
| Clear | Less than 0.1 |
| Moderately Turbid | 0.1-0.3 |
| Turbid | More than 0.3 |

Table 5

We obtained turbidity data from the NASA Ocean Biology Processing Group, which developed and maintains a global MODIS (Moderate Resolution Imaging Spectroradiometer) Diffuse Attenuation Coefficient at 490 nm (Kd490). The data are a measure of light penetration (attenuation) into the water column as a function of the concentration of organic and inorganic particles. The data are available as a 1 km spatial resolution global raster. The turbidity data are continuous data and were used as such in the analysis. For subsequent classification grouping and labeling purposes, we used three turbidity classes (table 5) from Shi and Wang.[43] Each coastline segment midpoint was attributed with a turbidity value from the raster cell whose center was closest to the segment midpoint.

## Land variables

### 6. Climate setting

Every location on Earth can be classified by its climate regime. The two most common and widely understood climate properties are the temperature regime and the moisture regime. The distribution of vegetation and terrestrial ecosystems in the coastal zone, as elsewhere in the terrestrial domain, is largely controlled by temperature and precipitation.[27,40,44] Integrated measurements of long-term temperature and precipitation describe a fundamental climate expression for a region. To characterize the coastal zone climate setting, we used an integrated measure of long-term average annual temperature and precipitation. The data are from a delineation of World Climate Regions.[45]

| Temperature Regime | Moisture Regime | Climate Region |
|---|---|---|
| Polar | Desert | Polar Desert |
| Polar | Dry | Polar Dry |
| Polar | Moist | Polar Moist |
| Boreal | Desert | Boreal Desert |
| Boreal | Dry | Boreal Dry |
| Boreal | Moist | Boreal Moist |
| Cold Temperate | Desert | Cold Temperate Desert |
| Cold Temperate | Dry | Cold Temperate Dry |
| Cold Temperate | Moist | Cold Temperate Moist |
| Warm Temperate | Desert | Warm Temperate Desert |
| Warm Temperate | Dry | Warm Temperate Dry |
| Warm Temperate | Moist | Warm Temperate Moist |
| Subtropical | Desert | Subtropical Desert |
| Subtropical | Dry | Subtropical Dry |
| Subtropical | Moist | Subtropical Moist |
| Tropical | Desert | Tropical Desert |
| Tropical | Dry | Tropical Dry |
| Tropical | Moist | Tropical Moist |

Table 6

The World Climate Regions analysis identified 6 temperature regime classes and 3 moisture regime classes for a total of 18 classes of integrated temperature and precipitation (table 6). The data are in raster format at a 250 m spatial resolution. The temperature and precipitation input data are from WorldClim version 2.0.[46] The input temperature and precipitation data are continuous, but the 18 resulting IPCC-compatible[47] World Climate Region classes represent categorical data. Each coastline segment midpoint was attributed with a climate region value from the raster cell whose center was closest to the segment midpoint.

## 7. Erodibility class

The effect of waves, tides, and rivers acting as hydrodynamic forces on substrates depends mostly on the erodibility of those substrates. Substrate erodibility is therefore an important element of the ecological setting of a coastal area. Relatively erodible substrates are the source of materials characteristic of depositional environments such as beaches and estuaries, whereas relatively inert substrates are associated with erosional environments such as rocky coasts.

| Lithological Class | Erodibility Class |
|---|---|
| Acid Plutonics, Acid Volcanics, Intermediate Plutonics, Metamorphics, Carbonate Sedimentary, Mixed Sedimentary | Low |
| Basic Plutonics, Basic Volcanics, Intermediate Volcanics, Siliciclastic Sedimentary, Evaporite | Medium |
| Unconsolidated Sedimentary, Pyroclastics | High |
| Water, Ice, Glacier, Other | Not Assigned |

Table 7

To include erodibility as one of our determinants of coastal ecological settings, we developed a simple erodibility index data layer using the Global Lithological Map (GLiM)[48] and the logic and definitions presented in Moosdorf et al.[49] We had used the GLiM previously in the development of the GEO-commissioned global ecological land units (ELUs),[50] but in that case we used lithology because it is an important driver of the distribution of vegetation assemblages due to differences in substrate chemistry.[51] We used a rasterized 250 m version of the GLiM that we developed previously when delineating the ELUs. We assigned a relative erodibility class of high, medium, and low to the 13 Level 1 classes in the set of GLiM attributes (table 7) using the logic and average global erodibility indices developed for the GEroID (global erodibility index) framework.[49] Four additional minor classes in the lithology dataset (water, ice, glacier, and other) were assigned an erodibility class of "Not Assigned". Our erodibility data layer is therefore represented by categorical data, and the erodibility class value assigned to a segment midpoint was from the raster cell whose center was closest to the segment midpoint.

## Coastline variables

### 8. Regional sinuosity

The sinuosity of a stretch of coastline is a measure of its geometric complexity and is defined as the ratio of the length of the actual, curvilinear coastline to the length of a straight line connecting both ends of the segment. Also known as the roughness index (RI), sinuosity is a geometric indicator that can provide information about the type of coastline structure.[14] Relatively smooth and straight coastlines with a low RI are likely to represent beaches, bluffs, or rocky headlands, depending on the erosional and depositional nature of the substrate. Stretches of coastline with a relatively high RI are more likely to be deltaic or estuarine in nature.

| Sinuosity Class | Sinuosity (unitless) |
|---|---|
| Straight | Less than 1.5 |
| Sinuous | 1.5-5.0 |
| Very Sinuous | More than 5.0 |

Table 8

Ecologically, coastline sinuosity has terrestrial and marine dimensions. Terrestrial impacts of coastline complexity include the distribution of freshwater, groundwater, nutrients, and sediments to the coastal zone. In the marine domain, sinuosity influences wave energy, water residence time, and protection or exposure of biotic communities.[52]

Our team calculated the RI of every 10 km stretch of coastline, rather than calculating the sinuosity of the individual 1 km segments, and as such, our index is more a measure of regional sinuosity. This decision was made in response to the observation of Nyberg and Howell[14] that calculating sinuosity from 5 km segments was inadequate for the "capture" of many landward-intruding, funnel-shaped coastline complexes. The RI values are continuous data and were used as such in the analysis. The regional sinuosity value calculated from the 10 km segment was attributed to each of the segment midpoints of the

10 1 km segments comprising the 10 km coastline length. For subsequent classification and description purposes, we identified three sinuosity classes (straight, sinuous, and very sinuous) based on ranges of the RI (table 8).

## 9. Slope profile

Coastal areas can contain steeply sloping mountains plunging into the ocean, low gradient mudflats with almost imperceptible sloping in a seaward direction, and everything in between. Slope is a determinant of the width of the littoral zone, which can range from narrow steep beaches to wide tidal flats.[53] The slope gradient at the coastline influences many aspects of wave energy and shoaling, swash zone morphodynamics, sediment deposition and erosion,[54] and associated differences in biotic distributions.[55] Slope gradient is a strong determinant of changing coastline position and is particularly important in the analysis of shoreline retreat from sea-level rise.[56] Interest in assessing change in coastline position has led to the development of methods (e.g., Doran et al.[57]) for calculating coastal slope gradient.

| Slope Class | Slope Range (%) |
|---|---|
| Flat | Less than 8.75 |
| Sloping | 8.75–57.3 |
| Steeply Sloping | 57.3–173.2 |
| Vertical | More than 173.2 |

Table 9

Our team developed a global coastline slope profile data layer by extending a perpendicular line from each segment midpoint 100 m in landward and seaward directions. The endpoints of this 200 m vector were attributed with elevation values from the corresponding raster cells that contained the 200 m perpendiculars. The elevation data source was the 15 arc seconds (~500 m) resolution GEBCO (General Bathymetric Chart of the Oceans) bathymetry and topography resource,[58] sharpened and gap-filled using data from an Airbus® global 12 m spatial resolution DEM. The slope values of the 200 m segments were attributed as continuous data to the segment midpoints. For subsequent classification and description purposes, we grouped the slope values into four categories: (flat, sloping, steeply sloping, and vertical (table 9) based on CMECS classes and value ranges.

## 10. River outflow index

Rivers and streams are the source of freshwater inputs and particulate matter to the coastal zone, structuring important ecosystems such as estuaries and deltas. Sediment discharge at the mouth of rivers along the coast is the source of most sediment in the coastal zone,[8] and river outflow influences coastal zone dilution processes, nutrient levels, sediment and particulate organic matter composition, pollutant and pathogen exposure levels, etc.[59] River-dominated systems are one of three fundamental depositional morphotypes in coastal areas, along with tide-dominated and wave-dominated systems.[17,18]

| Fluvial Importance | River Outflow Index (unitless) |
|---|---|
| Low | 0–.000012 |
| Moderate | .000013–.001128 |
| High | .001129–1.0 |

Table 10

Using data from approximately 160,000 rivers, we developed a global coastal river outflow index to capture the global distribution and magnitude of annual discharge of rivers at the coastline. The river outflow index was the most complex of the 10 ecological settings variables attributed to the coastline segments. Unlike the other nine variables, all of which represent physical measurements of the coastal environment, the river outflow index is a modeled value of the magnitude and extent of riverine influence. We first obtained global river mouth data from the MERIT (Multi-Error-Removed Improved Terrain) Hydro resource.[60] MERIT Hydro rivers are interpreted from a hydrologically conditioned 3 arc second (~ 90 m) global digital elevation model (DEM) and, unless they drain internally in an inland basin, terminate at a river mouth where the land meets the ocean. River mouth locations were obtained from the MERIT Hydro resource and subsequently "transferred" to corresponding locations on the GSV. MERIT Hydro rivers are associated with basin delineations, and these basin areas were used to approximate the average annual discharge for the rivers.

Using the WorldClim version 2.0 data,[46] the long-term (30 years) average annual precipitation figures for all 1 km cells containing river mouth locations were obtained. This river mouth precipitation value was treated as a uniform measure of the quantity of water falling in every cell in the basin, acknowledging that in reality there will be some level of spatial variation in precipitation input across the watershed. The precipitation quantity at the river mouth was multiplied by the total number of cells in the basin as an approximation of the average annual total amount of water falling in the watershed. This total watershed input quantity was used as a proxy for the discharge amount at the river mouth location in a "what pours in must spill out" sense, acknowledging that some of the input precipitation will be evapotranspirated or lost to groundwater flow and therefore will not arrive at the river mouth. The river discharge was then spatially distributed from the river mouth into the ocean using a statistical smoothing (kernel density) operation to identify a normalized spatial footprint and magnitude of river outflow.

When freshwater discharges into estuaries and the ocean, the spatial dynamics of the mixing waters are complex.[61] Plumes form in the mixing zone based on properties of the freshwater inputs (quantity, velocity, composition, etc.) and the receiving ocean environment (currents, tides, waves, obstructions, etc.).[59] Differences in the magnitude of freshwater influence in the coastal zone from one river to the next could indeed be assessed from a robust characterization of plume dynamics, but the spatial and temporal characterization of riverine discharge plumes globally is currently impractical and well beyond the scope of this exercise.

The team instead developed a standardized, spatial measure of freshwater "influence" using a quartic statistical kernel density algorithm[62] that spread the river discharge into the ocean as a probability smoothing function. This measure, which we call a river outflow index, describes the relative spatial distribution and magnitude of the riverine input and is in essence a potential spatial footprint of river influence in the coastal zone. Importantly, the spatial distribution and magnitude of the modeled river outflow is not intended to represent actual plume shapes and sizes; rather, it is a conceptual geospatial model of the relative influences of fluvial processes in the coastal zone. Essentially, we have modeled a standardized, potential river discharge footprint in the absence of currents or other directional energies or non-coastal barriers. This variable captures spatial differences in river outflows as a simple measure of land-to-sea influence.

A 1 km raster framework was established for the purpose of calculating a river outflow index for each segment using a moving neighborhood analysis window (NAW), such as is commonly used in DEM processing for calculations of terrain attributes. The NAW size was 1 decimal degree (~ 110 km). The spreading function was constrained seaward to a distance of approximately 55 km, the seaward limit of the NAW. Landward, the shoreline vector acted as a hard boundary preventing the spreading function from allowing the backward flow of water onto the land. The river outflow index data are continuous data in a 1 km raster grid and represent a NAW-derived measure of the magnitude (expressed as the spatial distribution) of the amount of precipitation/ riverine discharge that has been spread from the cell. Specifically, the river outflow index characterizes the size (expected number of pixels) of the spatial footprint created by the spread of discharge into the ocean. The coastline segment midpoints were attributed with the river outflow index value from the raster cell whose center was closest to the segment midpoint. For subsequent classification and description purposes, the river outflow index values were rescaled from zero to one using a minimum-maximum method and grouped into three levels (low, moderate, and high) of fluvial importance (table 10).

Geba River estuary, Guinea-Bissau.

# CLUSTERING AND CLASSIFICATION

In addition to the data development effort to create coastal segments and attribute them with values for the 10 variables, we sought to identify global groups of segments with similar aggregate ecological settings. We therefore performed a TwoStep[63] statistical clustering analysis in which all variables were included and given equal weightings, an "all-in and all-equal" clustering approach. However, the clustering was complex due to the number (10) and types (categorical and continuous) of variables, and the big data nature (4 million segments) of the inputs. Rather than operate in 10-dimensional data space, we sought to reduce dimensionality through statistical reduction of the complexity (variance) of the input data. We included all the data from the seven continuous variables in a t-stochastic neighbor embedding (t-sne) analysis,[64] reducing the number of continuous variables from seven to two. The t-sne reduction is similar to a principal components analysis (PCA) but more appropriate for use with non-orthogonal data. Sonnewald et al.[65] used t-sne to reduce dimensionality when clustering global plankton community structure and nutrient flux data for the delineation of marine ecological provinces.

Our final clustering routine included the three categorical variables and the two reduced continuous variables. Although the number of variables was reduced, it is important to note that all 10 variables were used to assess statistical variability of the aggregate ecological setting. None of the variables was dropped as unimportant or weighted less than any other variable. Our team did not constrain the clustering to output a desired number of classes and instead used the collective variability in the input data to identify an optimal number of clusters using a ratio of distance measure approach.[66]

## Results and discussion

### Coastal segment units (CSUs)

After attributing and classifying the 4 million segments, there were 22,534,848 total possible combinations (23 marine physical environment classes x 3 chlorophyll classes x 6 tidal range classes x 7 wave height range classes x 3 turbidity classes x 18 climate region classes x 4 erodibility classes x 3 sinuosity classes x 4 slope profile classes x 3 river outflow index classes), of which a total of 80,977 unique coastal segment units (CSUs) were actually identified. A summary name descriptor for each CSU was developed as a simple concatenation of the attribute classes in this order: slope, sinuosity, erodibility, temperature and moisture regime, river discharge, wave height, tidal range, marine physical environment, turbidity, and chlorophyll. An example CSU label follows:

*Steeply sloping, straight, medium erodibility, warm temperate dry, low river discharge, moderate wave energy, moderately tidal, euhaline-oxic-cool, clear, low chlorophyll.*

Any one CSU may differ markedly from another CSU with considerable differences in the classes of all or most of the 10 attributes. Similarly, any two CSUs might be almost identical, with only slight differences in the classes of one or a few of the 10 variables. The sheer number of CSUs precludes a rigorous analysis of their individual global distributions and comparisons of the differences between them. However, in a basic inventory sense, the CSU data may have great utility for managers at local scales as a comprehensive inventory of ecological properties for a 1 km stretch of coastline in a management area.

## Working with the data

*Clicking any coastal segment brings a pop-up with the query results. Shown here are the attribute queries from two 1 km segments from very different coastal environments.*

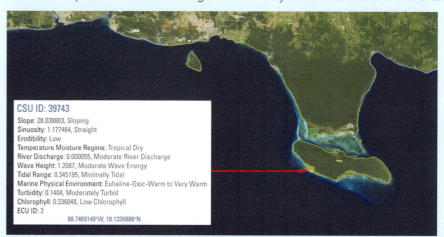

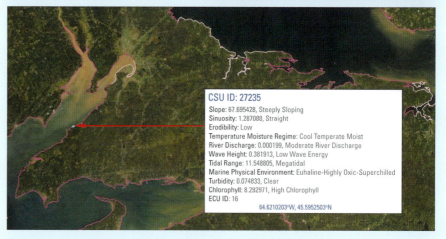

*The top panel shows the coastlines that surround Parque Nacional del Este in the southeast peninsula of the Dominican Republic. The bottom panel shows the coastal zone around the top end of the Bay of Fundy, New Brunswick, Canada. The Bay of Fundy has the largest tidal range in the world, and a query of its tidal range at this location reveals a range of 11.5 m, contrasted with a tidal range of 0.3 m in the southeast region of the Dominican Republic. As a tide-dominated system, the Bay of Fundy is a low-wave energy system (0.39 m mean significant wave height), whereas the segment from the Dominican Republic is in a moderate wave energy class (1.2 m). The Bay of Fundy is eutrophic at this location, where the primary productivity as indicated by the chlorophyll level (8.3 micrograms per liter, or μg/L) is high. In the Dominican Republic site, the chlorophyll level is relatively low (0.3 μg/L), indicating a lower productivity from an oligotrophic system. The climate regions obviously differ dramatically from Cool Temperate Moist (Bay of Fundy) to Tropical Dry (Parque Nacional del Este). The marine physical environment also differs from Euhaline-Highly Oxic-Superchilled (Bay of Fundy) to Euhaline-Oxic-Warm to Very Warm (Parque Nacional del Este). Neither segment is strongly turbid or strongly sloping, and their sinuosities are similar. Each segment has low erodibility. Importantly, fluvial importance in the Bay of Fundy segment is four times greater than the segment in the Dominican Republic. Note that the last attribute listed in the pop-up query box is the ECU/cluster to which the segment belongs.*

The intended practical value of the CSU data is that any 1 km segment on the global coastline can be queried in this manner, supporting the understanding of place-based coastal zone dynamics, and facilitating the comparison of stretches of coastline at local scales. Coastal classifications are used in many ways, including coastline vulnerability studies;[6,23,56] conservation planning;[4] ecosystem services assessments;[5] understanding of coastal ecosystem structure, function, and distributions;[24,27,28] design of marine protected area (MPA) networks;[15] and other applications. Our team hopes that the practical value of the CSU classification will be realized by its adoption and use in these and other applications.

## Ecological coastal units (ECUs)

The global assessment of optimal cluster numbers identified 16 clusters, shown in the world map of ECUs. Fewer than 16 clusters appeared largely responsive to latitudinal gradients, whereas more than 16 clusters did not identify new regional groupings, instead demonstrating successive partitioning of the parent clusters. The 16 clusters represent the identification of a preliminary set of standardized and replicable ECUs with potential utility for understanding global scale differences in coastal ecological settings and for global scale conservation assessments and priority setting. The world map below depicts the ECUs at a glance with maximum color separation. An easier interpretation of the individual ECU distributions is presented in the set of 16 maps on the next pages showing the global distributions of each ECU.

Several ECU distributions appear to have strong climate region associations, mainly driven by latitudinal temperature gradients. ECUs 1 and 4 are clearly tropical systems. ECU 2 has a more subtropical distribution with extensions into the warm temperate climate regions. ECU 6 is largely distributed in warm temperate regions. ECUs 7, 9, 15, and 16 are largely boreal. ECUs 8, 10, 11, 12, and 14 are largely distributed in polar regions, and of these, two (ECUs 8 and 12) are only distributed in the Northern Hemisphere, while ECUs 10, 11, and 14 are found in both hemispheres.

While none of the ECUs show a strong pole-to-pole distribution, ECUs 3 and 5 span multiple latitudes, indicating that the climate region influence in these clusters is low. ECU 5 exhibits a strong pattern of distribution along major estuaries/river mouths. ECU 16 shows an interesting distribution in coastal areas characterized by numerous coastal islands (Chile, Norway) and island chains.

The 10-dimensional nature of the dataset complicates visual identification of additional patterns in ECU distributions. The exact nature of each ECU, including the relative influences of the input variables driving their existence and distributions, can be quantitatively described, and the data are available. We summarize the main properties and descriptive statistics of the ECUs in the appendix. The team did not present a rigorous assessment and comparison of the ECU compositional properties herein, as our purpose is rather to describe the development of the ECUs

(continued on page 24)

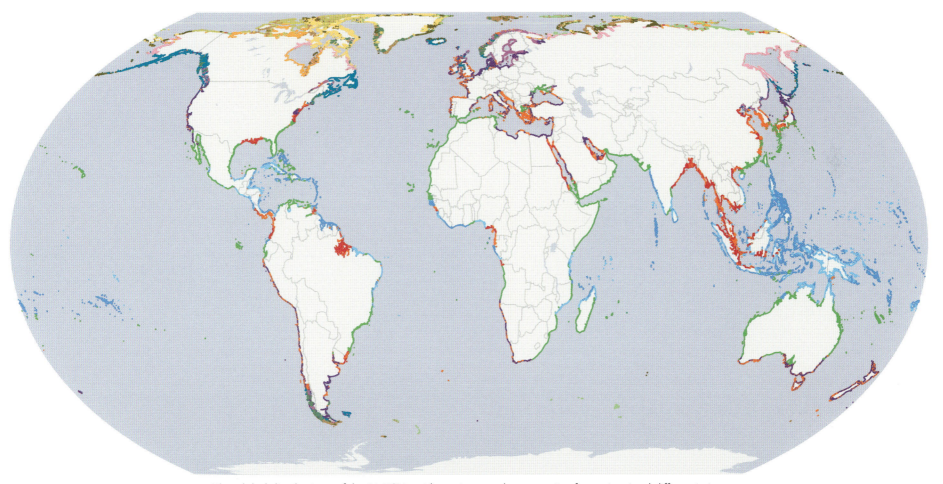

*The global distributions of the 16 ECUs, with maximum color separation for easier visual differentiation.*

# GLOBAL ECU DISTRIBUTION: UNITS 1–8

Shown in black for each of the 16 ecological coastal units

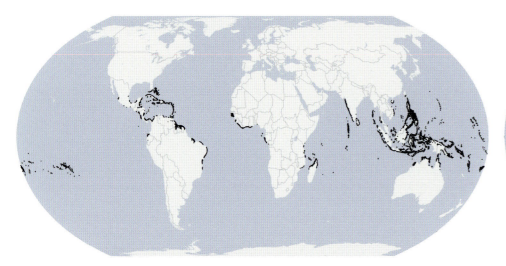

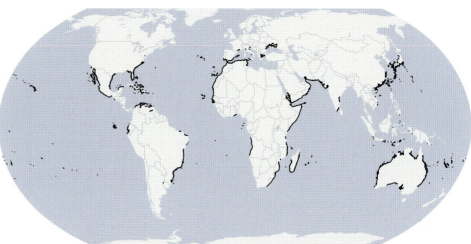

**1** ECU 1—Steeply sloping, sinuous, low erodibility, tropical moist, moderate river discharge, low wave energy, moderately tidal, euhaline-oxic-warm to very warm, moderately turbid, low chlorophyll.

**2** ECU 2—Steeply sloping, sinuous, high erodibility, tropical dry, moderate river discharge, low wave energy, moderately tidal, euhaline-oxic-warm to very warm, moderately turbid, moderate chlorophyll.

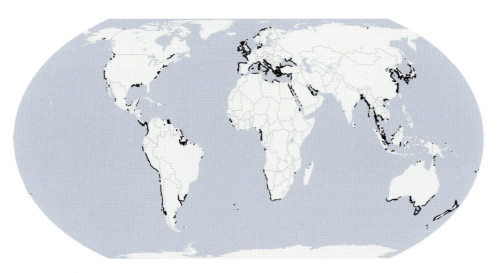

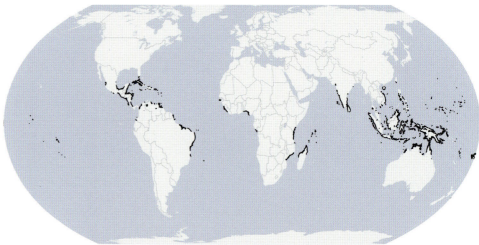

**3** ECU 3—Steeply sloping, sinuous, low erodibility, warm temperate moist, high river discharge, low wave energy, moderately tidal, euhaline-oxic-moderate, moderately turbid, moderate chlorophyll.

**4** ECU 4—Sloping, sinuous, high erodibility, tropical moist, high river discharge, low wave energy, moderately tidal, euhaline-oxic-warm to very warm, moderately turbid, moderate chlorophyll.

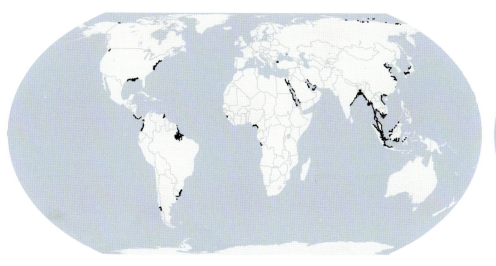

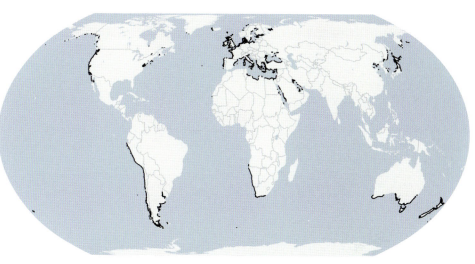

5 *ECU 5—Sloping, sinuous, high erodibility, tropical moist, high river discharge, low wave energy, moderately tidal, euhaline-oxic-moderate, moderately turbid, high chlorophyll.*

6 *ECU 6—Sloping, sinuous, high erodibility, cool temperate moist, moderate river discharge, moderate wave energy, moderately tidal, euhaline-oxic-moderate to cool, turbid, moderate chlorophyll.*

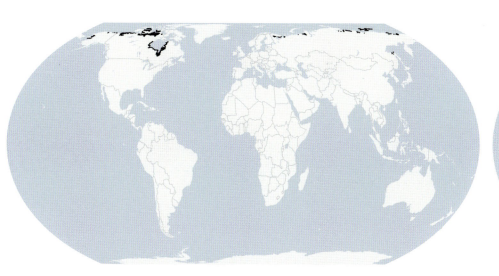

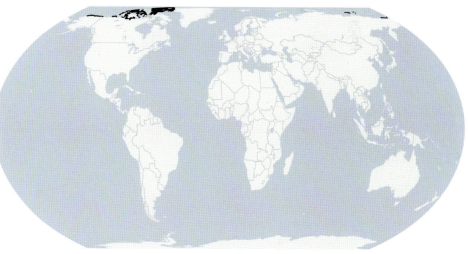

7 *ECU 7—Sloping, sinuous, low erodibility, polar dry, moderate river discharge, low wave energy, moderately tidal, polyhaline-anoxic-very cold, turbid, moderate chlorophyll.*

8 *ECU 8—Sloping, very sinuous, low erodibility, polar dry, low river discharge, low wave energy, minimally tidal, polyhaline-highly oxic-superchilled, turbid, low chlorophyll.*

# GLOBAL ECU DISTRIBUTION: UNITS 9–16

Shown in black for each of the 16 ecological coastal units

**9** *ECU 9—Sloping, sinuous, low erodibility, boreal moist, moderate river discharge, low wave energy, moderately tidal, euhaline-highly oxic-superchilled, turbid, moderate chlorophyll.*

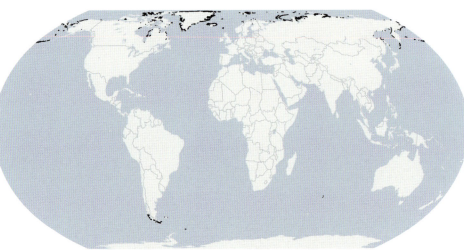

**10** *ECU 10—Sloping, sinuous, medium erodibility, polar moist, low river discharge, low wave energy, moderately tidal, euhaline-highly oxic-superchilled, turbid, moderate chlorophyll.*

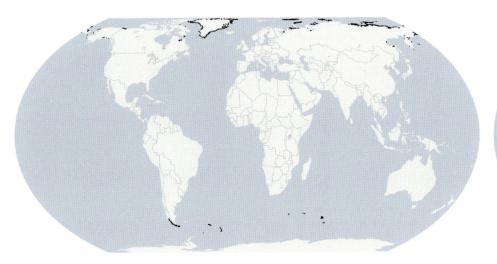

**11** *ECU 11—Sloping, sinuous, high erodibility, polar moist, moderate river discharge, low wave energy, minimally tidal, mesohaline-severely hypoxic-very cold, turbid, high chlorophyll.*

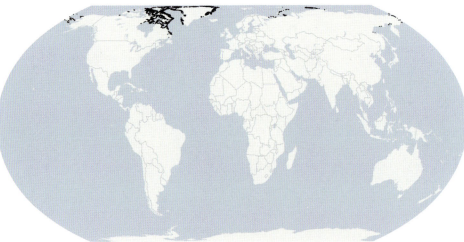

**12** *ECU 12—Sloping, sinuous, low erodibility, polar moist, low river discharge, low wave energy, moderately tidal, euhaline-highly oxic-superchilled, turbid, low chlorophyll.*

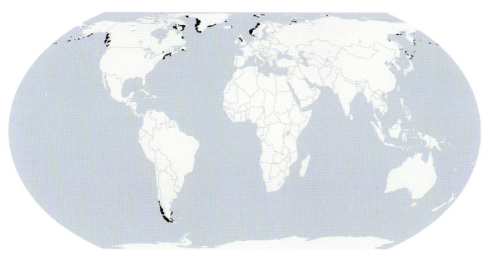

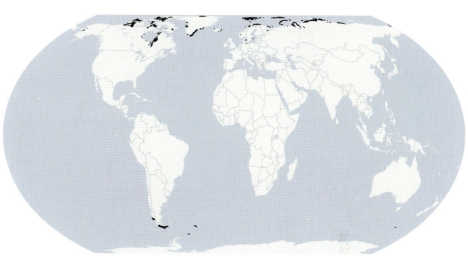

**13** *ECU 13—Steeply sloping, sinuous, low erodibility, cool temperate moist, low river discharge, moderate wave energy, moderately tidal, euhaline-oxic-very cold, turbid, low chlorophyll.*

**14** *ECU 14—Sloping, very sinuous, low erdodibility, polar moist, low river discharge, low wave energy, moderately tidal, polyhaline-highly oxic-superchilled, turbid, moderate chlorophyll.*

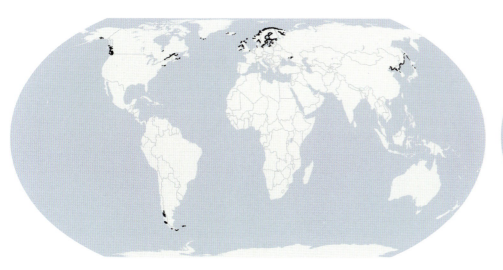

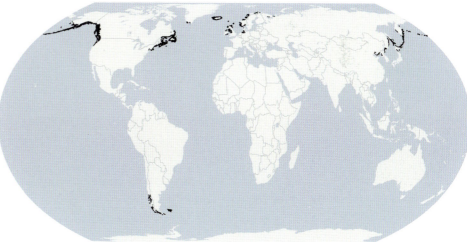

**15** *ECU 15—Steeply sloping, sinuous, low erodibility, cool temperate moist, moderate river discharge, low wave energy, moderately tidal, mesohaline-highly oxic-very cold, turbid, moderate chlorophyll.*

**16** *ECU 16—Steeply sloping, sinuous, medium erodibility, cool temperate moist, moderate river discharge, moderate wave energy, moderately tidal, euhaline-highly oxic-superchilled, turbid, moderate chlorophyll.*

(continued from page 19)

as a preliminary set of 16 macroscale global coastal ecosystems based on integrated information about the coastline and adjacent land and water environments. The ECUs are composed of many different coastal segment units (CSUs). Importantly, there is no hierarchical relationship or perfect spatial nesting between CSUs and ECUs.

One aspect in evaluating the accuracy and potential use of the ECUs relates to how well they line up with existing coastline segmentation and coastal ecosystem delineation efforts. This map of Australia shows the ECUs distributed along the Australian coast compared with the locations of Primary Coastal Compartments from a national sediment compartments assessment.[66] Five ECUs occur in mainland Australia, Tasmania, and territorial islands (a sixth ECU is found on an Australian island south of New Zealand not shown in the map). The distribution of the Primary Coastal Compartments is based on "major, usually distinctive structural features such as rocky headlands or major changes in orientation."[66]

The ECUs are globally distributed and are relatively few in number. They were developed to explore coastal variation at the global scale, and using them to attempt to explain local phenomena is not advised, in a "scale-inappropriate" sense. For example, it is common knowledge that Hawaiian islands have sandy beaches, rocky coasts, mangroves, and steep coastal cliffs, so an ECU user might ask, "Why are all Hawaiian islands just one ECU?" It would appear that the aggregate coastal variation among Hawaiian islands is less than the aggregate coastal variation between Hawaii and other locations. The CSU data, not the ECU data, should be considered the go-to resource for addressing these types of questions dealing with local and regional variation in coastal environments. But for global scale comparisons of similarity in coastal ecological setting, the ECU data are likely appropriate. Moreover, the ECU data might be used to frame new research collaborations between, say, Norwegians, Chileans, and Alaskans seeking to better understand their common ECU.

The ECUs do not contain any information on rocky headlands or coastal orientation, and as such would not necessarily be expected to predict the Primary Coastal Compartment geographies. Importantly, however, many of the locations along the coast where ECUs change are at or near a coastal compartment boundary. This may indicate that, as expected,[7] changes in the macro-level geology and orientation are associated with changes in other features of the general ecological setting that in turn are "captured" by ECUs.

Although ECUs often change at coastal compartment boundaries, they do not predict them. For example, one ECU in particular (ECU 2) has an extensive distribution along the Australian mainland, occurring on all four (eastern, western, northern, and southern) coasts. ECU 2 is subdivided into numerous coastal compartments. It would be interesting to see whether the much finer resolution CSUs would have stronger predictive value in identifying the coastal compartment locations.

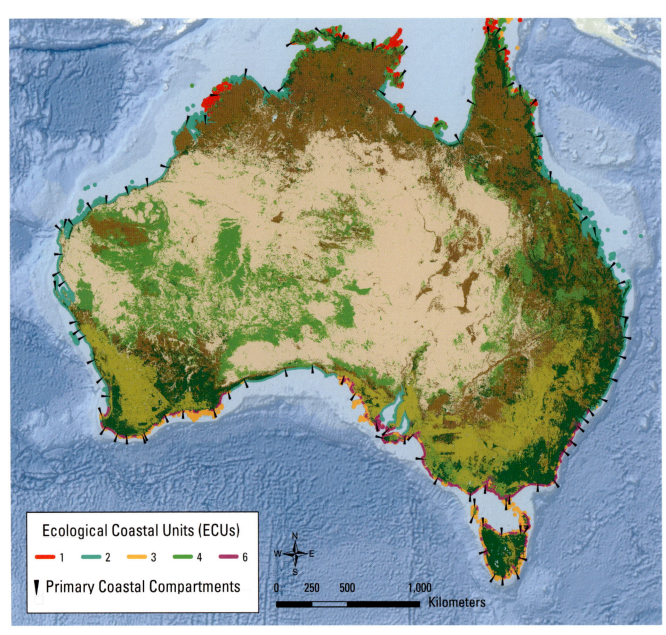

*A visual comparison of the distribution of ECUs along the Australian coast and the set of Primary Coastal Compartments from an Australian sediment compartments assessment.[66] The coastal compartments, bounded by black markers, were approximated from figure 1 in Thom et al.[66] and used with permission. The data for the interior represent the Australian distributions of World Terrestrial Ecosystems.[45] Bathymetry data are from Esri's Ocean Basemap®.*

**Ecological Coastal Units (ECUs)**

━ 1  ━ 2  ━ 3  ━ 4  ━ 6

🔻 **Primary Coastal Compartments**

N

W E

S

0   250   500   1,000

Kilometers

# CONCLUSION

The publication of this information introduces the availability of a detailed and comprehensive global dataset detailing 10 aspects of coastal ecological settings. This is a first-of-its-kind inventory of the properties of 4 million coastal segments at a management-appropriate spatial resolution (1 km). We classified those 4 million segments into 81,000 distinct coastal segment units (CSUs) using authoritative data and standards for describing the classes and their associated value ranges for each of the 10 variables. The data are freely available as an ArcGIS Online resource found at GISforScience.com. Our team plans to actively curate this data layer as an authoritative geospatial resource in ArcGIS Living Atlas of the World. The fully attributed, classified segment units data and the preliminary ECU data are available. The data will also be placed in the public domain after additional preparation and review, including making the data available in an open data format such as an OGC® (Open Geospatial Consortium) GeoPackage file or similar.

The CSU data permit spatial analysis and mapping of coastal ecological settings at any scale, anywhere on Earth. This comprehensive inventory of coastline environments should be useful for researchers, planners, managers, and policy makers. The community is invited to evaluate the data and provide feedback to improve the accuracy and utility of the data. Our team anticipates the development of a simple, web-based data visualization and query tool, the Global Coastline Explorer, which will allow anyone with an internet connection to access the data.

Ultimately, the utility of the classification will be judged by the end user who has assessed the CSU data and evaluated them for fitness of purpose. The classification has three major advantages in that regard: 1) it is exhaustive (every coast is classified), 2) the classes are mutually exclusive (every coast fits into only one class, but that class can be described in rich taxonomic detail), and 3) standardized comparisons of coasts across any oceanic or continental geography are enabled.

This presentation also includes a scientific assessment of the global similarity of these coastline environments. A statistical clustering of the 4 million segments identified 16 ecological coastal units (ECUs) that differ from each other based on their aggregate ecological setting as established by the 10 variables. The ECU work is intended for the scientific community and other interested parties seeking to understand global patterns of variation in coastal ecological settings. The ECU data may be useful in understanding global patterns of biodiversity distribution, helping biogeographers answer that enduring question, "Why are species distributed where they are?"

The ECU data are considered preliminary, and we anticipate feedback from the community on their veracity and utility. The statistical analysis was rigorous and defensible, but the selection of variables in any clustering exercise is always the subject of debate. Did we omit an important variable or variables, such as, for example, elevation change at the coast, important for understanding uplift and subsidence effects on coastal landforms?

Regardless of the research value of the current clustering results, the newly created set of granular, management-appropriate, and globally comprehensive coastline segments represents an advance in coastal ecology. It is our hope that this resource will contribute to better understanding and wise stewardship of the planet's many coastal environments.

## Appendix

| ECU | N (# pixels) | % of total N | Predominant Marine Physical Environment (% of total distribution of occurrences in cluster) | Average Chlorophyll Concentration (μ/l) | Average Tidal Range (m) | Average Significant Wave Height (m) | Average Turbidity (m-1) | Predominant Climate Region (% of total distribution of occurrences in cluster) | Predominant Erodibility Class (% of total distribution of occurrences in cluster) | Average Regional Sinuosity (unitless) | Average Slope (%) | Average Outflow Density (unitless) |
|---|---|---|---|---|---|---|---|---|---|---|---|---|
| 1 | 234,549 | 6 | Euhaline-Oxic-Warm to Very Warm (100) | 1.8 | 2.1 | 0.77 | 0.22 | Tropical Moist (100) | Low (67) | 2.57 | 97.87 | 0.0003 |
| 2 | 308,444 | 8 | Euhaline-Oxic-Warm to Very Warm (97) | 2.99 | 2.37 | 0.88 | 0.22 | Tropical Dry (33) | High (59) | 2.69 | 60.9 | 0.0006 |
| 3 | 263,112 | 7 | Euhaline-Oxic-Moderate (45) | 2.85 | 2.68 | 0.88 | 0.29 | Warm Temperate Moist (37) | Low (71) | 2.26 | 79.7 | 0.0015 |
| 4 | 206,534 | 5 | Euhaline-Oxic-Warm to Very Warm (100) | 3.44 | 2.48 | 0.7 | 0.24 | Tropical Moist (100) | High (100) | 2.64 | 53.48 | 0.0017 |
| 5 | 248,600 | 6 | Euhaline-Oxic-Moderate (100) | 6.14 | 2.41 | 0.58 | 0.29 | Tropical Moist (65) | High (100) | 2.27 | 29.48 | 0.0115 |
| 6 | 206,328 | 5 | Euhaline-Oxic-Moderate to Cool (49) | 3.54 | 1.98 | 1.02 | 0.36 | Cool Temperate Moist (43) | High (64) | 2.53 | 49.95 | 0.0007 |
| 7 | 192,253 | 5 | Polyhaline-Anoxic-Very Cold (33) | 3.57 | 1.06 | 0.38 | 0.46 | Polar Dry (47) | Low (94) | 2.89 | 20.87 | 0.0002 |
| 8 | 199,849 | 5 | Polyhaline-Highly Oxic-Superchilled (54) | 1.8 | 0.93 | 0.3 | 0.33 | Polar Dry (100) | Low (81) | 6.01 | 25.25 | 0 |
| 9 | 179,728 | 4 | Euhaline-Highly Oxic-Superchilled (51) | 4.98 | 3.06 | 0.58 | 0.55 | Boreal Moist (98) | Low (48) | 1.9 | 47.2 | 0.0008 |
| 10 | 183,614 | 5 | Euhaline-Highly Oxic-Superchilled (42) | 3.46 | 1.22 | 0.58 | 0.44 | Polar Moist (100) | Medium (100) | 1.87 | 47.35 | 0.0001 |
| 11 | 215,600 | 5 | Mesohaline-Severely Hypoxic-Very Cold (16) | 5.69 | 0.77 | 0.5 | 0.42 | Polar Moist (64) | High (83) | 1.96 | 27.01 | 0.0005 |
| 12 | 401,296 | 1 | Euhaline-Highly Oxic-Superchilled (100) | 1.69 | 3.26 | 0.42 | 0.43 | Polar Moist (100) | Low (100) | 2.58 | 49.28 | 0 |
| 13 | 341,997 | 9 | Euhaline-Oxic-Very Cold (100) | 1.88 | 2.95 | 1.27 | 0.43 | Cool Temperate Moist (53) | Low (100) | 2.63 | 110.51 | 0.0001 |
| 14 | 271,424 | 7 | Polyhaline-Highly Oxic-Superchilled (29) | 2.28 | 1.46 | 0.74 | 0.36 | Polar Moist (100) | Low (100) | 5.07 | 54.33 | 0.0001 |
| 15 | 285,583 | 7 | Mesohaline-Highly Oxic-Very Cold (33) | 3.15 | 1.54 | 0.88 | 0.58 | Cool Temperate Moist (100) | Low (100) | 2.25 | 66.02 | 0.0004 |
| 16 | 266,931 | 7 | Euhaline-Highly Oxic-Superchilled (49) | 2.66 | 3.53 | 1.29 | 0.43 | Cool Temperate Moist (100) | Medium (77) | 1.97 | 84.27 | 0.0002 |

*Descriptive statistics on the distributions and average compositional characteristics of the ecological coastal units (ECUs).*

# NOTES AND CREDITS

1. R. Sayre, S. Noble, S. Hamann, R. Smith, D. Wright, S. Breyer, K. Butler, et al., "A New 30 Meter Resolution Global Shoreline Vector and Associated Global Islands Database for the Development of Standardized Ecological Coastal Units," *Journal of Operational Oceanography*, 12:sup2 (2019): S47–S56, https://doi.org/10.1080/1755876X.2018.1529714.

2. R. Sayre, M. Martin, J. Cress, N. Holmes, O. McDermott Long, L. Weatherdon, D. Spatz, K. Van Graafeiland, and D. Will, "The Geography of Islands," in D. Wright and C. Harder, eds. *GIS for Science: Volume 2* (Redlands, California: Esri Press, 2020).

3. UN (United Nations), "World Population Prospects: the 2010 Revision, Highlights and Advance Tables," *Working Paper No ESA/P/WP220*. New York: Department of Economic and Social Affairs, Population Division (2010).

4. L. Burke, Y. Kura, K. Kassem, C. Revenga, M. Spalding, and D. McAllister, *Coastal Ecosystems* (World Resources Institute, 2000), https://www.wri.org/publication/pilot-analysis-global-ecosystems-coastal-ecosystems.

5. E. Barbier, S. Hacker, C. Kennedy, E. Koch, A. Stier, and B. Silliman, "The Value of Estuarine and Coastal Ecosystem Services," *Ecological Monographs* 81, no. 2 (2011): 169–193.

6. UNEP-WCMC (United Nations Environment Program—World Conservation Monitoring Center, "Environmental Sensitivity Mapping For Oil and Gas Development—A High-Level Review of Methodologies," (2019).

7. J. Davies, "Geographical Variation in Coastal Development," Geomorphology Text Series 4. (London: Longman Group Ltd, 1980).

8. R. Davis and J. Fitzgerald, *Beaches and Coasts*, 2nd edition (New York: John Wiley & Sons, Ltd., 2020), https://doi.org/10.1002/9781119334491.

9. D. Johnson, *Shore Processes and Shoreline Development* (New York: John Wiley and Sons, 2019).

10. D. Inman and C. Nordstrom, "On the Tectonic and Morphologic Classification of Coasts," *Journal of Geology* 79, no. 1 (1971):1–21.

11. J. Pethick, "Saltmarsh Geomorphology," in *Saltmarshes: Morphodynamics, Conservation and Engineering Significance*, Allen, J. and K. Pye, eds. (Cambridge: Cambridge University Press, 1992): 41–62.

12. S. Haslett, *Coastal Systems*, Routledge Introductions to Environment: Environmental Science Series (London: Routledge, 2000), 218 pp.

13. C. Woodroffe, *Coasts: Form, Process, and Evolution* (Cambridge: Cambridge University Press) .

14. B. Nyberg and J. Howell, "Global Distribution of Modern Shallow Marine Shorelines—Implications for Exploration and Reservoir Analogue Studies," *Marine Petroleum Geology* 71 (2016): 83–104.

15. B. Neilson and M. J. Costello, "The Relative Lengths of Seashore Substrata around the Coastline of Ireland as Determined by Digital Methods in a Geographical Information System," *Estuarine, Coastal and Shelf Science* 49 (1999): 501–508.

16. M.J. Costello and C. Emblow, "A Classification of Inshore Marine Biotopes," in *The Intertidal Ecosystem: The Value of Ireland's Shores*, J.G. Wilson ed. (Dublin: Royal Irish Academy, 2005), 25–35.

17. R. Boyd, R. Dalrymple, and B. Zaitlin, "Classification of Clastic Coastal Depositional Environments," *Sedimentary Geology* 80, no. 3 (1992): 139–150.

18. P. Harris, A. Heap, S. Bryce, R. Porter-Smith, D. Ryan, and D. Heggie, "Classification of Australian Clastic Coastal Depositional Environments Based upon a Quantitative Analysis of Wave, Tidal, and River Power," *Journal of Sedimentological Research* 72, no. 6 (2002): 858–870, https://doi.org/10.1306/040902720858.

19. J. Cooper and S. McLaughlin, "Contemporary Multidisciplinary Approaches to Coastal Classification and Environmental Risk Analysis," *Journal of Coastal Research* 14, no. 2 (1998): 512–524.

20. S. Jelgersma, M. Van Derzip, and R. Brinkman, "Sealevel Rise and the Coastal Lowlands in the Developing World," *Journal of Coastal Research* 9, no. 4 (1993): 958–972.

21. R. Sayre, D. Wright, S. Breyer, K. Butler, K. Van Graafeiland, M. Costello, P. Harris, et al., "A Three-dimensional Mapping of the Ocean Based on Environmental Data," *Oceanography* 30, no. 1 (2017): 90–103, https://doi.org/10.5670/oceanog.2017.116.

22. R. Davidson-Arnott, *An Introduction to Coastal Processes and Geomorphology* (New York: Cambridge University Press, 2010).

23. L. McFadden, R. Nicholls, A. Vafeidis, and R. Tol, "A Methodology for Modeling Coastal Space for Global Assessment," *Journal of Coastal Research* 23, no. 4 (2007): 911–920.

24. FGDC (Federal Geographic Data Committee), *Coastal and Marine Ecological Classification Standard Version 4.0. Report FGDC-STE-18-2012* (Reston, Virginia: Federal Geographic Data Committee Secretariat, U.S. Geological Survey, 2012).

25. R. MacArthur, *Geographical Ecology: Patterns in the Distribution of Species*, First Edition (New York: Harper & Row, 1972).

26. M. Lurgi, N. Galiana, B. Broitman, S. Kéfi, E. Wieters, and S. Navarrete, "Geographical Variation of Multiplex Ecological Networks in Marine Intertidal Communities," *Ecology* 101, no. 11 (November 2020): e03165. https://doi.org/10.1002/ecy.3165.

27. T. Gagné, G. Reygondeau, C. Jenkins, J. Sexton, S. Bograd, E. Hazen, and K. Van Houtan, "Towards a Global Understanding of the Drivers of Marine and Terrestrial Biodiversity," *PLoS ONE* 15, no. 2 (2020): e0228065. https://doi.org/10.1371/journal.pone.0228065.

28. K. Smyth and M. Elliott, "Effects of Changing Salinity on the Ecology of the Marine Environment," Chapter 9 in *Stressors in the Marine Environment*, eds. M. Solan and N. Whiteley (Oxford Scholarship Press, 2016), https://doi.org/10.1093/acprof:oso/9780198718826.003.0009.

29. H.U. Sverdrup, M.W. Johnson, and R.H. Fleming, *The Oceans: Their Physics, Chemistry and General Biology* (Englewood Cliffs: New Jersey, Prentice-Hall, 1942).

30. J. Steele and I. Baird, "Relations between Primary Production, Chlorophyll and Particulate Carbon," *Limnology and Oceanography* 6, no. 1 (1961): 68–78, https://doi.org/10.4319/lo.1961.6.1.0068.

31. S. Sathyendranath, R. Brewin, C. Brockmann, V. Brotas, B. Calton, A. Chuprin, P. Cipollini, et al., "An Ocean-Colour Time Series for Use in Climate Studies: The Experience of the Ocean-Colour Climate Change Initiative (OC-CCI)," *Sensors* 19, no. 19 (October 3, 2019): 4285. doi:10.3390/s19194285.

32. G. Jenks, "The Data Model Concept in Statistical Mapping," *International Yearbook of Cartography* 7 (1967): 186–190.

33. C. Crain, B. Silliman, S. Bertness, and M. Bertness, "Physical and Biotic Drivers of Plant Distribution across Estuarine Salinity Gradients," *Ecology* 85, no. 9 (2004): 2539–2549, https://www.jstor.org/stable/3450251.

34. J. Carstensen, M. Sanchez-Camacho, C. Duarte, D. Krause-Jensen, and N. Marba, "Connecting the Dots: Responses of Coastal Ecosystems to Changing Nutrient Concentrations," *Environmental Science and Technology* 45 (2011): 9122–9132.

35. C. Harley, "Tidal Dynamics, Topographic Orientation, and Temperature-Mediated Mass Mortalities on Rocky Shores," *Marine Ecology Progress Series* 371 (November 2008), 37–46, https://doi.org/103354/meps07711.

36. K.B. Moffett and S.M. Gorelick, "Relating Salt Marsh Pore Water Geochemistry Patterns to Vegetation Zones and Hydrologic Influences," *Water Resources Research* 52 (2016): https://doi.org/10.1002/ 2015WR017406.

37. L. Windham-Myers, F. Anderson, C. Sturtevant, and B. Bergamaschi, "Direct and Indirect Effects of Tides on Ecosystem-scale $CO_2$ Exchange in a Brackish Tidal Marsh in Northern California," *JGR Biosciences* 123, no. 3 (2018): 787–806, https://doi.org/10.1002/2017JG004048.

38. C. Bloch and B. Klingbeil, "Anthropogenic Factors and Habitat Complexity Influence Biodiversity but Wave Exposure Drives Species Turnover of a Subtropical Rocky Inter-tidal Metacommunity," *Marine Ecology* 37, no. 1 (2015): 64–76, https://doi.org/10.1111/maec.12250.

39. W. Odum, E. Odum, and H.T. Odum, "Nature's Pulsing Paradigm," *Estuaries* 18 (1995): 547, https://doi.org/10.2307/1352375.

40. E. Odum, *Fundamentals of Ecology* (Philadelphia: W. B. Saunders Company, 1953).

41. T. Burel, J. Grall, G. Schaal, M. Le Duff, and E. Ar Gall, "Wave Height vs. Elevation Effect on Macroalgal Dominated Shores: An Intercommunity Study," *Journal of Applied Phycology* 32 (2020): 2523–2534.

42. A. Austin, J. Hansen, S. Donadi, and J. Eklf, "Relationships Between Aquatic Vegetation and Water Turbidity: A Field Survey across Seasons and Spatial Scales," *PLoS ONE* 12, no. 8 (2017): e)181419, https://doi.org/10.1371/journal.pone.0181419.

43. W. Shi and M. Wang, "Characterization of Global Ocean Turbidity from Moderate Resolution Imaging Spectroradiometer Ocean Color Observations," *Journal of Geophysical Research*: Oceans 115, no. C11 (2010), https://doi.org/10.1029/2010JC006160.

44. R. Bailey, *Ecosystem Geography: From Ecoregions to Sites*, 2nd edition. (New York: Springer-Verlag, 2009).

45. R. Sayre, D. Karagulle, C. Frye, T. Boucher, N. Wolff, S. Breyer, D. Wright, et al., "An Assessment of the Representation of Ecosystems in Global Protected Areas Using New Maps of World Climate Regions and World Ecosystems," *Global Ecology and Conservation* 21, e00860 (2020): 2351–9894, https://doi.org/10.1016/j.gecco.2019.e00860.

46. S. Fick and R. Hijmans, "WorldClim 2: New 1 km Spatial Resolution Climate Surfaces for Global Land Areas," *International Journal of Climatology* (2017), https://doi.org/10.1002/joc.5086.

47. IPCC (Intergovernmental Panel on Climate Change), "2019 Refinement to the 2006 IPCC Guidelines for National Greenhouse Gas Inventories," Chapter 3, "Consistent Representation of Lands," in *Agriculture, Forestry, and Other Land Uses*, Volume 4 (2019), https://www.ipcc-nggip.iges.or.jp/public/2019rf/vol4.html.

48. J. Hartmann and N. Moosdorf, "The New Global Lithological Map Database GLiM: A Representation of Rock Properties at the Earth Surface," *Geochemistry, Geophysics, Geosystems* 13, no. 12 (2012): 1-37, https://doi.org/10.1029/2012GC004370.

49. N. Moosdorf, S. Cohen, and C. von Hagke, "A Global Erodibility Index to Represent Sediment Production Potential of Different Rock Types," *Applied Geography* 101 (2018), 36–44.

50. R. Sayre, J. Dangermond, C. Frye, R. Vaughan, P. Aniello, S. Breyer, D. Cribbs, et al., *A New Map of Global Ecological Land Units—An Ecophysiographic Stratification Approach* (Washington, DC: Association of American Geographers, 2014).

51. A. Kruckeberg, *Geology and Plant Life: The Effects of Landforms and Rock Types on Plants*, (Seattle: University of Washington Press, 2002).

52. J. Bartley, R. Buddemeier, and D. Bennett, "Coastline Complexity: A Parameter for Functional Classification of Coastal Environments," *Journal of Sea Research* 46, no. 2 (September, 2001): 87–97. https://doi.org/10.1016/S1385-1101(01)00073-9.

53. S. John, D. Brew, and R. Cottle, "Coastal Ecology and Geomorphology," in *Methods of Environmental and Social Impact Assessment*, 4th edition, R. Therivel and G. Wood, eds. (New York: Routledge, 2017), 234–296, https://doi.org/10.4324/9781315626932.

54. A. Short, "Australian Beach Systems—Nature and Distribution," *Journal of Coastal Research* 22, no. 1, 221 (2006): 11–27, 0749-0208, https://doi.org/10.2112/05A-0002.1.

55. O. Defeo, A. McLachlan, D. Schoeman, T. Schlacher, J. Dugan, A. Jones, M. Lastra, and F. Scapin, "Threats to Sand Beach Ecosystems: A Review," Estuarine, *Coastal and Shelf Science* 81, no. 1 (2009): 1–12, https://doi.org/10.1016/j.ecss.2008.09.022.

56. V.R. Burkett and M. Davidson, eds. *Coastal Impacts, Adaptation and Vulnerability: A Technical Input to the 2012 National Climate Assessment*, a National Climate Assessment Regional Technical Input Report Series (Washington, D.C.: Island Press, 2012).

57. K. Doran, J. Long, and J. Overbeck, "A Method for Determining Average Beach Slope and Beach Slope Variability for U.S. Sandy Coastlines," *U.S. Geological Survey Open File Report 2015–1053* (2015): 5 pp., http://dx.doi.org/10.3133/ofr20151053.

58. B. Tozer, D. Sandwell, W. Smith, C. Olson, J. Beale, and P. Wessel, "Global Bathymetry and Topography at 15 Arc Sec: SRTM15+," *Earth and Space Science* 6, no. 10 (2020): 1847–1864, https://doi.org/10.1029/2019EA000658.

59. A. Osadchiev and P. Zavialov, "Structure and Dynamics of Plumes Generated by Small Rivers," in *Estuaries and Coastal Zones—Dynamics and Response to Environmental Changes* (Rijeka, Croatia: IntechOpen, 2019), https://dx.doi.org/10.5772/intechopen.87843.

60. D. Yamazaki, D. Ikeshima, J. Sosa, P.D. Bates, G.H. Allen, T.M. Pavelsky, "MERIT Hydro: A High-Resolution Global Hydrography Map Based on Latest Topography Datasets," *Water Resources Research* 55 (2019): 5053–5073, https://doi.org/10.1029/2019WR024873.

61. R. Garvine, "A Steady State Model for Buoyant Surface Plume Hydrodynamics in Coastal Waters," *Tellus* 34, no. 3 (1982): 293–306, https://doi.org/10.3402/tellusa.v34i3.10813.

62. B. Silverman, *Density Estimation for Statistics and Data Analysis* (New York: Chapman & Hall, 1986), 175 pp., ISBN: 0412246201.

63. SPSS (Statistical Product and Service Solutions), *The SPSS Two-Step Cluster Component: A Scalable Component Enabling More Efficient Customer Segmentation, White Paper-Technical Report*, SCWP-0101 (Chicago: SPSS, 2001).

64. L. Van der Maaten and G. Hinton. "Visualizing Data Using t-SNE," *Journal of Machine Learning Research* 9, no. 86 (2008): 2579–2605.

65. M. Sonnewald, S. Dutkiewicz, C. Hill, and G. Forget, "Elucidating Ecological Complexity: Unsupervised Learning Determines Global Marine Eco-provinces," *Science Advances* 6 no. 22 (May 2020): eaay4740, https://doi.org/10.1126/sciadv.aay4740.

66. M. Norusis, SPSS 16.0 *Guide to Data Analysis* (Englewood Cliffs, New Jersey: Prentice Hall, 2008), ISBN 13: 9780136061366.

67. B. Thom, I. Eliot, M. Eliot, N. Harvey, D. Rissik, C. Sharples, A. Short, and C. Woodroffe, "National Sediment Compartment Framework for Australian Coastal Management," *Ocean and Coastal Management* 154 (2018): 103–120, https://doi.org/10.1016/j.ocecoaman.2018.01.00.

## Photo credits

Horseshoe Bay Beach, Bermuda, courtesy of Photographs by Joules (photographsbyjoules.com).

*Hic Sunt Dracones*—Latin for *Here Be Dragons*, from text on the circa1504 Hunt-Lenox Globe held at the New York Public Library. Graphic: Creative Commons CC0 1.0 Universal Public Domain.

Malibu, California, using DJI Mavic Pro drone, by Jason Collin (jasoncollinphotography.com).

Portland Head, Maine, by Rapidfire, CC BY-SA 3.0.

Sentinel Island, India, by Jesse Allen, using NASA Earth Observatory image.

The CSIRO *Sprightly* research vessel, courtesy of CSIRO, Creative Commons Attribution 3.0 Unported (CC BY 3.0) license [creativecommons.org].

Dover, England, by Immanuel Giel. CC BY-SA 3.0.

Jeju Island, South Korea, by Marcella Astrid. CC BY-SA 3.0.

Samoa, by the U.S. National Park Service (public domain)—National Park of American Samoa.

Asilomar State Beach, California, by docentjoyce. CC BY 2.0.

Hawaii Island, NOAA. Photographer: Kevin Lino, NOAA/NMFS/PIFSC.

Geba River estuary, Guinea-Bissau, contains modified Copernicus Sentinel data [2018], processed by Pierre Markuse.

## Acknowledgments

The authors appreciate the helpful reviews of Virginia Burkett and Sara Zeigler of the U.S. Geological Survey. Any use of trade, firm, or product names is for descriptive purposes only and does not imply endorsement by the U.S. Government.

# PREVENTING SPECIES EXTINCTIONS

Our natural world is under increasing threat from incompatible land use, pollution, and climate change. Scientists at NatureServe are using advanced GIS modeling approaches to map suitable habitat for imperiled species at high resolution, giving decision-makers the information they need to reduce conflicts between wildlife and humans and target conservation measures.

By Healy Hamilton and Regan Smyth, **NatureServe**

*Species from the Map of Biodiversity Importance. (See the photo credit section on page 41 for species names.)*

# PIONEERING THE USE OF DATA TO DOCUMENT BIODIVERSITY LOSS

Human transformation of our environment is causing widespread biodiversity loss. Destruction and degradation of natural habitats, overexploitation, pollution of terrestrial, freshwater, and marine ecosystems, and climate change are contributing to levels of extinctions not seen since an asteroid struck Earth 66 million years ago. Fraying the diverse web of life threatens our own well-being. Disease outbreaks, food and water insecurity, and catastrophic impacts of extreme climate events are just a few of the consequences. The footprint of human activities is now so pervasive that the fate of nearly all species and ecosystems hinges on our stewardship of the natural world.

A fundamental approach to managing biodiversity is to understand which individual species are most imperiled and then work to prevent their extinction. Many countries have codified the imperative of preventing extinction through endangered species legislation and policy. Effective species conservation requires three basic kinds of information: 1) taxonomic information that allows us to identify and communicate about a species, 2) distribution data that describe where a species occurs, and 3) conservation status information that describes the degree to which a species is imperiled.

The development and application of this foundational data to drive science-based biodiversity conservation is the mission and function of NatureServe, the world's first biodiversity information network. NatureServe was formed in the United States almost 50 years ago with the establishment of a network of programs to document the composition, distribution, and conservation status of rare and imperiled species and ecosystems. NatureServe is the hub of this network of biodiversity information programs that today operate in every U.S. state and Canadian province. The use of shared data standards and information management systems across the network has resulted in the most comprehensive database ever created describing which species are imperiled and where they are found. These data are used every day to inform and implement conservation plans, management actions, regulatory decisions, and public policy.

Since its inception, NatureServe has consistently applied new technology to biodiversity data management. Systems for managing shared spatial and tabular data, now in their eighth generation, allow seamless aggregation of observations recorded by field botanists, zoologists, and ecologists across state and national boundaries. These continually updated, range-wide datasets form the basis for generating the basic inputs for conservation action.

As technology has advanced, so has the power of NatureServe network data to inform biodiversity management and conservation decisions. This capability is nowhere more evident than with spatial biodiversity analyses that use the rapidly evolving capacity of GIS technology. From assessing strengths and gaps in our biodiversity observation data, developing species-habitat relationships, identifying the land management agencies with the greatest responsibility for biodiversity at risk, or evaluating the degree to which protected areas support imperiled species, GIS enables NatureServe to develop and deliver spatially explicit information to advance conservation.

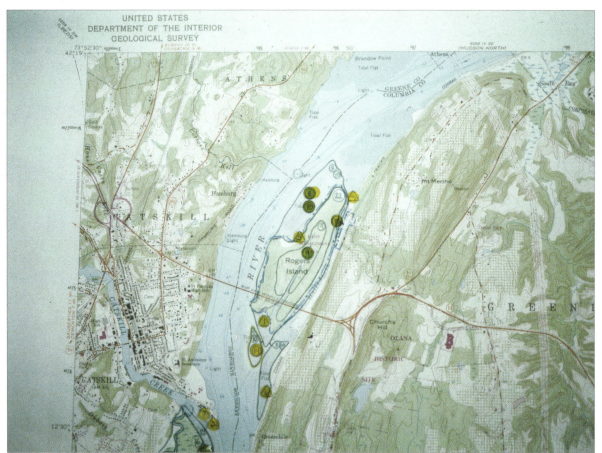

*Early efforts to map biodiversity used USGS topographic maps and hand drawn species occurrences.*
*Credit: New York Natural Heritage Program*

*Former NatureServe Chief Botanist Larry Morse computerizing biodiversity records with punch cards in 1975.*

# DISTRIBUTION DATA FOR BIODIVERSITY CONSERVATION

Transportation and energy infrastructure, forest harvests, grazing, pesticide use, and residential and commercial development take place in specific locations. A major goal of conservation is to ensure that these activities are carried out in areas away from the habitats where imperiled species occur. The more precisely the species distribution is known, the better we can avoid conflicts between human activities and species' needs. However, the complete distribution of many species is often challenging to determine with precision, especially for rare species. They can be hard to detect or identify, occur in remote areas, or undergo seasonal movements, making it challenging to document their distributions.

Traditionally, conservation practitioners have used two approaches to compiling distribution information. Field surveys conducted by qualified botanists and zoologists that produce confirmed records of species presence (and, sometimes, confirmed absence) offer the highest degree of spatial accuracy and confidence. But comprehensive biodiversity inventory is an immense challenge that has never been fully realized. The time and resources required for field surveys, the numbers of species involved, the inaccessibility of most privately owned land, and the vastness of natural landscapes combine to limit the numbers of records available for most species. As a result, field-based observations will almost always significantly underestimate species distributions.

The alternatives most often employed are coarse estimates of species ranges. For example, many U.S. Fish and Wildlife Service (FWS) range maps of threatened and endangered species include the counties where species are known or suspected to occur. Another example is coarse-scale range maps developed by experts by encircling the outer boundaries around known localities. These types of maps almost always overestimate species distributions, often encompassing significant area of unoccupied habitat (Hurlbert and Jetz 2007). Potential consequences are ill-informed land use decisions and unnecessary conflicts for areas where a species may not exist.

Today, NatureServe integrates GIS with cloud-based computational analyses, machine learning algorithms, and web-based tools for collaborative spatial science

to transform distribution information for the most vulnerable species in the United States. The foundation of this approach is habitat suitability modeling, also called species distribution modeling, an academically mature tool that combines verified records of species occurrences with environmental data layers to identify a suite of conditions that describe a species' habitat requirements (Guisan and Thuiller 2005). Machine learning algorithms then identify where else that suite of conditions occurs on the landscape. The output map is a probability distribution, with higher values indicating the closest match to the conditions at localities where the species is verified to occur.

NatureServe applies these tools to produce a dramatic refinement in our understanding of imperiled species distributions. For decades, NatureServe's network of botanists, zoologists, and ecologists have conducted field surveys using standardized approaches for documenting species occurrences and identifying the places and conditions most important for their persistence. Shared data management systems support the aggregation of these data into a national database of verified locations for a wide range of species, from ferns to frogs, mussels to mammals, and pollinators to vascular plants. This same network of experts also reviews model outputs, evaluating where they under- or overpredict, allowing models to be adjusted to produce more precise estimates of imperiled species distributions.

An example of the power of habitat models can be seen in the graphic below for the golden-cheeked warbler (*Setophaga chrysoparia*). This declining neotropical migratory bird is designated as endangered under the U.S. Endangered Species Act (ESA) and is considered globally imperiled (G2) by NatureServe's conservation status assessment. Distribution data provided by the FWS, which administers the ESA for terrestrial and freshwater plants and animals, shows that the species occurs across a swath of counties in central Texas. The Texas Department of Transportation (DOT) must consider this listed species for potential impacts from any planned road construction projects in these counties. But the habitat for this beautiful little songbird is not likely to occur throughout all these counties.

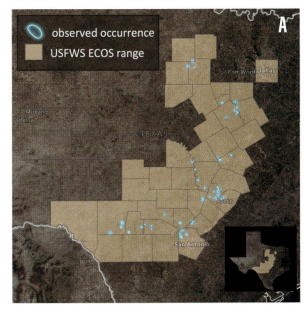
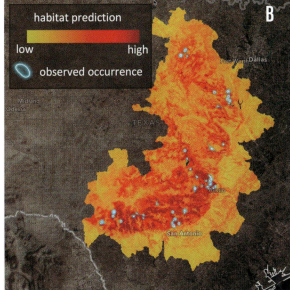
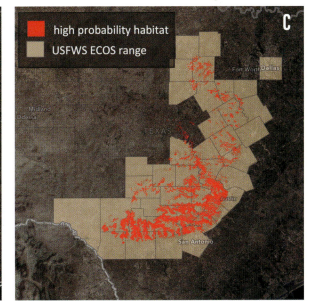

*Varying depictions of the distribution of the golden-cheeked warbler in Texas. Map A, county distribution provided by the FWS. Map B, modeled habitat probability. Map C, areas of highest probability where the warbler is most likely to occur.*

A habitat suitability model, based on verified warbler occurrences, offers a transparent, reproducible, high-resolution map of likely suitable habitat for the golden-cheeked warbler. This refined map can help direct survey efforts, as well as avoidance and mitigation actions in the places where warbler habitat overlaps with road construction. This map also greatly reduces the area where road construction and important warbler habitat are thought to overlap.

NatureServe has embarked on a major initiative to develop refined habitat models for species at risk, beginning with more than 2,200 of the most imperiled plants, pollinators, freshwater fish and invertebrates, birds, mammals, amphibians, and reptiles in the lower 48 United States. In 2020, NatureServe released the Map of Biodiversity Importance (MoBI) in ArcGIS Living Atlas of the World, a GIS analysis of the distributions of these species to identify the places that matter most for conserving our national natural heritage. In producing MoBI, NatureServe and its network of natural heritage programs also created an efficient, cloud-based engine for generating, collaboratively evaluating, revising, and sharing refined distribution maps for rare and imperiled species.

*The golden-cheeked warbler is the only bird species found in the U.S. whose population nests entirely in the state of Texas.*

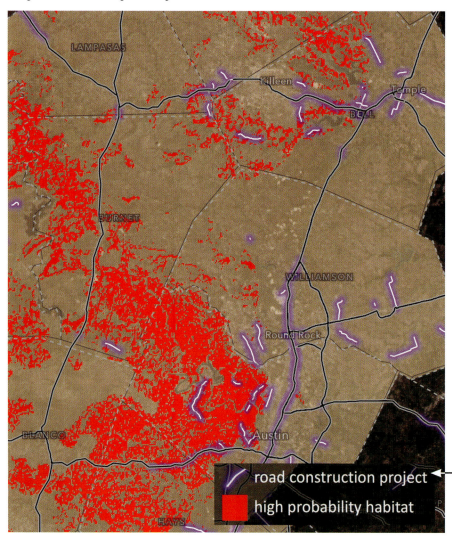

road construction project

high probability habitat

*Use of a habitat model greatly reduces the areas where road construction (purple lines) may interfere with golden-cheeked warbler habitat (red) compared with a county-level range map (tan polygons).*

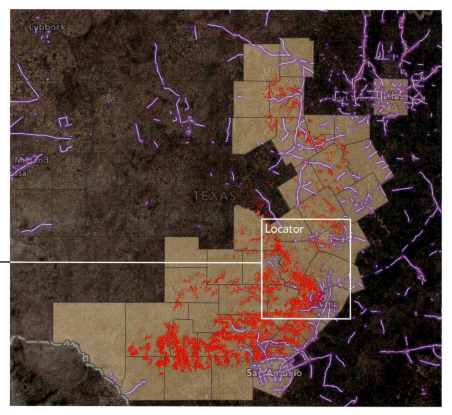

# MODELING HABITATS

With foundations in the mid-20th century, predictive approaches to developing habitat maps are not new. However, the science of habitat suitability modeling has matured, with advances in modeling algorithms and the increasing availability of fine-scale data characterizing the environmental factors to which species respond (Sofaer et al. 2019). The question today is not "Are good habitat models possible and useful?" but "How can habitat modeling efforts be brought to scale so that products are available and understandable to managers and decision-makers?"

To fulfill its mission to provide high-quality biodiversity location information for conservation actions, NatureServe has brought habitat modeling to scale using its accumulated inventory data and the biological and modeling expertise of its scientists. This approach moves habitat modeling out of the academic sphere and into more practical applications. Initial predictions of species habitat can be readily generated through a dynamic process and then, through review and iteration, refined over time. Through this process, the information available to guide decision-making is constantly improving.

## Habitat modeling process

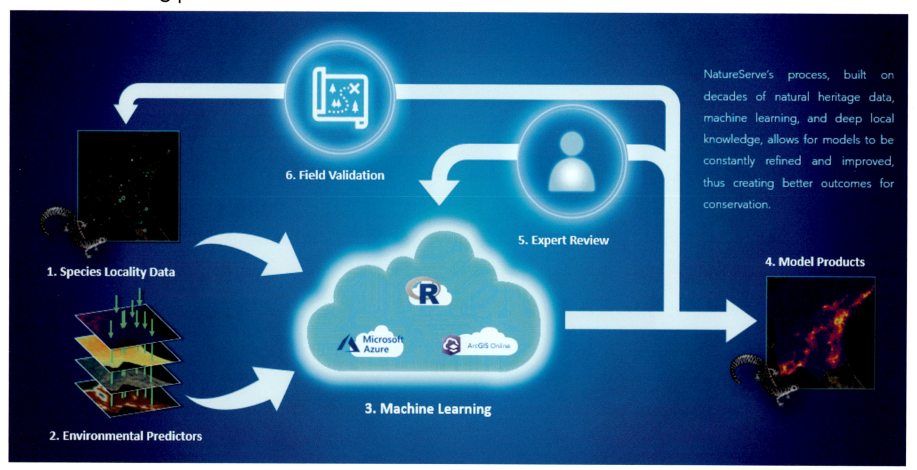

NatureServe's process, built on decades of natural heritage data, machine learning, and deep local knowledge, allows for models to be constantly refined and improved, thus creating better outcomes for conservation.

**1. Species Locality Data**

**2. Environmental Predictors**

**6. Field Validation**

**3. Machine Learning**

**5. Expert Review**

**4. Model Products**

1. **Species locality data**—*Records from nearly 50 years of biological inventory conducted by state Natural Heritage Program field biologists provide the expert verified occurrence data needed to train models. These data are often supplemented with occurrence data from additional sources, including natural history museum and citizen science records.*

2. **Environmental predictors**—*NatureServe has compiled a national library with more than 200 terrestrial and freshwater spatial environmental predictors on soils, climate, landform, land cover, hydrology, and other environmental factors. These high-quality environmental data layers make high-resolution modeling possible.*

3. **Machine learning**—*Using machine learning processes in a Microsoft cloud-based AI modeling engine, researchers process terabytes of data to intersect species occurrence data with environmental predictors. The goal is to characterize the relationship between these variables, build predictive models, and generate maps of potential habitat.*

4. **Model products**—*Predictive modeling products include continuous habitat suitability predictions from 0 (low–in purple) to 1 (high–in yellow) and classified maps distinguishing high, medium, and low probability habitat. Metadata provide information on environmental factors influencing the prediction, model performance, and appropriate uses for end products.*

5. **Expert review**—*Each model is uploaded into a GIS-based web application that allows experts familiar with the species to view the predictions and assess how well the models performed. Reviewers' comments help refine the models and communicate confidence in the predictions.*

6. **Field validation**—*Model products are also used to guide sampling and inventory efforts, providing new data that can be fed back into an iterative modeling process to improve products.*

# GOOD MODELS REQUIRE GOOD TRAINING DATA

The increased availability of remotely sensed data has advanced our understanding of the physical and biological processes that govern life on our planet. But making sense of that influx of data requires accurate and vetted field observations. The success of predictive models depends on the availability and quality of model training data in the form of verified species occurrences.

The primary source for training data used in the modeling effort presented here is NatureServe's biodiversity location database. Starting with circles drawn on topographic quad maps in the 1970s and progressing to today's sophisticated GPS-enabled field data collection systems, NatureServe network biologists have compiled millions of records representing areas of conservation significance for rare and imperiled species. These records are centrally managed in a web-enabled biodiversity information management system, which uses a unified taxonomy and consistent application of shared data standards and methodology. Range-wide point and polygon data are easy to extract for modeling.

NatureServe's biodiversity location data provide a foundation for building habitat models in the MoBI project. For species with insufficient data for modeling, we supplemented records from specimen and citizen science portals and academic researchers. (See GISforScience.com for the full list of data-gathering organizations employed.) Data from citizen science sources such as iNaturalist and eBird require careful screening for appropriate use as model training data, because they can have biases in where people record observations, large locational uncertainty, or misidentifications.

The quantity and quality of species occurrence records is one of the most important inputs into the modeling process. Model refinements may be made by obtaining species locality data from a wider variety of sources, developing methods for better filtering of citizen science data, and using the models themselves to guide additional field data collection. The latter is a particularly powerful means for improving model outcomes; model-guided field inventory is used to validate results and generate new presence and absence data that can be fed into an iterative model refinement process.

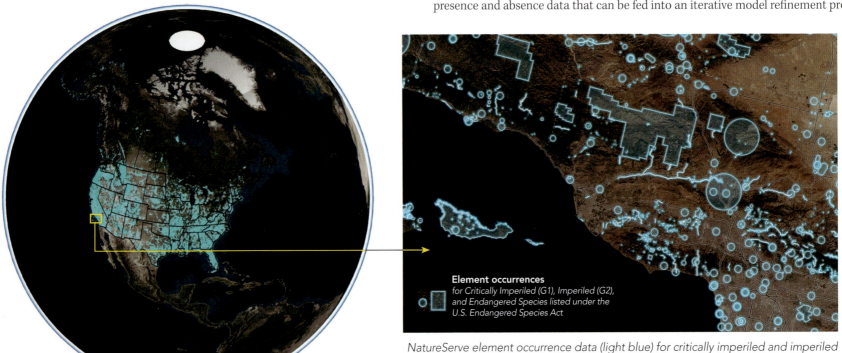

**Element occurrences**
*for Critically Imperiled (G1), Imperiled (G2), and Endangered Species listed under the U.S. Endangered Species Act*

*NatureServe element occurrence data (light blue) for critically imperiled and imperiled species, and other species listed as threatened or endangered under the ESA.*

## Application data case study: Golden-cheeked warbler

*The accuracy and relevance of species locality data used to train models of habitat suitability are an important determinant of modeling success. This map shows research-grade observations for the golden-cheeked warbler collected by citizen scientists using the iNaturalist application (yellow markers with black dots) in the area around Lake Travis, northwest of Austin, Texas. Although these sightings are likely to be legitimate observations of this easy-to-identify species, many records are located in suburban environments or over water where individuals may have flown by, but these locations may not represent areas important to the persistence of the species. Using these records to train a model could contribute to erroneous predictions. Element occurrence records collected by the Texas Natural Diversity Database are shown in blue, which represent areas of significance for the persistence of warbler populations. Drawing model training samples from within these areas increases the likelihood of successful modeling outcomes. Citizen science data can be valuable for modeling, but for the reasons illustrated here, they require careful screening for use as inputs to species habitat models.*

# IMPROVING MODELS THROUGH COLLABORATION

The machine learning tools used to produce predictive models are powerful. However, AI methods are not foolproof. When a child studying the American Revolution asks a smart device whether the patriots won, they just might be informed that the National Football League team lost 20 to 14 in overtime. When a modeler asks a machine learning algorithm where a species is likely to occur the results may be similarly misconstrued. Sometimes the algorithm passes internal tests very well (the Patriots lost!), but outside reviewers may know better.

To address this issue, NatureServe developed an interactive online Model Review Tool to facilitate expert evaluation of species habitat models developed for MoBI and other conservation applications. The information received allows modelers to identify where machine learning methods have fallen short and gain insight that can support iterative model refinement. These tools help bring together the promise of machine learning and the expertise of scientists who understand these species and their habitats.

The Model Review Tool is a web application built on an ArcGIS Online platform that allows species experts to view model outputs and associated metadata and provide feedback on how well the model performs. Reviewers can import their own data to evaluate the results, indicate areas where results are questionable, provide comments on how the model might be improved, and rate model performance.

The Model Review Tool for the MoBI project helped determine which models performed well, which required adjustments, and for which species we needed to pursue an alternative approach to mapping habitat. The reviews also provide insight about the promise of these predictive modeling approaches, including lessons about which groups of species resulted in the most defensible models. Models for plants tended to perform well, which is unsurprising given their lack of mobility and simpler relationships to the environmental factors represented in the predictor data. Successfully modeling pollinators proved to be more difficult, partly reflecting the complicated relationship between pollinators and their food plants.

Successful modeling of habitat for many pollinators may depend on first developing models for nectar plants. Rarer species, and those with small range sizes, tend to model best, reflecting the geophysical limitations often driving their rarity. While habitat modeling has many applications, these findings highlight its promising application for management of imperiled species.

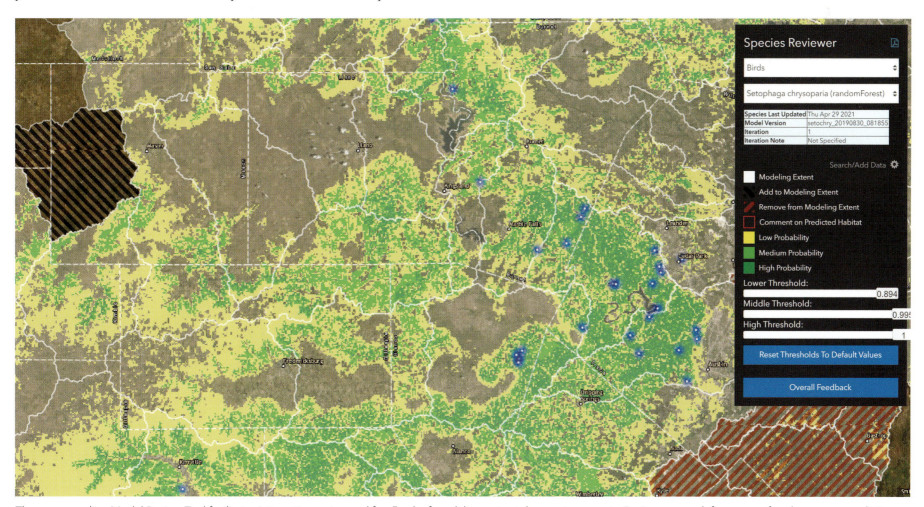

*The secure, online Model Review Tool facilitates interactive review and feedback of modeling outputs by species experts. Reviewers can define areas of under- or overprediction, adjust thresholds, drop their own occurrence records onto the model spatial extent, and provide written feedback. The review tool is a critical part of NatureServe's iterative modeling process that aims to continually improve and refine distribution data for species at risk.*

# THE MAP OF BIODIVERSITY IMPORTANCE

Individual habitat models can be used in many practical ways to advance species management and conservation, including guiding field inventories, identifying partners in shared stewardship of a species, avoiding conflicts in land use planning, and informing the sustainability practices of agriculture and forestry. When analyzed in aggregate, as for the MoBI project, habitat models can reveal hot spots of species imperilment, offering new insights into opportunities for conservation.

MoBI focused on species from an unprecedented diversity of taxonomic groups (vertebrates, vascular plants, freshwater mussels and crayfish, and pollinating bumblebees, butterflies, and skippers) that face a high degree of imperilment. The inclusion of invertebrates and plants beyond trees is significant, as they have previously been excluded from continental conservation prioritizations due to a lack of comprehensive data. MoBI is actually a series of maps, with different products designed to meet different information needs:

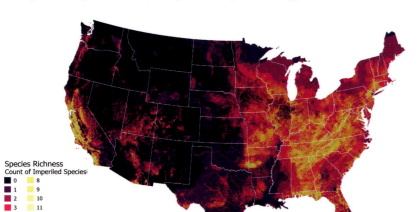

**Species Richness**
Count of Imperiled Species
| | |
|---|---|
| 0 | 8 |
| 1 | 9 |
| 2 | 10 |
| 3 | 11 |
| 4 | 12 |
| 5 | 13 |
| 6 | 14 |
| 7 | ≥ 15 (max = 31) |

**Species richness:** *Species richness is simply a count of the number of imperiled species with modeled habitat in a grid cell. Although intuitive, it can be misleading when used to identify areas of high conservation significance. By definition, the rarest species have small distributions and are thus unlikely to co-occur with other imperiled species. Species richness can identify where at-risk species are on the landscape, but high values do not necessarily correspond with areas where conservation need is greatest.*

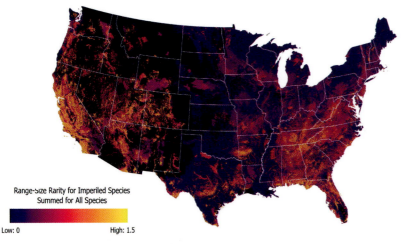

Range-Size Rarity for Imperiled Species
Summed for All Species

Low: 0      High: 1.5

**Range-size rarity:** *Range-size rarity (RSR) considers the overall extent of a species' range. This is a common metric in spatial conservation analyses. RSR recognizes the inverse relationship between range size and opportunities for conservation interventions. Each species is assigned an RSR score equal to the inverse of the total area mapped as suitable habitat, and those scores are summed for each species occurring in a grid cell. Higher values identify areas where species with small ranges (and thus fewer conservation opportunities) are likely to occur; the presence of multiple restricted-range species contributes to higher scores.*

Protection-Weighted Range-Size Rarity
for Imperiled Species
Summed for All Species

Low: 0      High: 1.5

Protected Areas Managed for Biodiversity
(Gap Status 1 & 2 in USGS PAD-US 2.0)

**Protection-weighted range-size rarity:** *Protection-weighted range-size rarity (PWRSR) maps combine information on range-size rarity and the degree to which habitat for the species is protected. We defined protected habitat as that occurring within protected areas managed for biodiversity (in other words, GAP Status 1 and 2 lands in the USGS Protected Areas Database; PAD-US 2.0). Each species was assigned a PWRSR score equal to the product of RSR and the percent of habitat that is unprotected. The PWRSR raster sums these scores for all species with habitat in a cell to identify areas of greatest conservation need. Higher scores depict areas where narrow-range imperiled species fall outside current protected areas; it is a road map to preventing extinctions.*

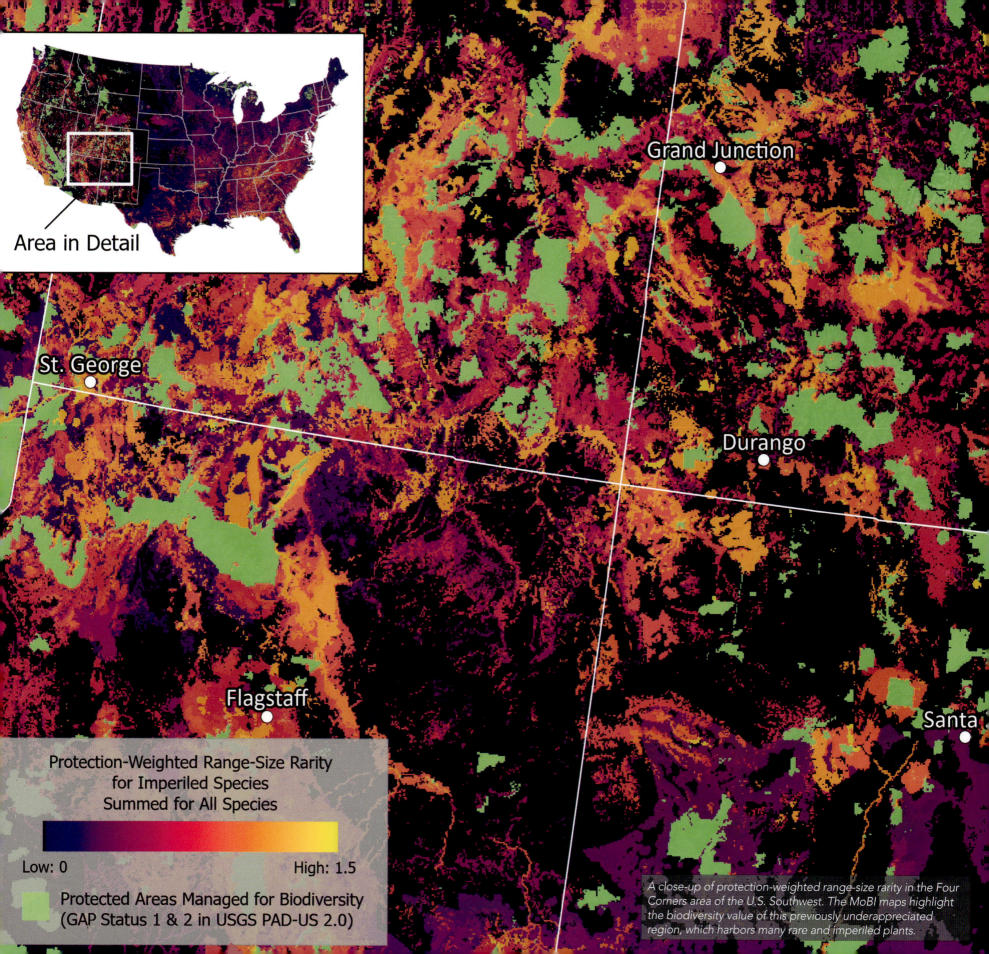

Grand Junction

St. George

Durango

Flagstaff

Santa

## Protection-Weighted Range-Size Rarity
### for Imperiled Species
### Summed for All Species

Low: 0                                    High: 1.5

Protected Areas Managed for Biodiversity
(GAP Status 1 & 2 in USGS PAD-US 2.0)

Area in Detail

*A close-up of protection-weighted range-size rarity in the Four Corners area of the U.S. Southwest. The MoBI maps highlight the biodiversity value of this previously underappreciated region, which harbors many rare and imperiled plants.*

The MoBI PWRSR map reconfirms some areas already recognized as conservation priorities, such as the California Floristic Province, Southeastern Coastal Plain, and Southern Appalachian Mountains. But the fine resolution of the input data and inclusion of an unprecedented diversity of taxa provides many novel insights. Smaller areas of ecological significance are revealed that can be lost in continental prioritizations: the Edwards Plateau in Texas, the Lake Wales Ridge in Florida, the Ouachita Mountains of Arkansas, and the White Mountains of New Hampshire are just a few examples (Hamilton, Smyth, Young et al., in press). The results show that areas of biodiversity significance are often local in nature. When we "thresholded" the map to identify the highest-scoring locations—places we call Areas of Unprotected Biodiversity Significance, or AUBIs, those locations were well dispersed across the country, occurring in almost every state.

Examining the results broken down by taxonomic groups helps reveal the drivers of the patterns observed, and can also help tailor effective conservation strategies. High scores on the Colorado Plateau and in California are driven largely by a great diversity of imperiled plants found in these topographically and ecologically diverse landscapes. High values in the Southeast are due in part to concentrations of imperiled freshwater species. Regional assessments consistently identify the southeastern United States as a hot spot of imperiled aquatic biodiversity (Elkins et al. 2019). But previous assessments have been based on species ranges mapped to watersheds, coarse geographic scales that are often challenging to translate into management and conservation actions on the ground. MoBI, for the first time, provides a fine-resolution picture of the freshwater reaches most critical for maintaining the globally significant freshwater biodiversity of the southeastern United States (see map on facing page).

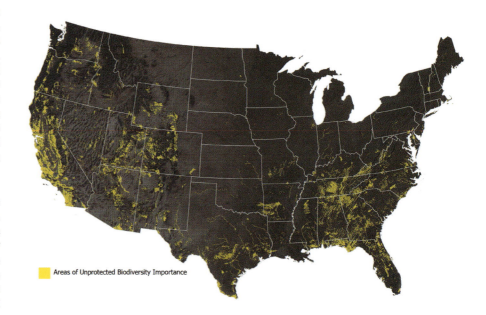

Areas of Unprotected Biodiversity Importance

*Areas of Unprotected Biodiversity Importance (AUBIs), where modeled habitat for imperiled species falls outside GAP Status 1 & 2 protected areas. From Hamilton, Smyth, Young et al. (in press).*

## Anatomy of the PWRSR score

Golden-cheeked warbler
Range-size: 40,898
Percent Protected: 2%
PWRSR: 0.00002

Tricolored bat
Range-size: 3,589,468 km²
Percent Protected: 2%
PWRSR: 0.0000003

Canyon mock orange
Range-size: 2,523 km²
Percent Protected: 11%
PWRSR: 0.0003

Bracted twistflower
Range-size: 570 km²
Percent Protected: 7%
PWRSR: 0.0016

Jollyville Plateau salamander
Range-size: 351 km²
Percent Protected: 10%
PWRSR: 0.0026

Barton Springs salamander
Range-size: 51 km²
Percent Protected: 20%
PWRSR: 0.015

Summed PWRSR: 0.0203

*What does a PWRSR score mean on the ground? Consider a single 990 m pixel in the Texas Hill Country. At this location southwest of Austin, with a rapidly growing population and little land area set aside for biodiversity conservation, there is predicted habitat for six imperiled species, including the golden-cheeked warbler. All have 20% or less of their habitat overlapping current protected areas. The tricolored bat (Perimyotis subflavus) is found across the eastern United States and is imperiled primarily due to disease. Because of its large range, it contributes little to the PWRSR score at this (or any) location. But this pixel also contains habitat for four species found only in the limestone hills of south-central Texas. The critically imperiled bracted twistflower (Stepptanthus bracteatus) has already seen several populations lost to housing developments. The Jollyville Plateau salamander (Eurycea tonkawae) and Barton Springs salamander (Eurycea sosorum) have extremely small ranges. Their survival is threatened by urban development and groundwater depletion. The combination of small range sizes and lack of protection for these species leads to a high PWRSR score at this location.*

*A view of the Texas Hill Country from a rural road in Hays County.*

# STEWARDSHIP RESPONSIBILITIES FOR CONSERVING IMPERILED SPECIES

Effective conservation requires an understanding of land ownership. Who has responsibility for conserving imperiled species, and where should land managers focus their efforts to prevent species extinctions? The MoBI maps provide biodiversity data for spatial overlays with land ownership that can guide stewardship of imperiled species on public and private lands and support more efficient and effective compliance with state and federal endangered species regulations.

Analysis of species distribution by land ownership category reveals patterns that inform strategies for protecting biodiversity. Reinforcing the findings of previous research (Jenkins et al. 2015), the extent to which private lands hold the fate of our nation's natural heritage emerges as a striking result. In all taxonomic groupings, more at-risk species have the greater part of their modeled distribution on private lands than in any other stewardship class.

This finding is overwhelmingly true for freshwater invertebrates, represented by crayfish and mussels. MoBI is the first analysis to produce comprehensive, high-resolution distribution maps for all imperiled species in these freshwater invertebrate groups. These often overlooked and poorly studied animals play essential roles in freshwater ecosystem services, purifying water and forming important links in aquatic food webs that support economically valuable recreational and commercial fisheries. The southeastern United States has globally significant diversity of these freshwater species, with richness that rivals that of tropical aquatic systems. Yet they are also the most imperiled group of organisms in the United States and have already suffered many extinctions.

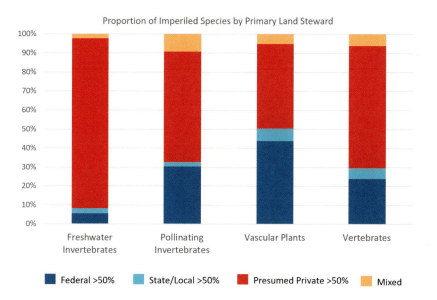

Conservation responsibility for species in major taxonomic groups. The graph shows the relative count of imperiled species in each group for which 50% or more of the species' modeled habitat overlaps federal, state, local, or privately owned lands.

The Center for Biological Diversity publishes regular videos on YouTube that make animals such as this freshwater mussel, the Arkansas fatmucket (Lampsilis powellii), more understandable to people. While not as cute and cuddly as, say, a black-footed ferret, these "out of sight, out of mind" animals are essential to healthy freshwater ecosystems and have experienced more documented extinctions than any other species group in the United States. (Video linked at GISforScience.com.)

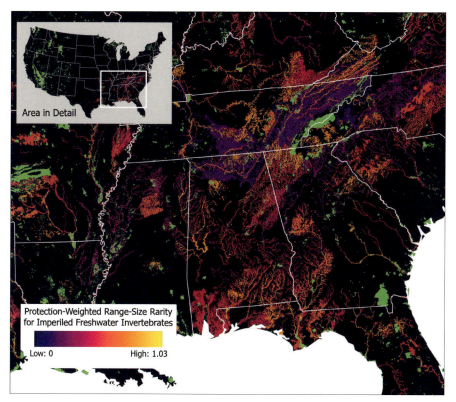

The MoBI analysis of PWRSR for freshwater invertebrates highlights stream reaches in the southeastern United States, shows where imperiled, range-restricted, unprotected mussel and crayfish species have likely suitable habitat. The analysis provides unprecedented spatial resolution to conservation efforts for these economically and ecologically important groups.

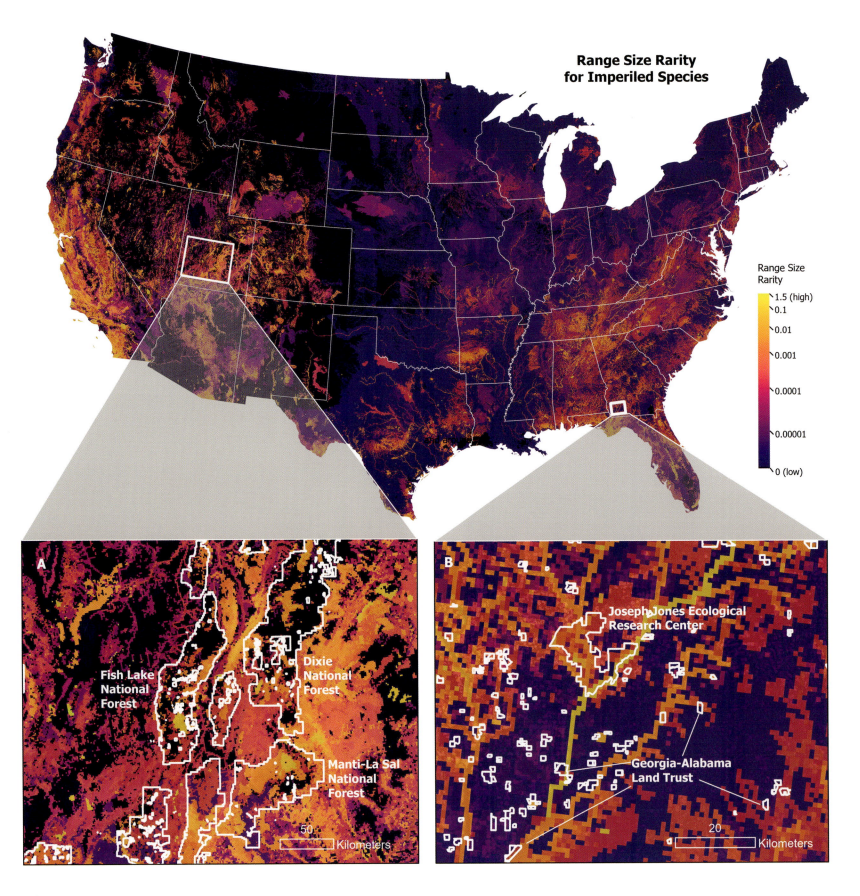

**Range Size Rarity
for Imperiled Species**

Range Size
Rarity

1.5 (high)
0.1
0.01
0.001

0.0001

0.00001

0 (low)

**A**

Fish Lake
National
Forest

Dixie
National
Forest

Manti-La Sal
National
Forest

50
Kilometers

**B**

Joseph Jones Ecological
Research Center

Georgia-Alabama
Land Trust

20
Kilometers

# CASE STUDY: FEDERAL LAND MANAGEMENT

The federal Bureau of Land Management (BLM), whose mission is "to sustain the health, diversity, and productivity of public lands for the use and enjoyment of present and future generations," manages almost 250 million acres, more than 10% of the surface lands of the United States, the vast majority of which are in 13 western states. These lands are largely designated as multiple use, which charges the agency with balancing grazing, mineral and fossil fuel extraction, hunting, fishing, and off-road vehicle use with the conservation of species and their habitats. To do so efficiently and effectively, the BLM must understand with high spatial precision where potential conflicts exist among these diverse resource management goals.

New Mexico is a case in point. Recent years have seen an enormous increase in requests for oil and gas leases in areas of the state that are also known to harbor rare and imperiled plant species whose geographic distributions are poorly known. BLM staff review these requests, balancing the need for national energy production against the importance of preventing decline of native ecosystems and the species that inhabit them.

To address this challenge, the New Mexico office of the BLM worked with NatureServe and the New Mexico Natural Heritage Program to develop habitat models for six plant species that could be impacted by oil and gas development, including a rare water-willow (*Justicia wrightii*), a federally endangered milkvetch (*Astragalus gypsodes*), and a critically imperiled species of flax (*Linum allredii*) that was only described as a new species in 2011. Existing distribution information for the plants was fragmentary, but it was known that their habitat might overlap with specific areas targeted for new oil and gas development.

Field surveys that confirm the presence or absence of the species are the most verifiable tool for decision-making—but they can also be time-consuming, needle-in-a-haystack efforts across vast desert expanses. Using documented observations for these species, NatureServe developed habitat models and used statistically defined thresholds to identify areas with low, medium, and high probability of suitable habitat. Using these 30 m resolution maps as a guide, botanists from the New Mexico Natural Heritage Program designed a structured field survey protocol and went on an imperiled plant species treasure hunt.

For several of these species, the hunt was successful in finding new, previously undocumented populations. Presence and absence data recorded during these field surveys were used together with refined environmental predictor inputs, including detailed soils maps, to develop a second-generation set of models for these rare plants.

This example highlights perhaps the most important and useful feature of habitat modeling as a tool for supporting species management and conservation. Beginning with available information on confirmed occurrence of a target species and high-resolution digital data on variables such as soils, climate, and topography, the modeling process produces a spatially explicit, testable hypothesis of the probability of suitable habitat. Choices regarding model parameters can be recorded in the metadata, so the process is transparent and reproducible. The model results can be used to guide field biologists to efficiently survey the areas of highest probability of suitable habitat. The results of field surveys in turn provide new presence and absence data for the next round of models, resulting in increased precision in model outputs. Approached this way, habitat modeling is a continuous, dynamic process, and the distribution information that guides decisions about the nation's most imperiled species represents the best available science.

*Justicia wrightii*

Documented occurences

*The imperiled* Justicia wrightii, *commonly referred to as Wright's water-willow, is known to occur only in a few locations in desert grasslands and shrublands of southeastern New Mexico and west-central Texas. Oil and gas development is extensive where it occurs on BLM lands in New Mexico.*

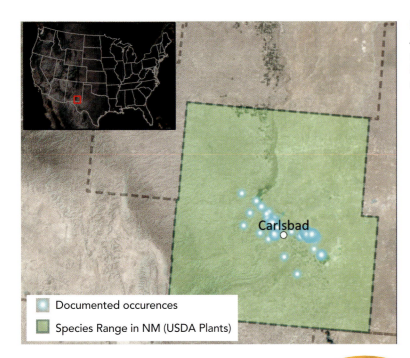

*Prior to habitat modeling efforts, information on the distribution of Justicia wrightii in New Mexico was limited to broad-range maps or field observations, like those maintained by New Mexico Natural Heritage. Without adequate information on the species distribution, assessing conservation status or mitigating threats was difficult.*

Documented occurences

Species Range in NM (USDA Plants)

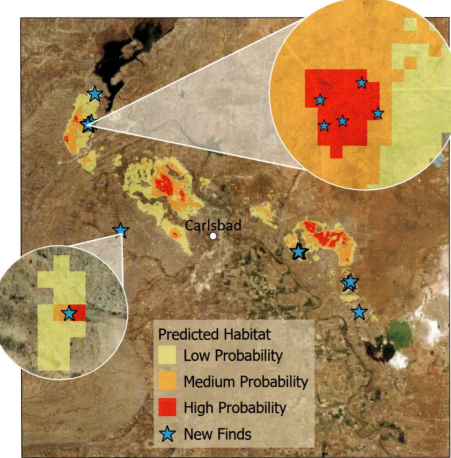

Predicted Habitat

Low Probability

Medium Probability

High Probability

★ New Finds

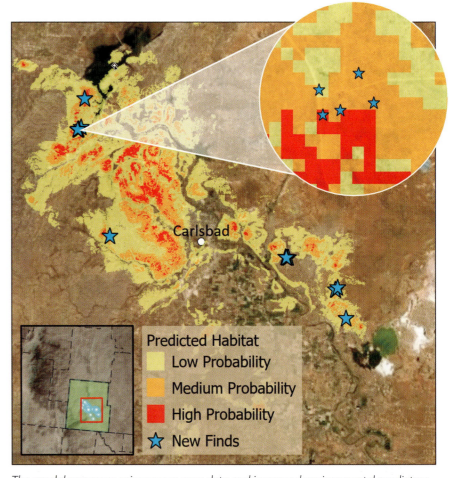

Predicted Habitat

Low Probability

Medium Probability

High Probability

★ New Finds

*To better understand where Justicia wrightii may occur, an initial habitat model was run based on readily available species locality data and a collection of national environmental predictor data. Field crews used the results to systematically survey for unknown populations; several new populations were identified.*

*The model was rerun using new survey data and improved environmental predictors, including highly refined local soil layers. The result is a significantly refined habitat map that highlights the areas most likely to support the species. With this new information, BLM is better able to better focus management.*

# CONCLUSION

Habitat suitability modeling has been deeply explored in academic literature for several decades. Yet still, more than 20 years into the 21st century, we are routinely using coarse, imprecise, incomplete, and often subjective information to answer one of the most important and consequential questions for the management and conservation of biodiversity: Where do species and their habitats occur? We have shown that this type of modeling is a mature science that, when combined with GIS analysis and tools to support a transparent, collaborative, and iterative process, can dramatically improve our understanding of the distribution of most species in ways that reduce conflict, save time and money, increase transparency in decision-making, and improve conservation outcomes.

The Map of Biodiversity Importance is a product and a process that can guide conservation in this biodiverse country. As a map product, MoBI provides an unprecedented view of hot spots of species imperilment for the most diverse suite of species yet examined. The MoBI data layers—freely available on the ArcGIS Living Atlas—allow viewers to zoom to their land trust, national forest, wildlife refuge, park, prospective conservation easement, wildlife corridor, development project, or any other location where better, higher-resolution information about the potential distribution of species of high conservation concern can benefit decision-making. In producing MoBI, we have also created a process for developing improved species distribution maps that is iterative and dynamic, and responsive to information changes. Vetted models can guide field inventory, which yields new data that feed new models in a continuous process of refinement. In a world committed to ongoing global change, this dynamic integration of technology, natural history data, and expertise is an important step forward for conserving the diversity of life on which our own well-being depends.

# REFERENCES

Elkins, D., S.C. Sweat, B.R. Kuhajda A.L. George, K.S. Hill, and S.J. Wenger. 2019. "Illuminating Hotspots of Imperiled Aquatic Biodiversity in the Southeastern U.S." *Global Ecology and Conservation* 19:e00654.

Graf, D.L. and K. S. Cummings, Review of the Systematics and Global Diversity of Freshwater Mussel Species (Bivalvia: Unionoida), *Journal of Molluscan Studies*, Volume 73, Issue 4, November 2007, Pages 291–314

Guisan, A., and W. Thuiller. 2005. "Predicting Species Distribution: Offering More Than Simple Habitat Models." *Ecology Letters* 8: 993–1009.

Hamilton, H., R. Smyth, B.E. Young, T.G. Howard, C. Tracey, S. Breyer, D.R. Cameron, A. Chazal, A.K. Conley, C. Frye, and C. Schloss. "Increasing Taxonomic Diversity and Spatial Resolution Clarifies Opportunities for Protecting Imperiled Species in the U.S." *Ecological Applications*, in press.

Hurlbert, A.H., and W. Jetz. 2007. "Species Richness, Hotspots, and the Scale Dependence of Range Maps in Ecology and Conservation." *Proceedings of the National Academy of Sciences* USA 104:13384–13389.

Jenkins, C.N., K. S. Van Houtan, S. L. Pimm, and J. O. Sexton. 2015. "U.S. Protected Lands Mismatch Biodiversity Priorities." *Proceedings of the National Academy of Sciences USA* 112:5081–5086.

Sofaer, H.R., C.S. Jarnevich, I.S. Pearse, R.L. Smyth, S. Auer, G.L. Cook, T.C. Edwards, Jr., G.F. Guala, T.G. Howard, J.T. Morisette, and H.H. Hamilton. 2019. "Development and Delivery of Species Distribution Models to Inform Decision-Making." *BioScience* 69:544–557.

Taylor, C.A., G. A. Schuster, J. E. Cooper, R. J. DiStefano, A. G. Eversole, P. Hamr, H. H. Hobbs III, C. E. Skelton, and R. F. Thoma. 2007. A Reassessment of the Conservation Status of Crayfishes in the United States and Canada after 10+ Years of Increased Awareness. *Fisheries* 32:372-389.

Referenced data sources: Biodiversity Data Serving Our Nation [BISON]; Butterflies and Moths of North America [BAMONA], Integrated Digitized Biocollections [iDigBio]; iNaturalist; Symbiota Collections of Arthropods Network [SCAN]; L. Richardson, unpublished bumble bee locality data.

| | | | | | |
|---|---|---|---|---|---|
| Fireback Crayfish, Dale Jackson | Florida scrub-jay, Mike Carlo, USFWS | Dakota skipper, Jessica Petersen | Barton Springs salamander, Ryan Hagerty, USFWS | Rusty-patched bumble bee, Jessica Petersen | Bonytail chub, Brian Gratwicke |
| Blair's fencing crayfish, Dustin Lynch, AR Natural Heritage Comm. | Golden-cheeked warbler, USFWS | Venus flytrap, Misty Buchanan, NC Natural Heritage Program | Agassiz's desert tortoise, Beth Jackson, USFWS | Bracted twistflower, Steven Schwartzman, Lady Bird Johnson Wildflower Center | Kirtland's warbler, Joel Trick, USFWS |
| Yellowcheek darter, Dustin Lynch, AR Natural Heritage Comm. | Clubshell, Craig Stihler, USFWS | California tiger Salamander, Robert Fletcher, Ohlone Preserve | Caddo mountain salamander, Dustin Lynch | Eastern Massasauga, Richard Staffen | Black-footed ferret, Kimberly Fraser, USDA |
| Canyon mock orange, Melody Lytle, Lady Bird Johnson Wildflower Center | Boston Mountains crayfish, Dustin Lynch, AR Natural Heritage Comm. | Chiricahua leopard frog, Jim Rorabaugh, USFWS | Jollyville Plateau salamander, Piershendrie | Tricolored bat, Missouri Department of Conservation | Maguire's primrose, Larry England, USFWS |

Page 28-29 photos

*Species common name, Photographer*

# MAPPING HALF-EARTH

Spatial biodiversity knowledge is vital for effective conservation planning. The Half-Earth Project has created a comprehensive map of our planet's biodiversity to inform and track conservation efforts and ensure that no species is driven to extinction from lack of knowledge.

By D. Scott Rinnan, **Yale University;** Greta C. Vega, Estefanía Casal, Camellia Williams, **Vizzuality;** and Joel Johnson and Chris Heltne, **E.O. Wilson Biodiversity Foundation**

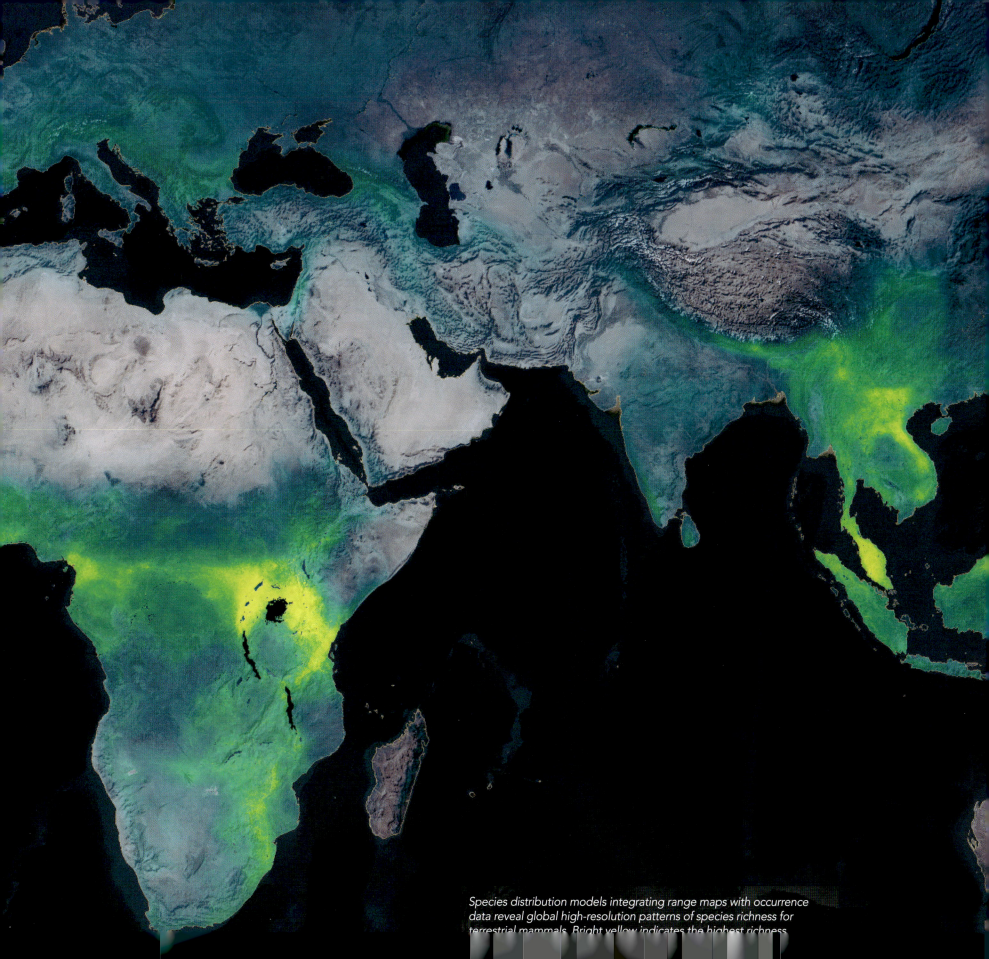

Species distribution models integrating range maps with occurrence data reveal global high-resolution patterns of species richness for terrestrial mammals. Bright yellow indicates the highest richness.

# THE NEED FOR SPATIAL BIODIVERSITY DATA

"We are drowning in information, while starving for wisdom. The world henceforth will be run by synthesizers, people able to put together the right information at the right time, think critically about it, and make important choices wisely."

—Edward O. Wilson

When people are introduced to the concept of Half-Earth, two questions invariably arise: "Why half?" and "Which half?" The answer to the first question is derived from the principles of island biogeography, which explain the relationship between the amount of habitat and the number of species that habitat can sustain. The curve predicted by this relationship indicates that if we protect half the land and sea, we can safeguard the bulk of the biodiversity. At its core, Half-Earth is about raising our conservation ambition to an easily understood goal that inspires collaborative action and addresses the urgent need for comprehensive, global biodiversity protection to sustain the health of our planet.

Identifying which half will sustainably support the bulk of biodiversity requires a clear picture of how and where species are distributed across the globe, their habitat needs, how they move, and how they depend on one another. These are some of the most fundamental questions in ecology, but for all the simplicity in stating them, the answers remain astonishingly elusive. By some estimates, more than 85% of the planet's eukaryotic species (in other words, plants, animals, fungi, and protists) have yet to be scientifically described. Moreover, for only a small percentage of species described so far do we have the most rudimentary descriptions of their spatial distributions.

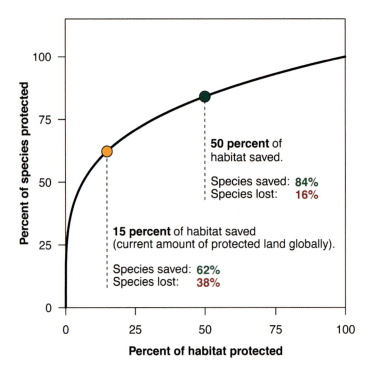

The island biogeographic principles behind the Half-Earth concept. Protecting 15% of the land could be enough to sustain roughly 60% of terrestrial species, whereas protecting 50% of the land could sustain the bulk of biodiversity.

Paradoxically, the large gaps in spatial and taxonomic coverage have persisted through the recent explosion of available spatial biodiversity data that resulted from the proliferation and increasing accessibility of citizen science tools coupled with the ubiquity of portable technology. This proliferation is partly due to the underlying sampling effort of the data itself: in 2020, the Global Biodiversity Information Facility (GBIF) added almost 28 million user-contributed species occurrence records, but some geographic locations and taxa are much better represented in these data than others. Central Park in New York City, for example, contained almost 3 times the number of reported observations as the entirety of Madagascar, and almost 10 times the number of animal observations. What is needed, then, is not simply more data, but rather a systematic approach to cataloguing our planet's life and synthesizing available data into knowledge useful for guiding conservation decisions.

Spatial occurrence data of plant (red) and animal (blue) species contributed to GBIF in 2020. Central Park in Manhattan (left) contains 9,104 unique observations in an area of only 3.4 km², compared with just 3,846 unique observations in the entirety of Madagascar (right), with an area of 587,041 km².

# THE HALF-EARTH PROJECT

Inspired by E.O. Wilson's sweeping call to action in his book *Half-Earth: Our Planet's Fight for Life*, the E.O. Wilson Biodiversity Foundation launched the Half-Earth Project© in 2016. Together with its partners, the Half-Earth Project is driving research to better understand the species of our planet and how they interact with their ecosystems. The project provides conservation management leadership by mapping biodiversity, identifying the best opportunities to protect the most species and engaging with people globally to care for our planet, with the goal of protecting Earth's biodiversity.

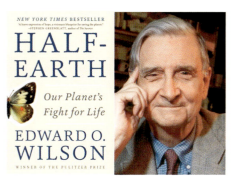

A pillar of the Half-Earth Project is the *Half-Earth Project Map*, a tool for scientific communication and planning that is collaboratively designed and maintained by four core organizations. Map of Life—the flagship project of Yale University's Center for Biodiversity and Global Change—leads the scientific research, contributing the information needed for informed conservation planning. Vizzuality—a company of scientists, developers, and data visualization specialists—leads the user-centered design and development aspects, along with Esri, which provides additional cartographic basemap design, spatial analysis, and data management functionality. And finally, the E.O. Wilson Biodiversity Foundation provides the leadership and vision for the map and other programs such as educational initiatives and the annual Half-Earth Day, bringing the focus and voice of E.O. Wilson to this endeavor.

"The foundation for a new way of understanding the beautiful intricacy of our planet and how we can best steward its enduring stability is science," says Dr. Paula Ehrlich, president and CEO of the E.O. Wilson Biodiversity Foundation. "When E.O. Wilson conceived of Half-Earth, he imagined that we would bring together our scholarship in many walks of life, many areas of expertise and experience, and work together within the spirit of a moonshot. He imagined that by driving significant scientific

*Science teacher and Half-Earth Project Educator Ambassador Lucretia Smith leads a group of middle-school students in a biodiversity mapping exercise.*

innovation, we would provide leadership regarding the most effective path forward for protection of endangered species and endangered ecosystems."

As Edward O. Wilson noted, the Half-Earth solution does not place biodiversity protection at odds with human activity. Rather, Half-Earth reminds us that if we lose species, we lose the ecosystems that sustain nature and sustain us as part of nature. Effective global conservation strategies will necessarily comprise many approaches and strategies tailored to the needs of different people, landscapes, activities, and interests.

The science of the Half-Earth Project places species as the core unit of conservation concern. "Species are the absolute key in all of this," says Dr. Walter Jetz, scientific chair of the E.O. Wilson Biodiversity Foundation and lead principal investigator of Map of Life. "They are the critical elements underpinning the ecosystems that constitute our landscapes. They're the nodes on this very intricate web of life that are ultimately behind nature's benefits to people." Ensuring that species are represented in our characterizations of the planet's biodiversity is a necessary first step in safeguarding them from extinction. The Half-Earth Project tracks conservation progress at the species level and aggregates this information to identify places where additional conservation actions will best contribute to the preservation of biodiversity. One of this project's primary goals is to provide a globally and taxonomically comprehensive mapping of species distributions for use in conservation planning.

*The* Half-Earth Project Map, *an interactive tool for exploring global biodiversity data, priority areas for conservation, and various biodiversity indicators.*

# COLLECTING SPECIES DATA

With more than 1,200 terrestrial vertebrate species (including 22 that are uniquely endemic), Mozambique is historically rich in biodiversity. But from 1977 to 1992, a civil war ravaged the country, destroying critical infrastructure, killing more than a million people through fighting and starvation, and taking a toll on species populations as well. Inside Gorongosa National Park—the jewel of Mozambique's incredible landscape—populations of large animals decreased by 90% as people turned to hunting bushmeat to survive. Fortunately, Mozambique's wild places remained relatively intact.

After the war ended, the Gorongosa Restoration Project began its efforts to restore the park's biodiversity, led by a team of local conservationists and philanthropist Greg Carr. The E.O. Wilson Biodiversity Laboratory at Gorongosa National Park was established to train a new cadre of Mozambican biologists and conservationists to support restoration efforts and to carry out comprehensive surveys of biological diversity in the park. The laboratory is directed by Dr. Piotr Naskrecki, entomologist, conservation biologist, author, photographer, and Half-Earth Chair, based at the Museum of Comparative Zoology at Harvard University.

*Gorongosa National Park in Mozambique is one of Africa's most ambitious wildlife restoration stories.*

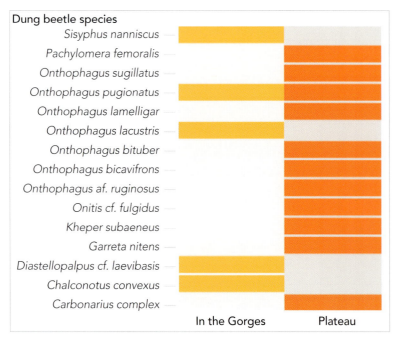

*The sampled composition of dung beetle species between the gorge and its adjacent plateau was almost disjoint, with only one species found in common between both habitats. Many of the individuals gathered have not yet been identified to the species level.*

*Dung beetles in Gorongosa National Park.*

Today, Gorongosa National Park is a spectacular 6,700 km² preserve located at the southern end of the Great East African Rift Valley. Gorongosa encompasses almost all types of habitat found in southern Africa, with a variety of microclimates and environments created by unique biogeographical features and a range of altitudes. Fed by rivers originating on Mount Gorongosa, the floodplain of Gorongosa National Park supports some of the densest wildlife populations in all of Africa, including charismatic carnivores, herbivores, and 475 bird species. Populations of many species of large mammals such as waterbuck (*Kobus ellipsiprymnus*), impala (*Aepyceros melampus*), and sable antelope (*Hippotragus niger*) have either returned to or exceeded pre-war levels, and other species such as the African bush elephant (*Loxodonta africana*) and African buffalo (*Syncerus caffer*) are quickly approaching them. Other species have been successfully reintroduced, such as African wild dogs (*Lycaon pictus*). Yet there is much still to be learned about many of the park's lesser-known species that build and support the ecosystems in which these high-profile examples of charismatic megafauna reside.

In March 2013, an expedition led by Dr. Naskrecki set out to conduct a survey of the Nhagutua Gorge (accessible only by helicopter), with a focus on detecting differences in species composition across an elevational gradient in smaller species not detectable by aerial surveys. Equipped with Sherman traps, mist nets, and pitfall traps, Dr. Naskrecki and his team gathered samples of rodents, bats, and dung beetles. The dung beetles alone comprised hundreds of specimens belonging to 12 separate genera, illustrating the astonishing diversity that individual regions can hold.

This single population survey exemplifies the extraordinary amount of coordination and effort to collect species data and the diverse historical contexts that surround each datum. Mapping the planet's biodiversity is only possible because of the blood, sweat, and tears of the countless thousands of individuals who have dedicated their time in gathering data, observations, and samples in the field, and make that information available to others. You can read more about the E.O. Wilson Biodiversity Laboratory and Gorongosa's community-based natural resource management efforts in this volume's online resources at GISforScience.com.

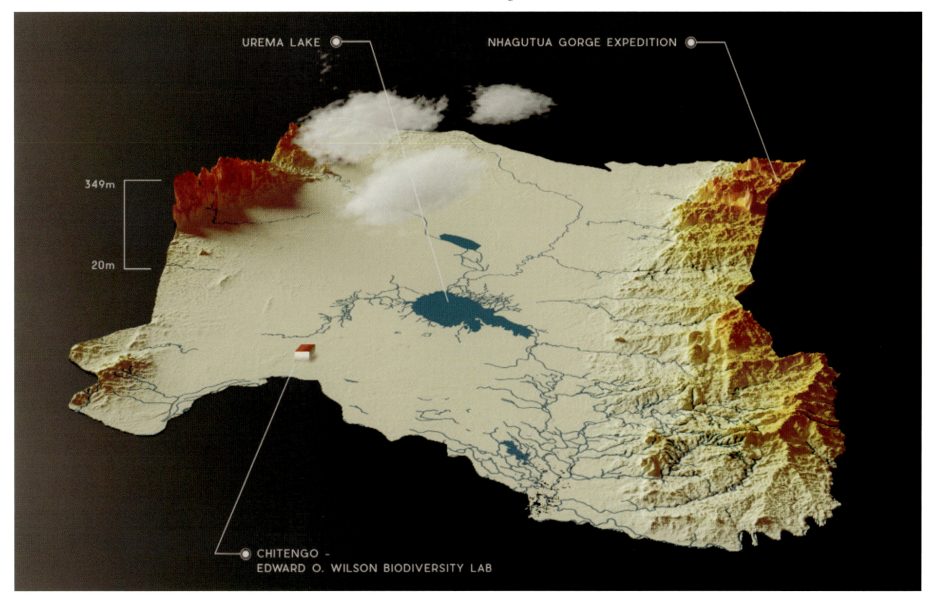

UREMA LAKE

NHAGUTUA GORGE EXPEDITION

349m

20m

CHITENGO - EDWARD O. WILSON BIODIVERSITY LAB

*Located at the southern end of the Great East African Rift Valley in the heart of central Mozambique in Southeast Africa, Gorongosa National Park includes more than 4,000 km² (1,500 sq mi) of protected park on the valley floor and parts of surrounding plateaus. Rivers flowing from nearby Mount Gorongosa irrigate the plain.*

# AGGREGATING AND INTEGRATING DATA

Spatial biodiversity data comes in many forms: reported observations of individuals from citizen scientists and wilderness enthusiasts through popular apps such as Map of Life and iNaturalist; presence/absence records of individuals from scientific surveys such as Dr. Naskrecki's work in Gorongosa; museum records; range maps that delineate general habitat preferences of species; inventory lists of a geographic area generated by "BioBlitz" activities; and larger regional checklists, often at the national level. Each of these data varieties vary in spatial accuracy, the amount and type of information that can be gleaned from them, and the value and utility of this information to inform conservation. Furthermore, within each of these categories, biodiversity data can vary in its availability (bird enthusiasts outnumber ant enthusiasts by at least an order of magnitude), amount (in New York City versus Madagascar), confidence (was that a chipping sparrow or an American tree sparrow?), accuracy (how accurate are my GPS coordinates?), and precision (did I see two chipping sparrows, or was it the same one twice?).

It's not surprising, then, to speak of the challenges of aggregating these disparate data types and integrating them to yield an accurate portrait of species distributions in space and time. Data aggregation is a continually ongoing cycle of identifying new data sources and potential collaborators, building partnerships, creating data-sharing agreements, ingesting new data, identifying what is usable, harmonizing taxonomies, updating datasets, and managing databases. This work is the nuts-and-bolts basis of building a comprehensive picture of biodiversity and often slips by underappreciated in the shadow of splashy, high-profile scientific publications that focus on the results of this work. Informatics underpins virtually every aspect of biodiversity research, and it would be folly to understate the importance of its role in facilitating our understanding of the biosphere.

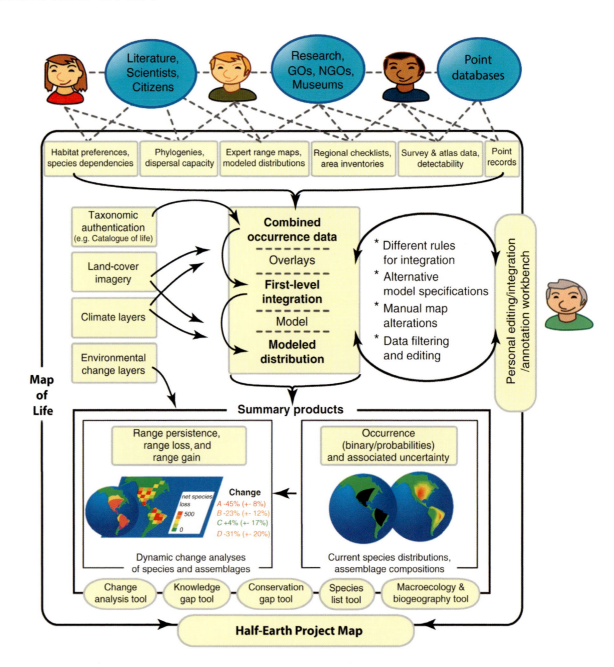

Schematic diagram (adapted from Jetz et al. 2012), showing Map of Life's data integration process. Map of Life facilitates the uploading of species distribution information from many organizations and sources, including data on habitat preferences, point occurrences, and expert range maps. The infrastructure stores this data and provides a workbench for integrating it for one or many species. The data compiled, resulting summary information such as binary and probabilistic occurrence maps, and products from analysis tools, including ArcGIS Pro and ArcGIS API for Python, are then used for various types of modeling. Model outputs are displayed in the Half-Earth Project Map.

## Methodology

As one of the Half-Earth Project's core teams, Map of Life (www.mol.org) leads the biodiversity informatics research that informs the *Half-Earth Project Map*. Map of Life comprises a group of more than 25 data scientists, ecologists, taxonomic experts, postdoctoral associates, technicians, and students led by Jetz, and Robert Guralnick, a scientist at the Florida Museum of Natural History. "Our role in the Half-Earth Project is to deliver the science and information for effective conservation decision-making, to ensure that species are not unknowingly left behind," Jetz said.

Map of Life has developed a complex, sophisticated process for integrating different datasets into a common modeling framework. The outputs of this framework provide a type of space-time-species data array known as an essential biodiversity variable (EBV), in which each datum specifies the occurrence probability of a given species in a given location at a given point of time. These EBVs can be used to infer a variety of ecological patterns such as species richness and change in community composition over time, and are what ultimately provide the high-resolution spatial information needed to guide conservation efforts.

### Three measures of biodiversity

**1** **Species richness** is a measure of the number of different species in a given region. This quantity can be summarized by distinct geographic regions such as countries or protected areas or by equal-area grids to reveal global patterns.

**2** **Species endemism** is the proportion of the distribution of a species found in a given region, summed across all species in that region. This term is also known as **total range-size rarity, rarity score, or weighted endemism**.

**3** **Species rarity** is simply species endemism divided by species richness, and is a measure of average geographic range-restrictedness of species in a given region. This is also known as **average range-size rarity** or simply **range-size rarity**.

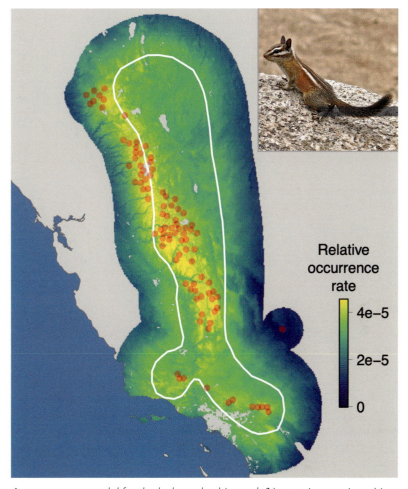

Relative occurrence rate

4e–5

2e–5

0

*An occurrence model for the lodgepole chipmunk (Neotamias speciosus) in California. This model uses Map of Life's infrastructure for data integration to combine range map (white outline) and observation data (orange points) with environmental variables and remotely sensed land-cover products to predict where the lodgepole chipmunk is most likely to occur.*

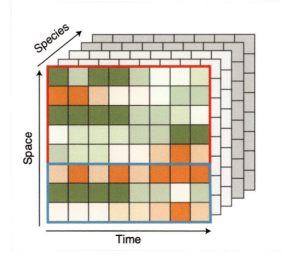

Species · Space · Time

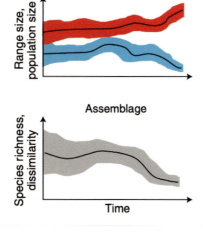

Single species

Range size, population size

Assemblage

Species richness, dissimilarity

Time

*A space-time-species essential biodiversity variable (EBV) of occurrence probabilities. When data is aggregated for single cells, the EBV informs about community change in, for example, species richness or compositional similarity, or—through ancillary data—functional or phylogenetic turnover. Adapted from Jetz et al. (2019).*

# TAXONOMIC REPRESENTATION AND DATA GAPS

To date, the *Half-Earth Project Map* includes global patterns of richness and rarity for all known species of amphibians, birds, mammals, reptiles, cacti, and conifers. These groups are displayed because of their comprehensive representation, i.e., every known species is accounted for. The Half-Earth Project is engaged in expanding the taxonomic coverage of the map to other groups such as ants, bees, butterflies, dragonflies, vascular plants, marine fishes, and crustaceans.

Because the *Half-Earth Project Map* is intended to inform and guide conservation planning, its patterns of richness and rarity must account for any spatial biases that may be present in the species data. At the continental scale, for example, the extent of the genus *Bombus* (bumblebees) is fairly well known, but sizable disparities can be observed when mapping individual species at finer spatial resolutions: while most North American and European distribution data is readily available and comprehensive, Asia and South America are currently relatively data poor and incomplete by comparison. As such, using the higher-resolution but taxonomically incomplete map to guide management decisions may lead to erroneous conclusions about the distribution and importance of bumblebee habitat.

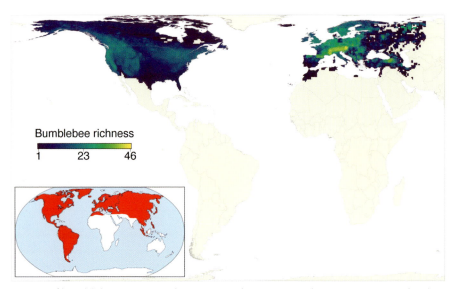

**Bumblebee richness**

1     23     46

*A map of bumblebee species richness in North America and Europe, compared with the global extent of bumblebees (inset). Biodiversity patterns inferred from taxonomically unrepresentative, incomplete, or spatially biased data may lead to erroneous conclusions about the conservation importance and value of different regions.*

Although large, under-sampled portions of the world remain for bumblebees, many other taxonomic groups show more diffuse and scattered data gaps. Many of these gaps are best understood by acknowledging the socioeconomic differences between countries and the political histories associated with different regions. Data coverage maps provide information about where more data and sampling efforts are needed. The key to filling these knowledge gaps resides in international collaboration and capacity-building between organizations linking local communities. Until these gaps are filled, other modeling approaches are needed to map biodiversity patterns that can be used in global planning, such as constructing representative subsets of species that are sampled across genera.

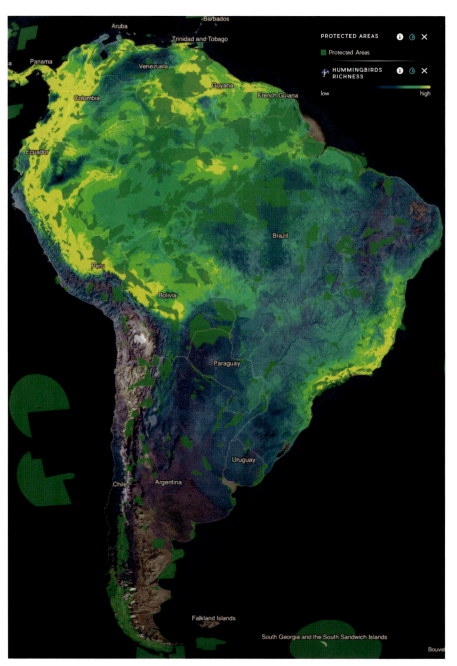

*The* Half-Earth Project Map *is a high-resolution, dynamic world map and decision-support tool that guides where place-based species conservation activities are needed the most to save the bulk of Earth's species, including humans. This view shows taxonomically complete rarity patterns (high rarity in yellow) for the world's 276 known species of hummingbirds, overlaid with currently protected areas (dark green areas).*

# PRIORITY AREAS FOR CONSERVATION: WHICH HALF?

Once equipped with the necessary species data, the project uses spatial conservation planning tools to answer how much habitat is needed for global biodiversity protection and where to direct conservation efforts. Spatial conservation planning describes the process of converting spatial data into a mathematical problem, using an optimization algorithm to solve this problem, and then translating the solution back into a spatial conservation network. When used effectively, solutions adhere to the four specific principles of conservation planning: comprehensive, adequate, representative, efficient solutions (CARE).

## CARE principles of spatial conservation planning

**Comprehensiveness**—Solutions should comprise as many facets of biodiversity as possible (e.g., habitat diversity, species composition, and ecological function).

**Adequacy**—Solutions should ensure the persistence of species through time.

**Representativeness**—Solutions should sample across the full range of variation for each species (e.g., nesting, breeding, and foraging habitat).

**Efficiency**—Solutions should achieve conservation objectives at minimal cost. Cost can reflect acquisition costs, operational costs, total area, opportunities lost (e.g., commercial or industrial activity), sociopolitical values, or any number of other characterizations.

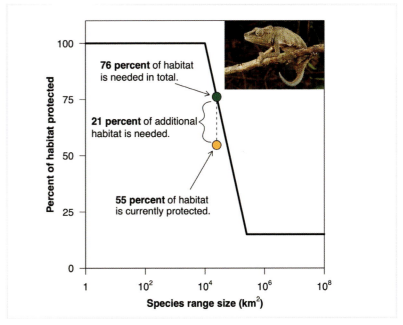

*An example of how species conservation targets are calculated for spatial conservation planning. The Gorongosa pygmy chameleon (Rhampholeon gorongosae) is a species endemic to Mozambique, with a range area of about 25,000 km². Of this range, 55% is already protected, and an additional 21% of its habitat is needed to ensure that its area-based target (black line) is met.*

What amount of habitat is adequate to ensure population persistence? A variety of methods determine the habitat needs of species, but one common method expresses areal conservation targets as a simple function of a species' range size that specifies up to 100% of habitat protected for species with smaller ranges, and 15% of habitat protected for the most common and widespread species. While habitat quantity alone is insufficient to guarantee persistence, it is a necessary baseline condition for species to thrive and a useful proxy that can be inferred for any species with distributional data.

Guided by these CARE principles, the Half-Earth Project employs spatial planning to explore various configurations of the areas needed to achieve the goal of comprehensive biodiversity conservation. Beginning with currently protected regions, these models minimize the amount of additional area needed to meet species conservation targets, while prioritizing intact habitat wherever possible. The *Half-Earth Project Map* features layers in the terrestrial and marine realms that illustrate one possible configuration of a global conservation network, in addition to the supporting layers of human impacts and protected areas. This featured network provides habitat for all species of amphibians, birds, mammals (terrestrial and marine), reptiles, and marine fish.

Once the conservation network is identified, this information is then aggregated in several ways to yield further insights crucial for decision-making. Most critically, the amount of area needed within different political regions, biomes, and ecoregions provides our first estimates of differential conservation needs that reflect the heterogeneous distribution of the planet's biodiversity. These individualized targets exemplify one of the core principles of international conservation policies such as the Convention on Biological Diversity, currently in negotiation for 2020–2030 and beyond.

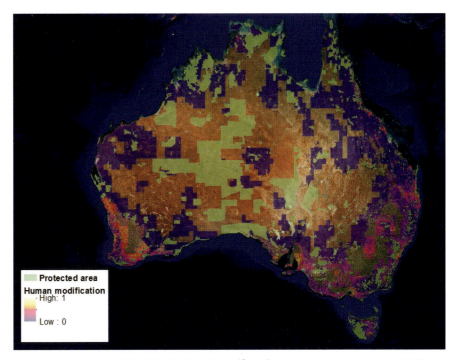

*A conservation network highlighted in the Half-Earth Project Map, comprising 56% of Australia's land, including the 19.5% that is currently protected. This network ensures that species targets are efficiently met while minimizing the amount of human modification in the additional selected areas.*

Likewise, aggregating across species reveals differences in conservation area priorities between taxonomic groups. These patterns can also be explored in the *Half-Earth Project Map*.

## Degree of human modification

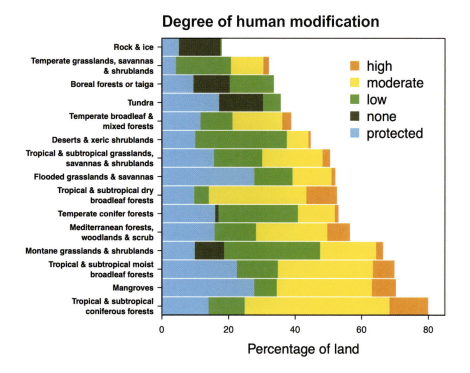

Composition of global conservation network shows the percentage of each biome needed to meet species conservation targets, the amount of human modification contained in the network, and the amount currently protected.

## Takeaways

These results reveal two key takeaways. First, established protected areas do not adequately safeguard global biodiversity to the extent predicted by island biogeographic theory. This result may be discouraging, although not surprising. Because biodiversity positively correlates with resource availability, many protected areas were historically placed in regions that did not inhibit resource exploitation and economic interests, which results in areas with less biodiversity protected. In contrast, the second key takeaway should be quite encouraging: by offering a systematic, strategic approach to global biodiversity conservation, spatial planning can help us drastically outperform the expectations of island biogeography. In our terrestrial example, 47.4% of land was needed to meet conservation targets for all species modeled.

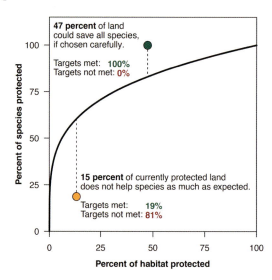

*Spatial conservation planning helps us to outperform the expectations of island biogeographic theory. Although our current protection of the planet's biodiversity is inefficient and insufficient, rapid gains in comprehensive conservation of the biosphere are possible with a strategic, global approach. The results shown here are derived from a model that accounts for all terrestrial vertebrate species.*

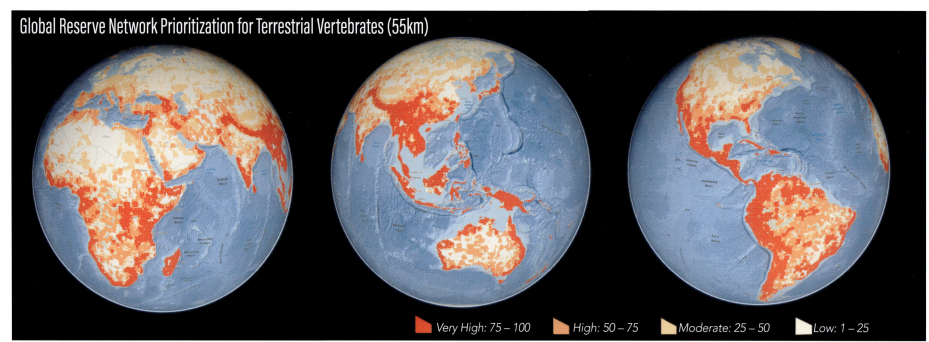

This feature layer shows global priority areas of conservation importance for all terrestrial vertebrate groups. Values are summarized within an equal-area grid, with a grid cell area of ~3,025 km² (approximately 55 km x 55 km in the tropics). This cell size represents the finest resolution at which currently available range map data can be used to accurately infer species presence without further habitat modeling.

# BIODIVERSITY INDICATORS

Although spatial conservation planning identifies the amount of protection needed and where, there is a big difference between identifying a mathematically optimal solution and turning that solution into action through conservation policies and resource management. This process is slow and messy (in the best of times) due to myriad additional considerations that may not have been accounted for in the modeling process, such as budgets, time horizons, conflicts with existing policies, and competing sociopolitical interests. Even with unanimous agreement on a united path toward global biodiversity conservation, it would take years to implement the policy needed to close the gap between where we are at today and what is needed to achieve Half-Earth. Consequently, it's unknown whether the path taken today would accomplish the same goals in the future. To complement spatial planning, we turn to methods for tracking conservation progress through time and global change.

Biodiversity indicators are measurements derived from biodiversity data that enable us to study, report, and manage biodiversity change. One prominent example featured on the Half-Earth Project Map's National Report Cards is the Species Protection Index (SPI). Map of Life developed this metric to quantify the extent of species habitat conserved by protected areas. When measured at the national level, the SPI reflects the average amount of area-based conservation targets met across all indigenous species within a given country in a given year, weighted by the country's stewardship. With a range of 0–100, the SPI is based on the amount and location of currently protected land and the number and location of species found inside and outside the protected areas. An SPI of 100, for example, reflects a country practicing good stewardship and promoting equitable conservation efforts within its borders.

As a biodiversity indicator, the SPI helps ensure that our conservation actions reflect and achieve our conservation goals over time by prioritizing areas where biodiversity protection is most needed. The SPI can be updated regularly to reflect additions to protected area networks and enhancements in our understanding of species distributions. Additionally, the SPI can be aggregated at different spatial scales (e.g., globally or by country) and for different taxonomic groups. For countries with low SPI values, the layers of priority areas for conservation show where efforts can be directed to make the most rapid gains in species protection.

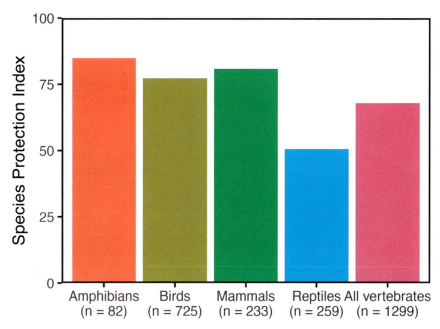

*National SPI of Mozambique aggregated by taxonomic group and calculated for 2019, illustrating disparities in current protection levels between species groups.*

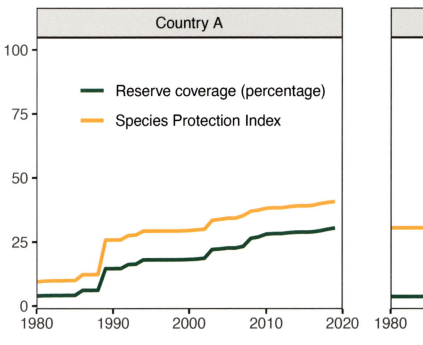

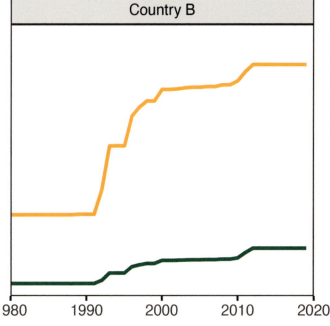

*Two example countries that provide stewardship for a similar number of terrestrial vertebrate species (A = 488, B = 521) and have similar percentages of protected area, yet country B's SPI is much higher than country A's due to the location of its protected areas in areas of biodiversity value and the extent to which they provide habitat for country B's species.*

# NATIONAL REPORT CARDS

This chapter has explored how the Half-Earth Project integrates local scale knowledge into a global portrait of the biosphere and uses this information to coordinate a global conservation strategy. To facilitate local conservation action, however, these results must be interpretable and meaningful at the scale in which policy and decision-making is implemented, which means translating them back into local scale knowledge. The National Report Cards in the *Half-Earth Project Map* summarize various aspects of conservation efforts at the national level. They can be used to explore various national indicators measuring conservation needs and progress and to understand different challenges faced by each country. Once a country is selected on the map for exploration, the rest of the world map falls away and exposes an interactive 3D map.

In addition to details about national SPI values and priority areas for conservation, the report cards feature information about each country's species composition, including downloadable tables of indigenous species and various species-specific metrics such as stewardship and a Species Protection Score (SPS). The species stewardship element of the National Report Card scales up the concept of joint responsibility for a species by considering all of the land vertebrates in each country. Through this approach, it's possible to see the number of countries that share the stewardship of a species. The SPS goes deeper into that concept by providing an assessment of the protection accomplished per species, per country.

The SPS differs from the SPI in that it reflects the level of protection an individual species receives within a given country. An SPS value indicates how close a country is to meeting a species' conservation target relative to the amount of species habitat it has stewardship over. A single species will therefore have a unique SPS for each country that overlaps with its global range. SPS values are presented as ranges (e.g., 75–100) to reflect some of the spatial uncertainty associated with species distributions.

The Challenges panel of the National Report Cards explores the relationships between the SPI and the various sociopolitical and biodiversity indicators of different nations. Scatterplots can be filtered to emphasize similarities between countries and the social challenges they face in ensuring equitable global biodiversity conservation. By grouping countries by their similarities, this feature of the National Report Cards could make it easier for countries to learn from one another and replicate each other's successes. Countries can be filtered by stewardship to reveal the 10 countries with the greatest number of species in common. This capability provides insight into which countries could work together to give to the largest number of species the best level of protection possible. Because individual species are often found in many countries, the entire global population of each species needs protection wherever they are found.

The Ranking panel of the National Report Cards provides a concise overview of each country's species composition, human modification, protection status, and SPI ranking, further facilitating comparisons between countries.

As the National Report Cards continue to be updated regularly, the Half-Earth Project plans to expand the concept to summarize regional areas such as states and provinces.

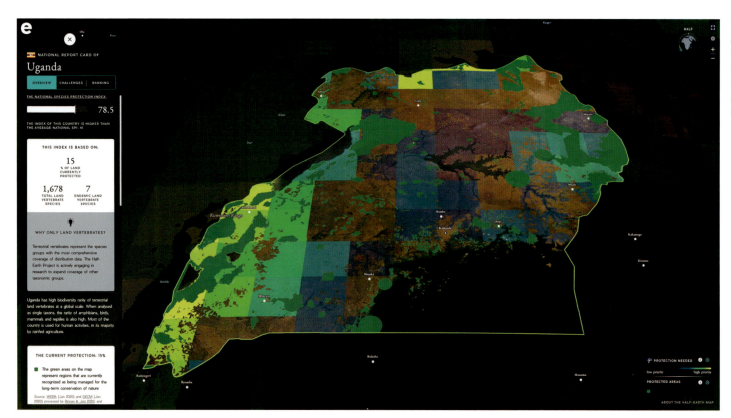

*The National Report Card (Uganda shown here) is a user-friendly interactive dashboard application that presents various indicators measuring conservation status and needs at the national level for every country on the globe.*

This vibrant representation of our planet created as part of the Half-Earth Project had the goal of invoking a reminder that our shared planet is something truly beautiful, unique, and worthy of sound stewardship.

# CONCLUSION

The Half-Earth Project recognizes the urgent need for comprehensive, global biodiversity protection to sustain the health of our planet and uses the best available science to help guide a coordinated global response. With the help of countless biologists and conservationists working around the world, GIS powers a synthesis of information that advances our understanding of biodiversity. This knowledge is used to identify the places best suited to safeguard species, prioritize conservation actions, promote equitable decision-making, and track our conservation progress.

The *Half-Earth Project Map* provides tools for use by international organizations such as UN Environment Programme World Conservation Monitoring Centre, The Group on Earth Observations Biodiversity Observation Network, and The Intergovernmental Science-Policy Platform on Biodiversity and Ecosystem Services to characterize the current state of biodiversity conservation and to facilitate collaborative, coordinated global action plans through mechanisms such as the Convention on Biodiversity. By synthesizing and making more accessible the scientific evidence obtained through the joint effort of many researchers and conservationists, the *Half-Earth Project Map* also provides information to help citizens hold leaders accountable for their promises and push for further action. While ambitious, the goal of Half-Earth is achievable because of our determination to succeed and the science to guide us.

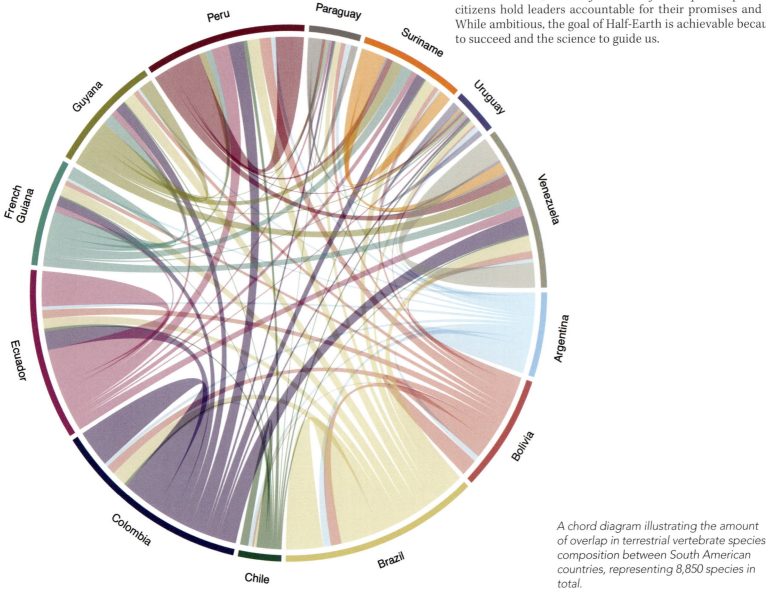

*A chord diagram illustrating the amount of overlap in terrestrial vertebrate species composition between South American countries, representing 8,850 species in total.*

# REFERENCES

GBIF: The Global Biodiversity Information Facility. "What is GBIF?" 2020. Available from https://www.gbif.org/what-is-gbif.

Jetz, Walter, Jana M. McPherson, and Robert P. Guralnick. "Integrating Biodiversity Distribution Knowledge: Toward a Global Map of Life." 2012. *Trends in Ecology & Evolution* 27, no. 3 (March 1): 151–59. https://doi.org/10.1016/j.tree.2011.09.007.

Jetz, Walter, Melodie A. McGeoch, Robert Guralnick, Simon Ferrier, Jan Beck, Mark J. Costello, Miguel Fernandez, et al. 2019. "Essential Biodiversity Variables for Mapping and Monitoring Species Populations." *Nature Ecology & Evolution* 3, no. 4 (April): 539–51. https://doi.org/10.1038/s41559-019-0826-1.

Kennedy, Christina M., James R. Oakleaf, David M. Theobald, Sharon Baruch-Mordo, and Joseph Kiesecker. 2019. "Managing the Middle: a Shift in Conservation Priorities Based on the Global Human Modification Gradient." *Global Change Biology* 25, no. 3: 811–26. https://doi.org/10.1111/gcb.14549.

Kukkala, Aija S., and Atte Moilanen. 2013. "Core Concepts of Spatial Prioritisation in Systematic Conservation Planning." *Biological Reviews* 88, no. 2 (May): 443–64. https://doi.org/10.1111/brv.12008.

Mora, Camilo, Derek P. Tittensor, Sina Adl, Alastair G. B. Simpson, and Boris Worm. 2011. "How Many Species Are There on Earth and In the Ocean?" *PLOS Biology* 9, no. 8 (August 23): e1001127. https://doi.org/10.1371/journal.pbio.1001127.

Possingham, H. P., K. A. Wilson, S. J. Andelman, C. H. Vynne, M. J. Groom, G. K. Meffe, and C. R. Carroll. 2006. "Principles of Conservation Biology." In *Protected Areas: Goals, Limitations and Design*, edited by M.J. Groom, G.K. Meffe, and R. Carroll, 509–552. Sunderland (MA): Sinauer Associates, Inc.

Rinnan, D. Scott, and Walter Jetz 2020. "Terrestrial Conservation Opportunities and Inequities Revealed by Global Multi-scale Prioritization." *bioRxiv*, (February 7). 2020.02.05.936047. https://doi.org/10.1101/2020.02.05.936047.

Rinnan, D. Scott, Gabriel Reygondeau, Jennifer McGowan, Vicky Lam, Rashid Sumaila, Ajay Ranipeta, Kristin Kaschner, Cristina Garilao, William L. Cheung, and Walter Jetz. 2021. "Targeted, Collaborative Biodiversity Conservation in the Global Ocean Can Benefit Fisheries Economies." *bioRxiv*, 2021.04.23.441004. https://doi.org/10.1101/2021.04.23.441004.

Rodrigues, Ana S. L., H. Resit Akçakaya, Sandy J. Andelman, Mohamed I. Bakarr, Luigi Boitani, Thomas M. Brooks, Janice S. Chanson, et al. 2004. "Global Gap Analysis: Priority Regions for Expanding the Global Protected-area Network." *BioScience* 54, no. 12: 1092. https://doi.org/10.1641/0006-3568(2004)054[1092:GGAPRF]2.0.CO;2.

UNEP-WCMC and IUCN. 2020. Protected Planet: The World Database on Protected Areas (WDPA), Cambridge, UK: UNEP-WCMC and IUCN. Available from www.protectedplanet.net. Accessed 2020-12-15.

Williams, Paul H. 1998. "An Annotated Checklist of Bumble Bees with an Analysis of Patterns of Description (Hymenoptera: Apidae, Bombini)." *Bulletin-Natural History Museum Entomology Series* 67: 79–152.

Wilson, Edward O. 2016. *Half-Earth: Our Planet's Fight for Life*. New York: W.W. Norton & Company.

Photo credits: Charles Marsh, Dennis Liu, Piotr Naskrecki, Peter Schoen, and Lisa Tanner.

The authors would like to thank Ajay Ranipeta, Yanina Sica, John Wilshire, and Charles Marsh for their data contributions to this chapter, and Walter Jetz and Paula Ehrlich for their helpful reviews.

Puzzle image by Sergio Cerrato from Pixabay.

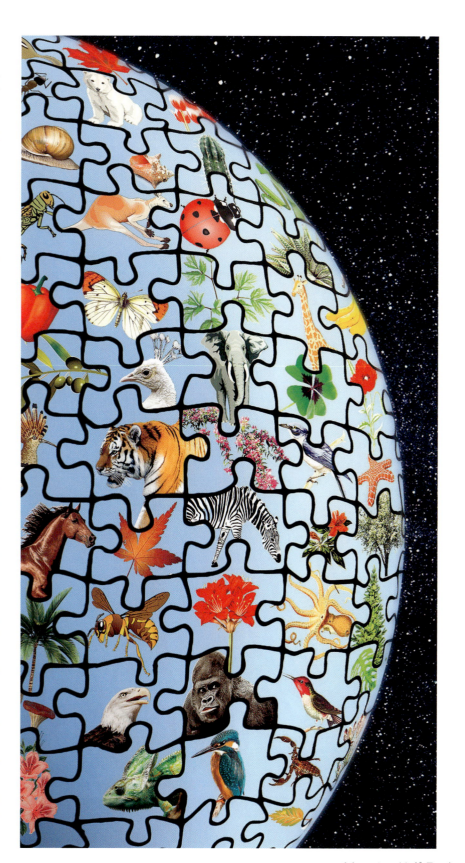

# PART 2
# HOW EARTH LOOKS

How Earth looks is essentially how we as humans change Earth's appearance and function, as illuminated by linkages between natural science and social science, in science partnerships that work across disciplines, geographies, and organizations. Here, we often use GIS interactively and iteratively to create and evaluate alternative (geo)designs to make better decisions, especially with land cover for land-use planning, green infrastructure planning, urban planning, and sustainability science.

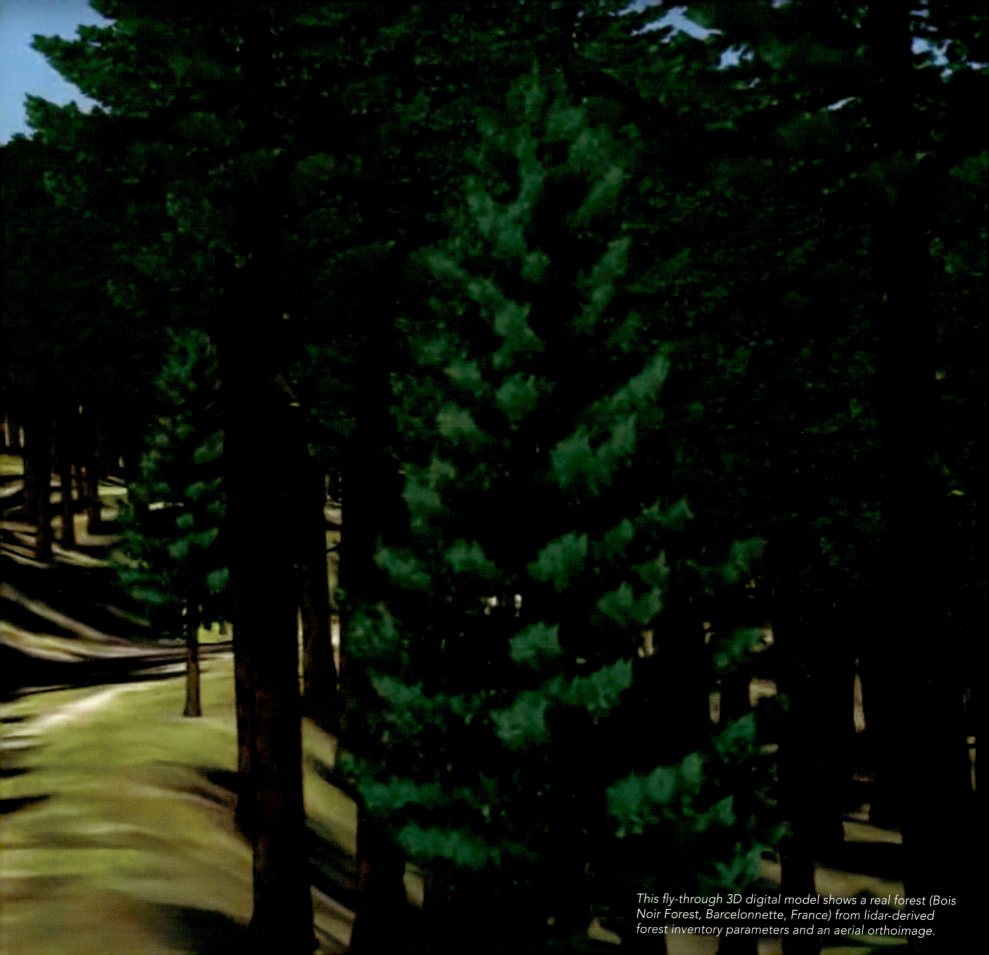

This fly-through 3D digital model shows a real forest (Bois Noir Forest, Barcelonnette, France) from lidar-derived forest inventory parameters and an aerial orthoimage.

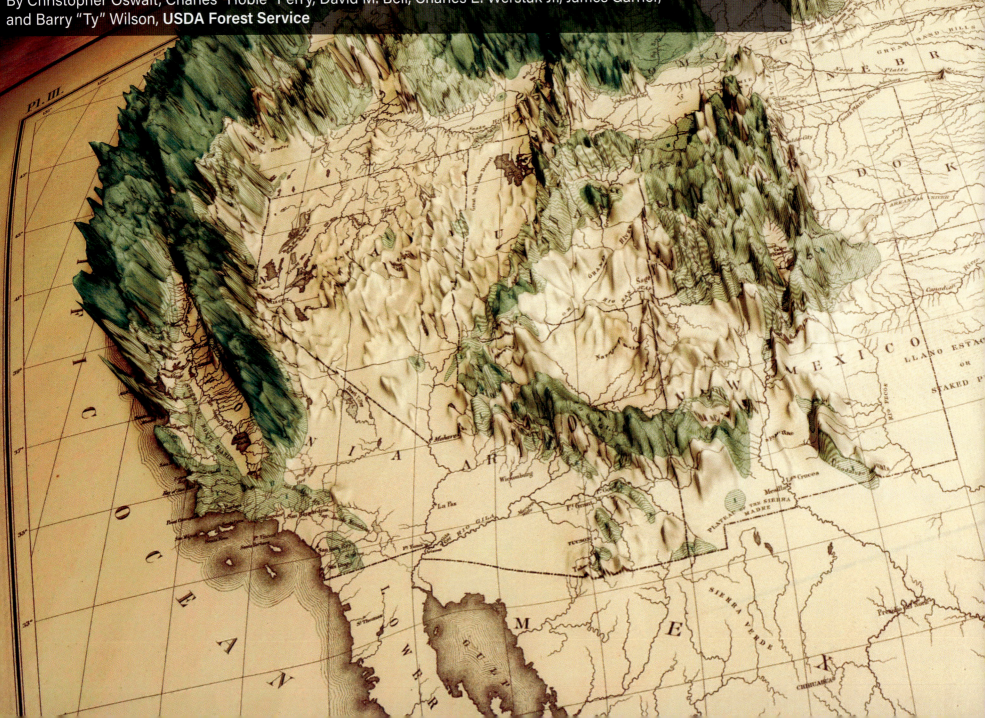

# FROM PLOTS TO PIXELS

Forest Inventory and Analysis (FIA), a USDA Forest Service program since 1930, collects data on more than 355,000 forest plots on public and private lands. FIA's latest data, deployed in the cloud with sophisticated GIS analysis tools, remains a mission-critical tool for the agency.

By Christopher Oswalt, Charles "Hobie" Perry, David M. Bell, Charles E. Werstak Jr., James Garner, and Barry "Ty" Wilson, **USDA Forest Service**

MAP
WING IN FIVE DEGREES OF DENSITY
THE DISTRIBUTION OF

An 1874 map of woodland density, updated with 3D elevation based on modern FIA carbon data.

# THE ANALYSIS OF FOREST INFORMATION

Monitoring the current status, health, and trends affecting forests is critical at a time when serious challenges confront us: globalization, climate change, and population growth, among others. The search for solutions creates a hunger for data and information on a variety of scales. The USDA Forest Service has a proud history of conservation and management, and through its Forest Inventory and Analysis (FIA) program, the agency takes the lead in monitoring the nation's forest resource. The federal program works with partners in states, universities, private industry, and other organizations to collect and provide state-of-the-art data, information, and knowledge. A timeline of the program shows various intersecting milestones in performing a seemingly impossible mission: inventorying the forest resources in one of the world's geographically largest nations.

In 1928, Congress directed the Secretary of Agriculture to conduct a comprehensive survey of the nation's "present and prospective requirements for timber and other forest products." It's important to note the economic focus of this instruction. Early inventories focused on timberland, those lands that can produce at least 20 cubic feet of wood per acre annually and are not excluded from management. These landscapes produce valuable forest products such as wood and paper.

As the FIA program matured, our collective understanding of the value of trees and forests has evolved to include much more than traditional forest products. Some people collect non-timber forest products (e.g., edible mushrooms, fish and game), and others use the forest for other purposes such as camping, canoeing, and biking. To inventory resources relevant to these broader needs and monitor long-term trends in forest health and productivity, the program had to change. The population of interest changed from timberland to forestland.

**Mission** FIA has been mandated by Congress for more than 90 years, and invested more than $1 billion since 1990, to provide authoritative data across three themes...

**1** A field plot network monitoring the ecology and management of forests and woodlands

**2** A census of the economic dynamics of the forest products industry

**3** A survey of the motivations and objectives of the nation's forestland owners

The FIA program maintains three primary streams of data that align with the three missions of the program:

1. Curate a network of field plots (the sampling areas where actual trees are counted) to measure the current status of the nation's forests.

2. Assess the nation's timber products industry.

3. Survey forest landowners to better understand their motivations, activities, and challenges.

Each component of the FIA program presents geospatial challenges. To address these challenges, the USDA Forest Service has long been at the forefront of applying the latest GIS technology. This chapter will explore these different dimensions of the FIA mandate, and its strategy of combining traditional observation and observation methods—boots on the ground—with modern technology to accomplish its goals.

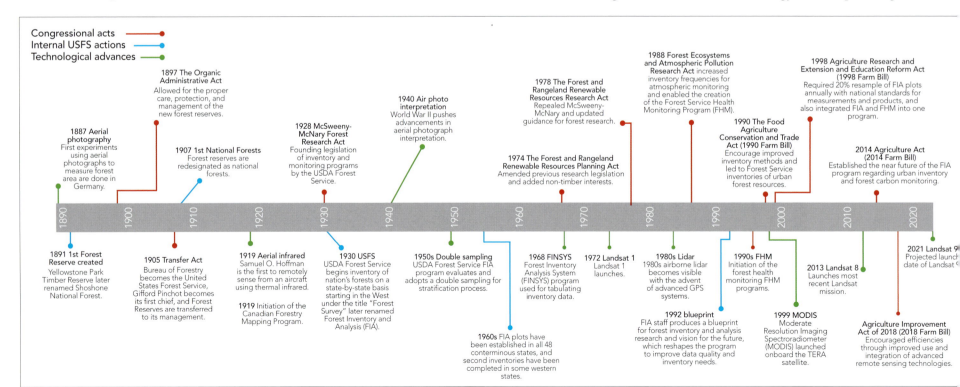

*A timeline of the congressional acts, USDA Forest Service actions, and technological advances that have organized and advanced the systematic collection of data about U.S. forestlands. (Timeline adapted from Bechtold and Patterson 2005; Landsat 2016; Shaw 2008).*

# BOOTS ON THE GROUND

Since 2000, the FIA program has inventoried most states annually, applying a nationally consistent sampling protocol using a quasi-systematic design covering all landownership in the United States. Fixed-area plots are installed in locations with accessible forestland cover, and field crews collect data on more than 300 variables, including landownership, forest type, tree species, tree size, tree condition, and other site attributes (e.g., slope, aspect, disturbance, and land use). The program collects and maintains a set of core national variables and uses a flexible approach that allows regional collection of variables.

The FIA program tracks and reports on the use of wood harvested from America's forests. Wood may be harvested for industrial purposes, such as the production of lumber or paper, or it may be removed for nonindustrial purposes, such as firewood. In either case, monitoring the removal and processing of wood provides information about a significant component of the U.S. economy. Timber Products Output (TPO) studies include two important components: mill surveys and use studies. Together, TPO data provides significant insight into the forest products economy, its impact on the larger U.S. economic situation, and how wood moves across America.

Approximately half of America's forests are privately owned, either by individuals or by corporations. Private landowners therefore have a huge impact on the state of America's forests. The FIA program conducts periodic surveys of private forest owners (the National Woodland Owner Survey) to assess their ownership objectives, expected benefits, harvest intentions, and management plans. These surveys are completed through voluntary questionnaires sent to private forestland owners. Responses are kept confidential to protect landowner privacy. Information is summarized at the state and regional level to provide information on status and trend in forestland ownership. As is the case with forest and timber product output data efforts, it is becoming increasingly important that the FIA program develops strategies to automate data and science delivery applications.

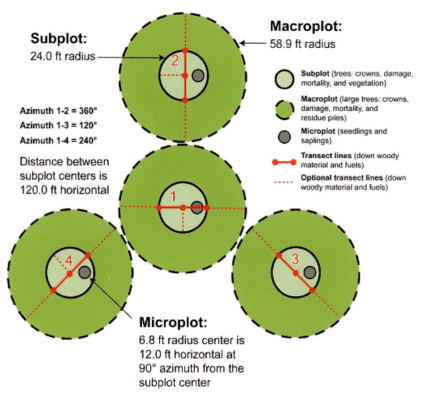

**Subplot:** 24.0 ft radius

**Macroplot:** 58.9 ft radius

Azimuth 1-2 = 360°
Azimuth 1-3 = 120°
Azimuth 1-4 = 240°

Distance between subplot centers is 120.0 ft horizontal

**Microplot:** 6.8 ft radius center is 12.0 ft horizontal at 90° azimuth from the subplot center

Subplot (trees: crowns, damage, mortality, and vegetation)

Macroplot (large trees: crowns, damage, mortality, and residue piles)

Microplot (seedlings and saplings)

Transect lines (down woody material and fuels)

Optional transect lines (down woody material and fuels)

*The FIA sampling protocol uses groundplots designed to cover 1-acre sample areas. Not all trees on the acre are measured. Groundplots may be new plots that have never been measured or remeasurement plots that were measured during one or more previous inventories.*

## Distribution of Six Forest Ownership Types in the Conterminous United States

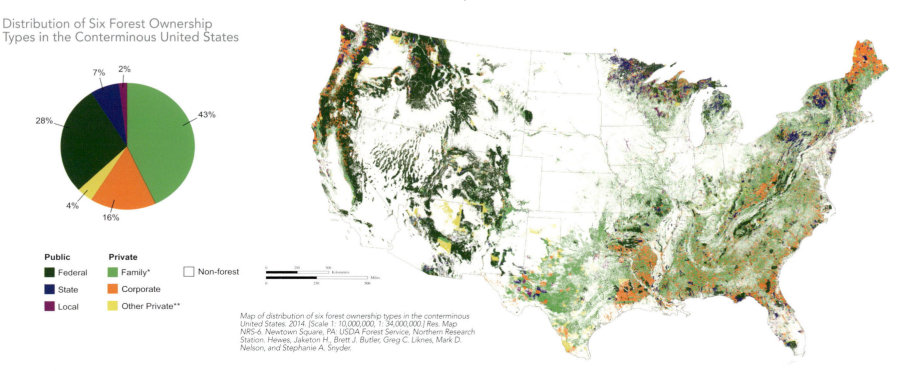

7% | 2%
28%
43%
4%
16%

**Public**
Federal
State
Local

**Private**
Family*
Corporate
Other Private**

Non-forest

250 500 Kilometers
250 500 Miles

*Map of distribution of six forest ownership types in the conterminous United States. 2014. [Scale 1: 10,000,000, 1: 34,000,000.] Res. Map NRS-6. Newtown Square, PA: USDA Forest Service, Northern Research Station. Hewes, Jaketon H., Brett J. Butler, Greg C. Liknes, Mark D. Nelson, and Stephanie A. Snyder.*

# METHODOLOGY

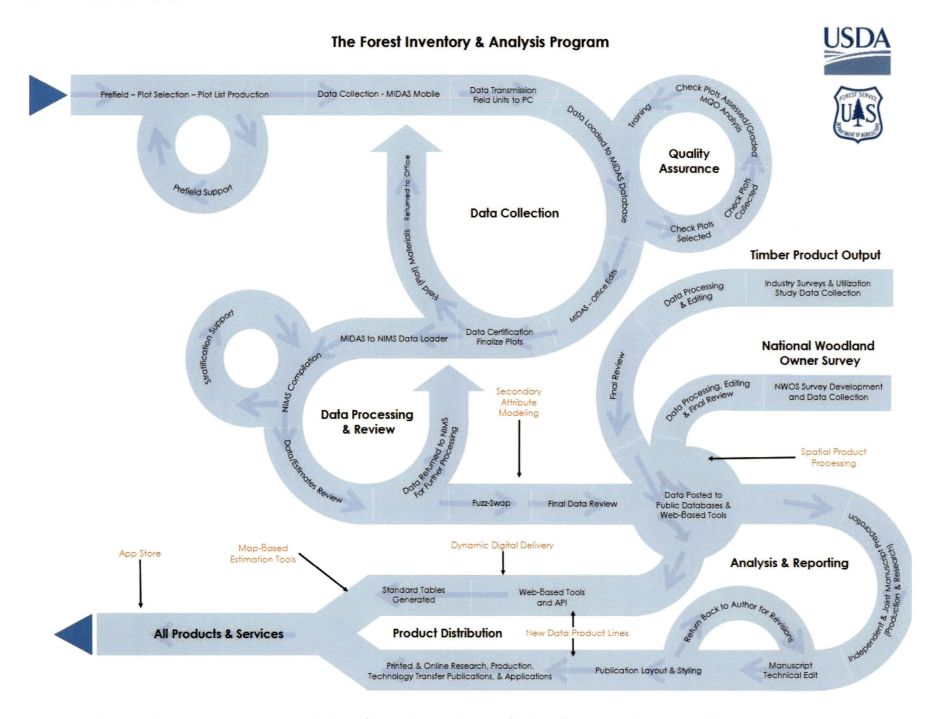

The Forest Inventory & Analysis Program

*From a technology standpoint, maintaining an accurate database of the condition and context of millions of trees is complex, as this workflow shows.*

# THE FOUNDATION

Each component of the FIA program—forest monitoring, the National Woodland Owner Survey, and TPO—faces similar challenges. Each part must continue to support traditional customers and data users while growing the relevance of the program by transforming FIA's product delivery machinery to serve an even larger number of groups.

The objective is to share data, information, and knowledge about the nation's forest resources with the public. To that end, the agency creates reports. Traditionally, the reporting had centered on hard-copy, paper reports that limited their capacity to build new audiences. Because printed reports increasingly are an outdated medium, it became imperative that the agency continue evolving its efforts to reach as many people as possible.

In the past and present, we have relied on the statistical design of our inventory to calculate official estimates of various forest attributes. While scientifically robust, this practice denies us the opportunity to use our full understanding of the forest ecosystem, collected data, and ancillary products such as remotely sensed imagery.

This heavy statistical underpinning also creates a formidable technical barrier for many of the data users who might not understand complex statistics. While maps have always been an integral part of the information products, the agency is moving into a future that places a greater reliance on map-based estimation. Thoughtfully implemented, this capability will increase the security of the FIA plot network and facilitate spatially resolute estimates at finer scales than currently possible.

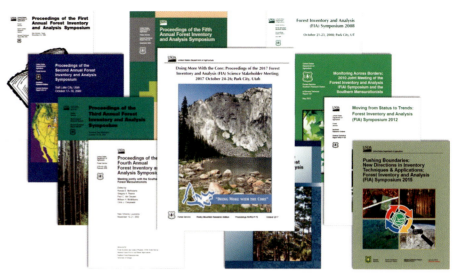

*The USDA Forest Service traditionally released information in the form of printed reports. Digital visualization tools and mapping platforms are changing this longtime practice and creating new generations of digital users of forest information.*

## From print to digital

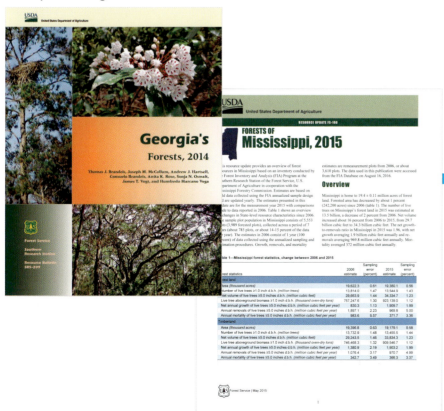

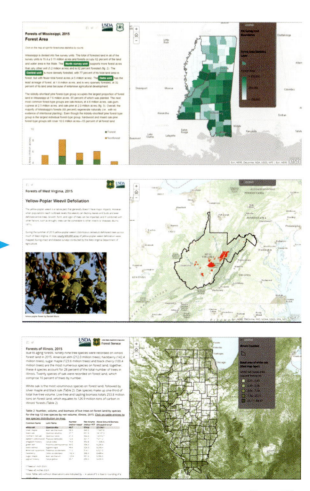

# DIGITAL INNOVATIONS

We are now designing a system to provide improved access to FIA data and associated analytical skills in an environment where the issue of "plot locations" is moot. Consider a future in which questions could be passed to the data and answers retrieved without exposing the underlying confidential content (or data). Reaching a broader spectrum of users requires a more powerful and flexible computing environment than existed just a short time ago within the agency. This environment also allows greater use of imagery and remote sensing without compromising owner confidentiality.

The current inventory program focuses on data collected in the field to generate authoritative estimates of forest attributes (e.g., forest area, species, and woodland density). Going forward, the FIA program has an opportunity to evolve and estimate other tree attributes that reflect the emergent thinking about forests, such as ecosystem services (e.g., wildlife habitat and carbon sequestration). This effort will produce a more valuable version of our existing database for users of this information.

Creating useful data products requires translating them into spatial data, and in time, published maps. Tables tend to hide patterns that spatial data can reveal through visualizations. The rapidly progressing fields of data visualization and digital geographic analysis tools provide FIA with easy-to-use solutions that help identify meaningful information from voluminous datasets. For example, interactive maps published in the cloud allow FIA to share data packaged with appropriate analytic tools that allow things such as summarization on the fly, simplifying an otherwise complicated workflow.

Cloud and GIS technology advancements allowed the FIA program to convert data publication tools and databases into digital data streams and an application programming interface (API). This capability allows the agency and its customers to submit queries from any tool that can write an HTML string, resulting in new public-facing applications that can be built using FIA data in an API-packaged approach. Moving forward, the agency anticipates internal and collaborative application development using the API with a future app store to distribute agency and partner content focused on specific needs.

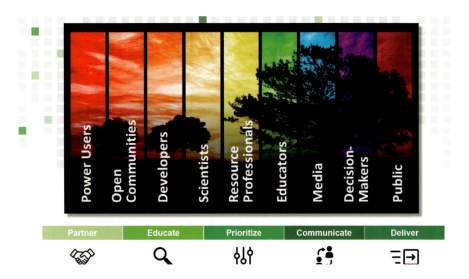

*Stakeholders in the FIA network.*

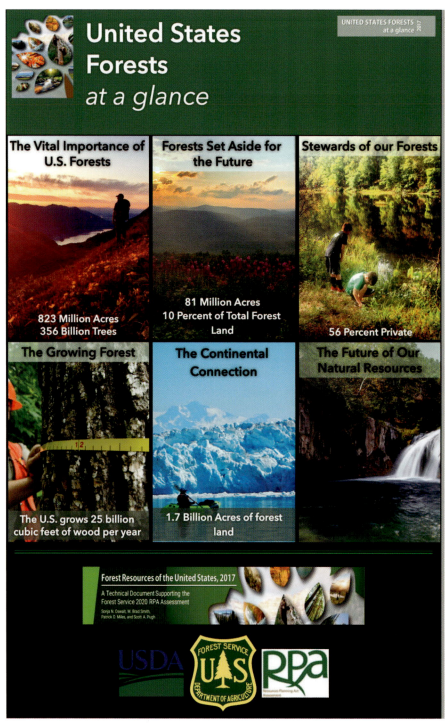

*This website (linked at GISforScience.com) explains the vital importance of U.S. forests to the nation's physical, mental, and economic survival, now and in the future.*

# THE FUTURE IS NOW

The digital transformation at FIA has been underway for some time and has included the creation of online tools such as EVALIDator, RPA DataWiz, and DATIM, and the adoption of tree canopy cover (TCC) for stratification. Now, deeper levels of integration have created even more opportunities across the workflows. Going forward, the transformation will require the adoption of national standards, because the tools must interact to be effective. The key is to build tools using best-of-class, commercial off-the-shelf solutions that are more likely to persist over time and customize only when necessary. Automating analytical, science, and reporting products will allow our scientists to add more value to other aspects of the program, develop new data product lines, integrate the Forest Inventory and Analysis Database (FIADB) with other databases, and ultimately populate an FIA application warehouse with programs that can help maximize the use of our data.

Digital transformation includes embracing the agency's current contracts, using open source and commercial digital tools, and collaborating with related partnerships where they are strongest. Transforming workflows when opportunities arise throughout the program will create the efficiencies needed to reinvest in other program areas.

Standing still in today's context means falling behind. If this is true, the FIA program can define its impact and utility going forward by the value it adds to the data. The application of meaningful science can improve the program in data collection techniques, biomass estimation, and producing statistically valid estimates at increasingly finer scales. As a result, the FIA program will have the capacity to turn data and information into knowledge. The FIA program must also consider ways to maintain and even improve its standing as the nation's authoritative source for knowledge about the forests and trees, their owners, and the associated economy. Digital transformation is about using technology while changing our focus from products to data and analytical services.

The next section describes four specific applications of FIA data in scientific contexts: mapping tree species distributions, forest carbon monitoring and accounting, mapping to support timber product outputs estimation, and tracking tornado damage in Wisconsin.

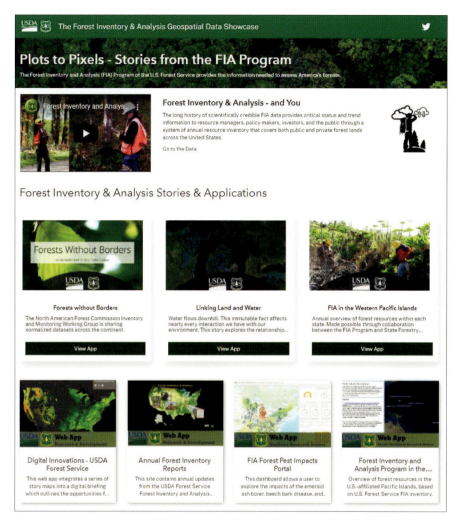

*Stories created using ArcGIS StoryMaps and apps allow users of all skill levels to interact with many of the tools and applications developed from the FIA program.*

# MAPPING TREE SPECIES DISTRIBUTIONS

Certain tree species are often found together in forest communities, as seen in the map of forest type groups in southern New England. These tree communities are themselves related to other plant communities (e.g., shrubs and grasses) and wildlife species. These communities may be sensitive to broadscale environmental drivers, such as climate, and local drivers, such as topography and disturbance history. Therefore, understanding local forest community patterns can support landscape planning, but this understanding increasingly depends on the fine-scale mapping of tree species distributions.

Knowing where certain types of forests, composed of differing combinations of tree species, are distributed across the United States is essential information for understanding how forests might respond, for example, to changing climate. These species may all have differing sensitivities to heat, drought, snowfall, disturbance, etc., meaning that forest vulnerability to change could be dependent on the assemblage of individual tree species within a given forest.

*Comparing forest type groups based on MODIS and Landsat satellite imagery.*

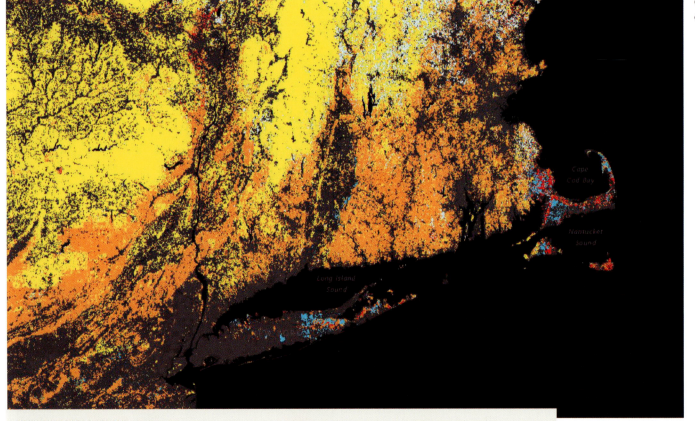

*Modified from the Forest Atlas of the United States, USDA Forest Service, this map of forest type groups for southern New England highlights the latitudinal transition from forests dominated by oak (orange) to maple (yellow).*

**FOREST-TYPE GROUPS**

| | | |
|---|---|---|
| White/red/jack pine | Western white pine | California mixed conifer | Aspen/birch |
| Spruce/fir | Fir/spruce/mountain hemlock | Exotic softwoods | Alder/maple |
| Longleaf/slash pine | Lodgepole pine | Oak/pine | Western oak |
| Loblolly/shortleaf pine | Hemlock/Sitka spruce | Oak/hickory | Tanoak/laurel |
| Pinyon/juniper | Western larch | Oak/gum/cypress | Other western hardwoods |
| Douglas-fir | Redwood | Elm/ash/cottonwood | Tropical hardwoods |
| Ponderosa pine | Other western softwoods | Maple/beech/birch | Exotic hardwoods |

# FOREST CARBON MONITORING AND ACCOUNTING

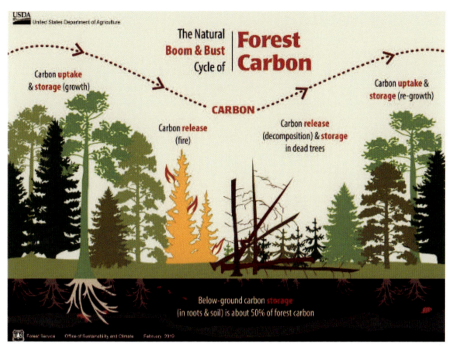

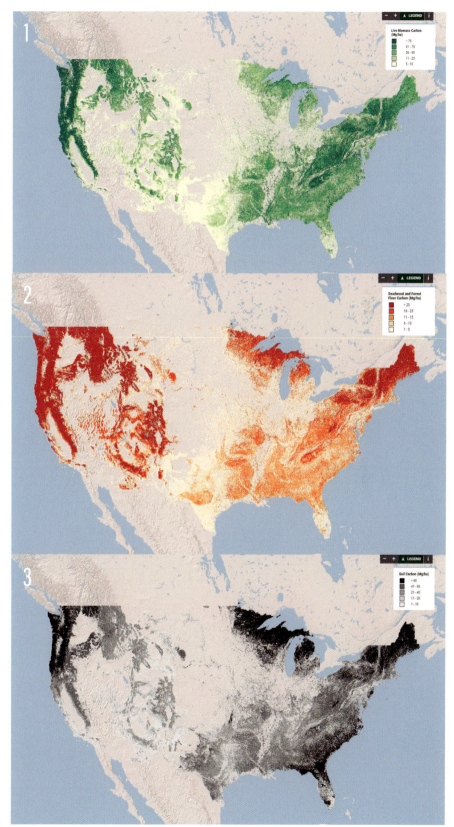

Concerns about monitoring and mitigating climate change often focus upon carbon as it is a component of carbon dioxide ($CO_2$) and methane ($CH_4$), two primary greenhouse gases. Carbon is released to the atmosphere through a variety of processes; major contributors include combustion of fossil fuels, non-energy uses of fuels, livestock production, and landfill decay. Forests contribute carbon to the atmosphere as they decay or when they burn in wildland fires. At the same time, forests are critical for their role in sequestering carbon from the atmosphere through tree growth. The balance of carbon emission and sequestration determines whether forests will act as a carbon source to or sink from the atmosphere.

The United Nations Framework Convention on Climate Change (UNFCCC) has the objective of stabilizing greenhouse gases in the atmosphere at levels that don't dangerously alter the globe's climate. To make this a reality, policy makers and managers need high quality information to better understand and improve the role forests play in carbon cycling. It's also important to track progress towards the goal, so the U.S. Environmental Protection Agency (EPA) leads an annual National Greenhouse Gas Inventory (NGHGI) with the Forest Service providing valuable input by monitoring the status and trends of carbon stocks in forested ecosystems.

Research on forest carbon monitoring focuses on improving the measurement and estimation of forest carbon pools, including standing trees, leaf litter, downed deadwood, and soils. Given the importance of providing a comprehensive assessment which includes all forests, additional research is conducted to inventory less accessible landscapes through a fusion of field inventory and remote sensing. This research by the FIA program to develop and improve upon methods of characterizing, monitoring, and communicating forest carbon pools is an important collaboration with and contribution to the international community.

*National mapping efforts provide estimates of forest carbon pools—1) live tree, 2) deadwood and litter carbon, and 3) soil carbon, which are synthesized and summarized for the public through web maps and related applications. Adapted from FIA Forest Carbon Accounting web application (linked at GISforScience.com).*

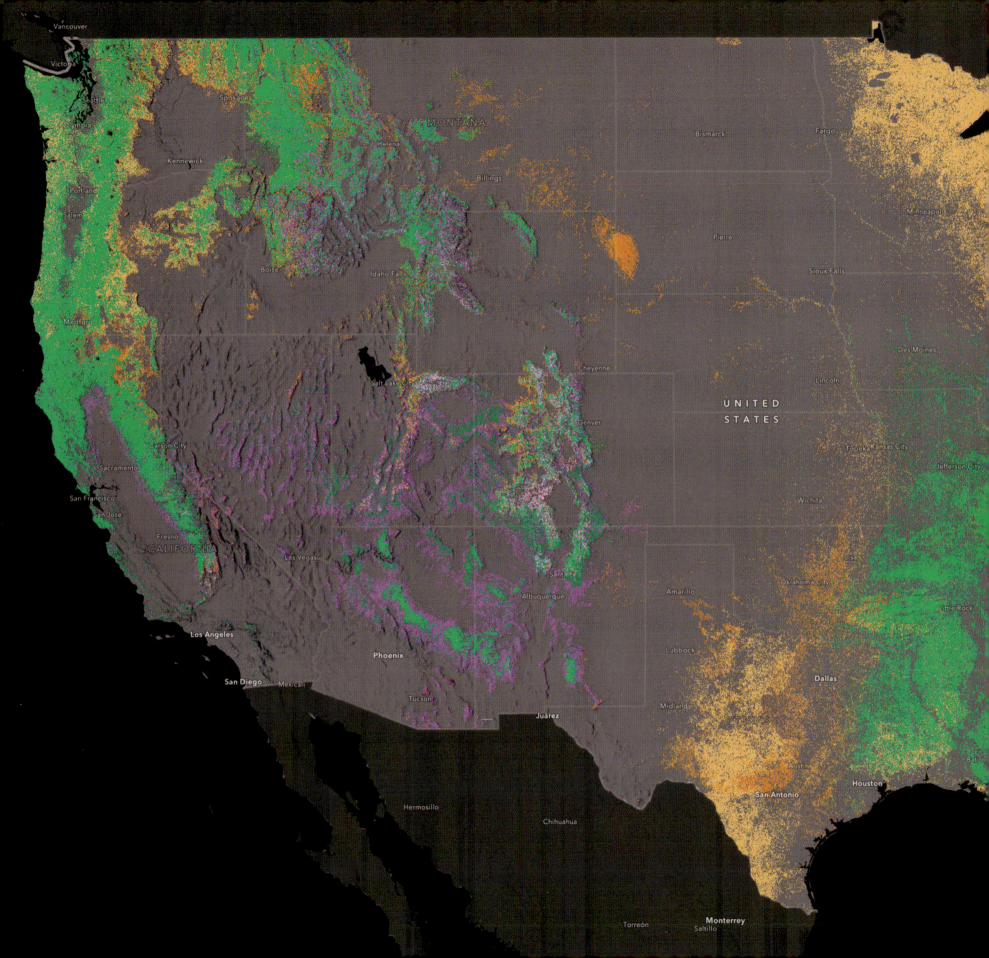

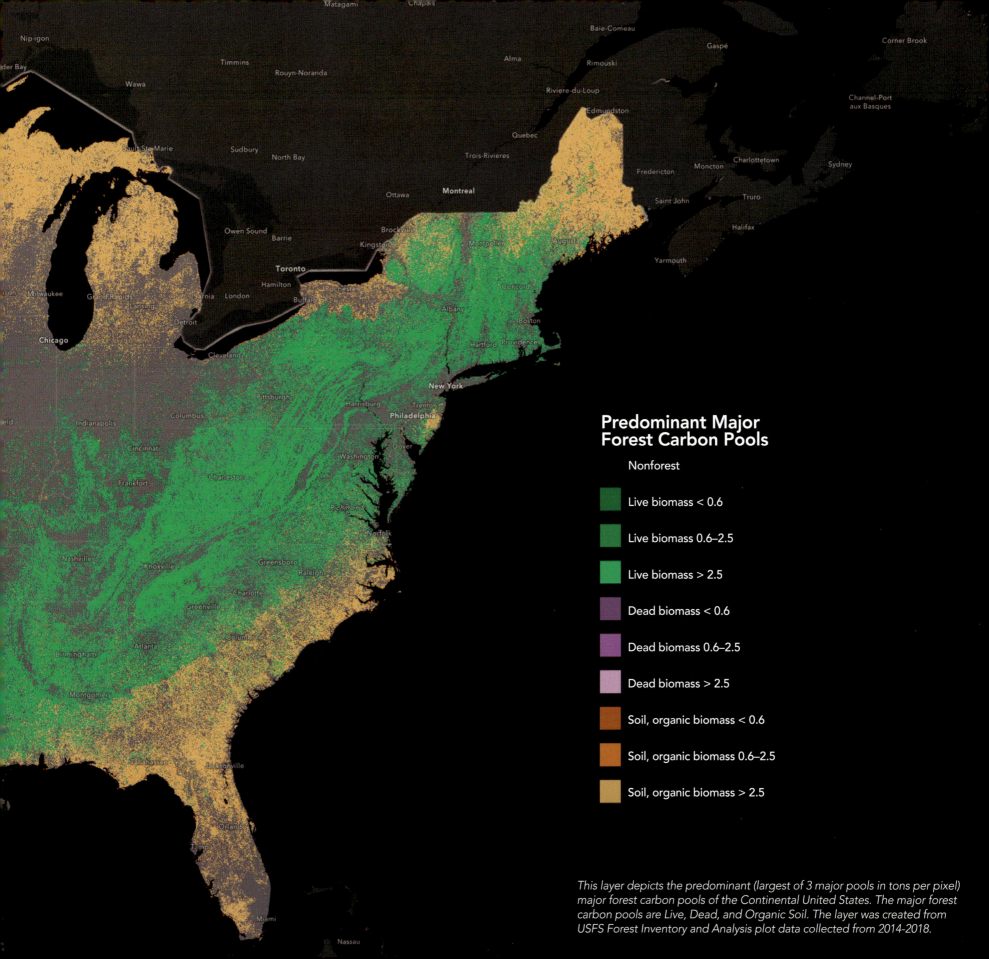

## Predominant Major Forest Carbon Pools

Nonforest

Live biomass < 0.6

Live biomass 0.6–2.5

Live biomass > 2.5

Dead biomass < 0.6

Dead biomass 0.6–2.5

Dead biomass > 2.5

Soil, organic biomass < 0.6

Soil, organic biomass 0.6–2.5

Soil, organic biomass > 2.5

*This layer depicts the predominant (largest of 3 major pools in tons per pixel) major forest carbon pools of the Continental United States. The major forest carbon pools are Live, Dead, and Organic Soil. The layer was created from USFS Forest Inventory and Analysis plot data collected from 2014-2018.*

# MAPPING TO SUPPORT FOREST PRODUCT PRODUCERS

White oak is an extremely important tree species ecologically and because it helps maintain an economic engine worth billions of dollars. Domestic and global demand for white oak is surging due to its use in making barrels to support a booming distilled spirits industry. Although there may not be a current shortage of white oak trees, there is a shortage in the infrastructure required to harvest the wood and turn the materials into barrels. Thus, there's a need to develop tools to promote and sustain white oak forests and support the economic and ecosystem benefits that those forests provide.

In 2010, an economic downturn depressed the forest products industry and accelerated mill closings and job losses in small towns across the United States. One bright spot was white oak cooperage (barrel-making). The FIA program worked with a leading spirits manufacturer to bring good news to some hard-hit communities by helping locate sites for new mills through data-intensive analytics and processing.

Now, with products such as the Bourbon Barrel Oak Finder, that work can be done with a click of a button anywhere on a map. With tools such as ArcGIS GeoPlanner℠ and the FIA API (along with some JavaScript programming), we can provide a set of metrics for a user-defined woodshed. These kinds of tools will help accelerate analyses and put the information resources directly in the hands of our users.

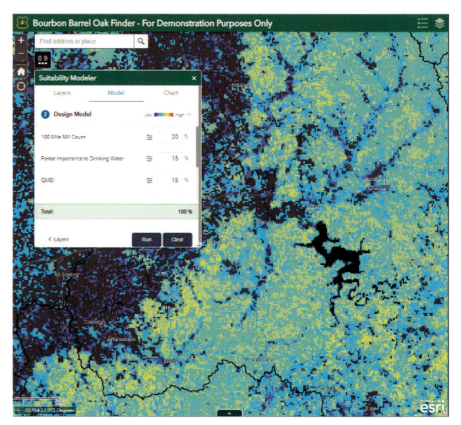

*Forest suitability for providing timber products (oak) that support bourbon barrel production within 50 miles of Olympia, Kentucky.*

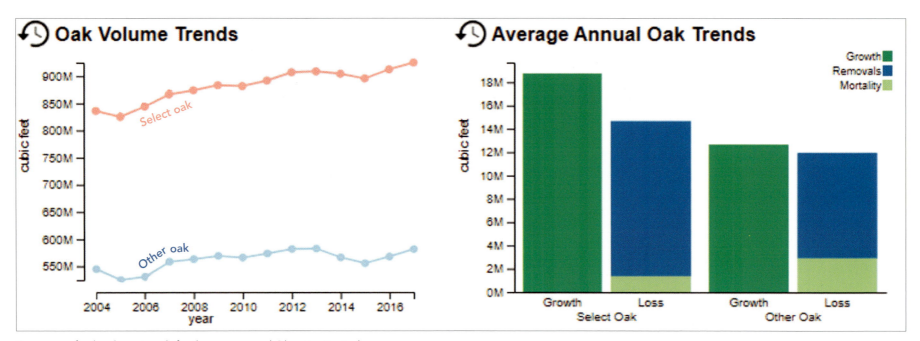

*Summary of oak volume trends for the area around Olympia, Kentucky.*

# TRACKING TORNADO DAMAGE IN WISCONSIN

Forests change in a variety of different ways from year to year. Typically, these changes involves year-to-year variation in forest dynamics, such as tree recruitment, growth, and death. However, some changes occur in a relatively short time, such as during forest disturbances including wildfire, insects and disease, and tornadoes. These kinds of disturbances, while naturally occurring in forests, substantially impact forest ecosystems and communities, more specifically, wildlife, timber, and recreation.

Tornadoes are hyper-local and highly destructive disturbances, felling trees, damaging property, and threatening lives in many parts of the United States, For example, a tornado swept through a portion of Wisconsin (Map 1) that is characterized by a diverse mix of forests and agricultural lands. The tornadoes destructive winds converted live trees to snags and down wood in the course of just a few minutes. Given the relatively small footprint of many tornadoes (tens of hectares), few if any FIA plots may overlap tornado tracks, requiring a mapping approach quantifying the magnitude of tornado impacts, such as mapping the changes in live tree biomass or carbon (Map 2). Even if some plots overlap tornado tracks, it is likely that too few plots will be available for traditional estimation (mean and variance), requiring small area estimation approaches that use plot data and maps to generate model-based estimates. Small area estimation can allow land managers, property owners, and policy makers to visualize the consequences of a disturbance, and its uncertainty, to help them plan for the future (Map 3).

*Wisconsin landscape shown in pre-tornado condition (with the path of destruction track shown).*

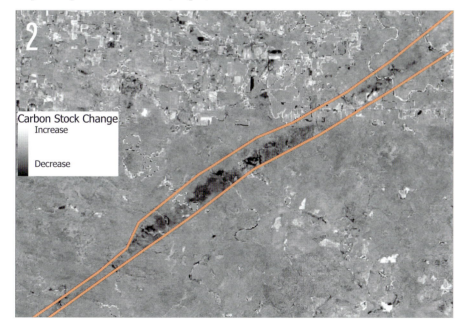

*Post-tornado, the damage scar (characterized by a decrease in carbon stock) is shown as dark patches.*

Carbon Stock Change
Increase

Decrease

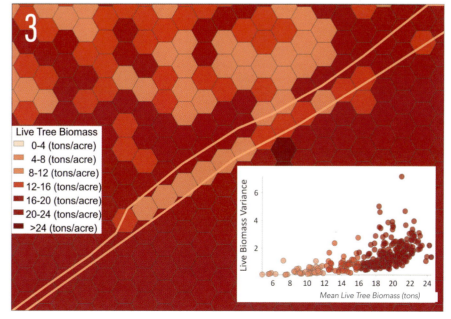

Live Tree Biomass
- 0-4 (tons/acre)
- 4-8 (tons/acre)
- 8-12 (tons/acre)
- 12-16 (tons/acre)
- 16-20 (tons/acre)
- 20-24 (tons/acre)
- >24 (tons/acre)

*Mean FIA estimates of post-tornado live tree biomass at 250-ha scales. Variance (inset), a measure of estimate precision, can itself vary with the mean biomass in forests, highlighting the degree of uncertainty in FIA estimates.*

# REMOTE SENSING IS PART OF THE FIA MANDATE

Since its earliest days, the forest inventory program has adapted and evolved in response to customer needs. In 1991, a panel of experts, stakeholders, and partners developed a national vision and strategy for the FIA program. This group recognized that the quality of FIA data would be enhanced by shorter inventory cycles (e.g., five years). The panel also recommended that FIA "develop and implement new technology, such as remote sensing, to accomplish the inventory in a more cost-effective and efficient manner." These efficiencies would facilitate more timely completion of the inventory.

Challenges with meeting the first panel's recommendations led to the convening of a second panel in 1997 and 1998. This group reiterated that "the development of new and expanded remote sensing technologies and modeling capabilities" is required for an effective forest inventory. These technologies should be considered strategically, with a vision for how they would be used in an evolving context. This panel understood that providing full interactive access to FIA data and analysis programs and improving Forest Service research and expertise in GIS would quickly give customers more timely information and improved accuracy of our estimates. To

that end, remote sensing was specifically acknowledged in the 2001 memorandum of understanding between the Forest Service and the National Association of State Foresters as essential for stratifying the sample to improve our estimates.

Congress has also provided clear direction to the FIA program. Along with mandating the transition to an annual inventory, the Agricultural Research, Extension, and Education Reform Act of 1998 directed FIA to employ remote sensing and other advanced technologies to carry out the program. The 2014 and 2018 Farm Bills included continued Congressional direction to integrate and improve the use of spatial analysis techniques and advanced remote sensing technologies. Currently, aerial imagery is used in plot selection, remotely sensed imagery is integrated with the field observations to provide additional insight, and operational remote sensing and geospatial modeling yields novel map products. Even with the development of national maps fusing FIA and remotely sensed data (e.g., forest type group, forest biomass, and tree species ranges), our task is not complete. Users require maps of other forest attributes, such as forest stocking, along with updates of extant maps using more timely and higher resolution remote sensing. These continuing and emerging needs motivate continued FIA research and development.

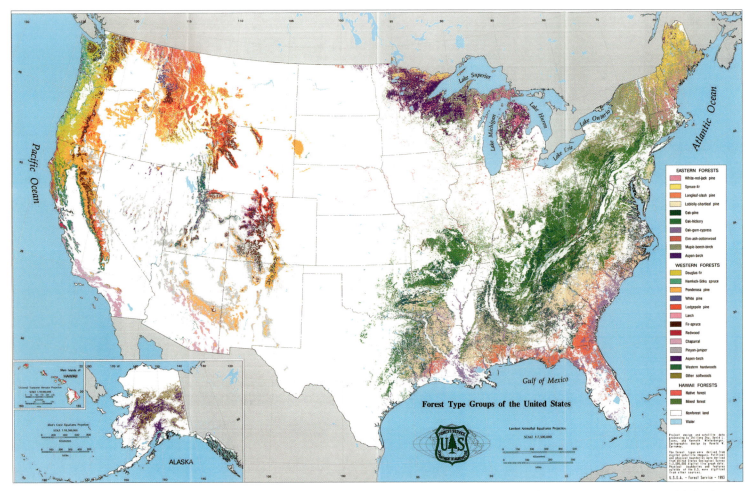

*The first national map of forest type groups based upon the integration of remotely sensed imagery with forest inventory data was published in 1993 as part of a Resource Planning Act (RPA) Assessment Update, Source: Powell, Douglas S.; Faulkner, Joanne L.; Darr, David R.; Zhu, Zhiliang; MacCleery, Douglas W. 1993. Forest resources of the United States, 1992. Gen. Tech. Rep. RM-GTR-234, Rev. 1994. Fort Collins, CO: USDA Forest Service, Rocky Mountain Forest and Range Experiment Station.*

# MODELING AND MAPPING: FOREST INVENTORY AND IMAGERY IN HARMONY

In earlier mapping projects, FIA used a spectro-temporal approach with moderate spatial resolution satellite imagery to produce raster datasets of numerous forest attributes across the continental United States. Today, with ready access to imagery from a variety of sensors and powerful cloud computing environments, the program has shifted focus to do this work using dense time series of finer resolution imagery.

$$\sum_{j=0}^{m} a_j cos(j2\pi t/n) + b_j sin(j2\pi t/n)$$

*As m increases, additional harmonics are added to the series.*

The fitted Fourier series coefficients describing vegetation phenology, along with other auxiliary data such as climate and topography, are potentially powerful features for modeling forests. Through ecological ordination techniques such as canonical correspondence analysis, these predictors can be combined with data about tree species collected on forest inventory plots to order them along environmental gradients. The location of forest inventory plots along these gradients can be used with machine learning algorithms such as k-nearest neighbors. This approach works by imputing to each pixel in the raster a "bucket of plots" based on their proximity to the pixel as measured in the space of features derived from the predictors.

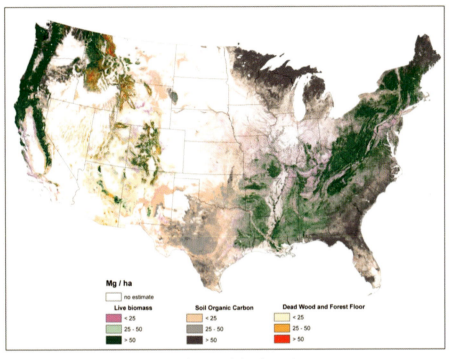

*Map of live biomass, soil organic carbon, and dead wood.*

By tracking changes over time in spectral indices derived from Landsat imagery, scientists can monitor the beginning and end of the growing season (vegetation phenology) across large geographic areas. From this information, they infer the presence of various types of forest, their composition of tree species, and overall forest conditions. One of the techniques used for analyzing these spectro-temporal profiles is harmonic regression (shown as the blue curve fitted to the green "greenness index" values) which characterizes the shape of such profiles as a Fourier series, a simple sum of a series of cosine and sine functions.

**Transformed feature space** → **Imputed plot map**

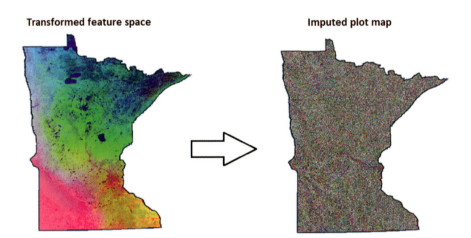

*Each bucket of plots represents a distribution of attribute values, stored in FIADB, the FIA database, that can be used as the machinery to make pixel-level predictions and maps of forest resources.*

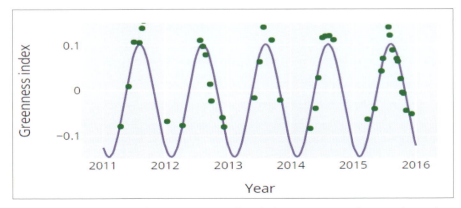

*Example of Fourier series fit to time series of tassled cap greenness for a single pixel.*

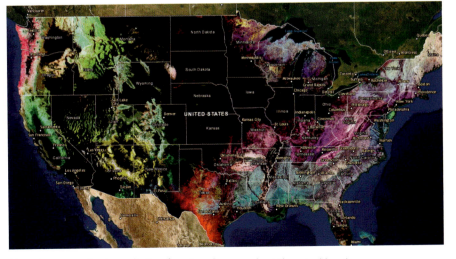

*Sample map output predicting forest carbon pools at the pixel level.*

# REACHING FOR THE CLOUD: THE BIG WHY?

National forest inventories (NFIs), such as the ones conducted by FIA, provide a gold standard for the estimation of forest status and trends of forest attributes useful for planning and decision-making at multiple scales. While FIA's collection of content is well managed, these resources are not suitable for providing information and knowledge to a wide spectrum of users. The FIA estimation tools require an understanding of relational databases to create relevant estimates. Additionally, users increasingly require information at finer spatial and temporal scales for which NFIs were not designed.

The internet has demonstrated the public's growing competency with maps, and FIA has a long history of exploring map-based approaches to sharing information and knowledge. Modeling frameworks that integrate FIA plot data with ancillary geospatial data, such as satellite imagery, can also play an important role in FIA estimation. Artificial intelligence and machine learning (AI and ML, respectively) algorithms show promise in accounting for complex relationships between forest attributes and ancillary data. In particular, deep learning techniques could provide important advances in FIA estimation procedures for small areas and change analyses.

To meet the user's needs in a timely manner, FIA must dramatically increase the pace and scale of computation. It currently takes months to convert field observations into information and estimates. We need to publish future map-based content—likely dozens if not hundreds of geospatial data layers—simultaneously with the release of tables to the public. These maps will include current estimates accompanied by geospatial information that addresses other agency priorities, such as the 1990 baseline for carbon accounting.

Given the focus on delivering map-based content, our ideal computing environment is tuned to facilitate parallel processing of geospatial data through efficient scripting and implementation of AI/ML techniques. This capability will likely require a transformation of our business models away from a traditional focus on desktop modeling and estimation into new approaches (and languages) that efficiently use the potential of the cloud to implement deep learning applications.

The maps derived from the integration of FIA data and ancillary predictors are peer-reviewed, transparent, and shareable modeling approaches, and the models are constructed to simultaneously protect plot security and provide reliable estimates of various forest attributes across a range of spatial scales. These ancillary predictors are truly massive datasets that are proven to improve the precision of estimates based on FIA plot data. For example, incorporating structural information related to the height of the forest canopy adds value to these models, but this work cannot be accomplished in desktop environments.

At the same time, the resulting data and information are hosted and maintained in cost-effective ways that empower FIA to create the broadest range of products imagined. The goal is to create and host the data in a manner that reduces data migration and related concerns about keeping content current and error-free. This goal is particularly important because it relates to the various systems of record across the USDA Forest Service in multiple disciplines. Central to the goal is identifying appropriate points of simple yet consistent integration into enterprise information systems to maximize the value of the FIA content. The result is increased capacity and understanding about how to maximize FIA's use of geospatial technology and data. Maps of individual forest attributes can then be combined easily with other content to provide integrative assessments (developed by the agency and its partners) relevant to resource managers, policy makers, and the public that holds them accountable.

The future of FIA's reporting and decision-support requires access to new, computing-intensive resources to fully integrate USDA Forest Service data with the wide varieties of other content relevant to create authoritative models and maps of forest attributes at the scale of continents. This collaborative work capitalizes on recent innovations in GIS and remote sensing technologies to integrate a growing archive of satellite-based earth-observation platforms such as Landsat—the longest continuous record of our changing Earth—and to enable efficient processing and modeling of these massive archives at the national scale, to geo-enable one of the largest, most detailed archives of forest inventory information in the world.

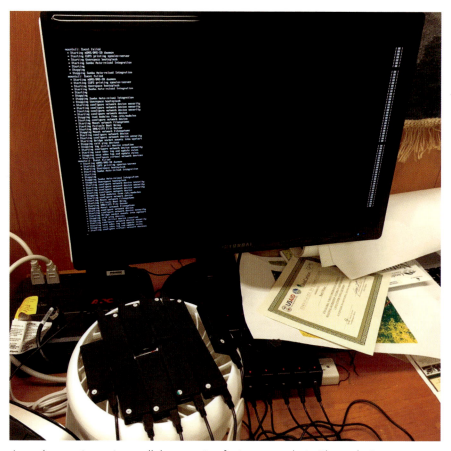

*An early experiment in parallel processing for image analysis. These devices were originally for Android-based video streaming, designed to be plugged into the HDMI port of a television set. Researchers at the USDA Forest Service installed Linux and the R statistical package on each of the devices and set them up on a detached Wi-Fi network where they shared a USB drive for storage. To maximize performance and run them at the fastest clock speed, they placed a fan beneath the CPUs. Although the system worked, it wasn't an optimal solution and was ultimately deployed using raster analytics in ArcGIS® Enterprise.*

This 1874 map of woodland density is given modern 3D relief relating to current rates of forest carbon. This mashup of vintage cartography with current inventory data was created in ArcGIS Pro by first georeferencing the vintage map then assigning a USDA Forest Service image service of carbon rates as its 3D elevation surface.

# BLUEPRINT FOR A BETTER FUTURE

The nearly 200 member states of the United Nations aim to end poverty, protect the planet, and promote peaceful and inclusive societies by 2030. This chapter introduces the framework for these aspirational goals and examines how GIS is helping to move them forward.

By Maryam Rabiee, **United Nations Sustainable Development Solutions Network**, and Ismini Ethridge, **Columbia University**

The many intersections of the United Nations Sustainable Development Goals in the modern world.

# THE SUSTAINABLE DEVELOPMENT GOALS

"We have reached a defining moment in human history. The people of the world have asked us to shine a light on a future of promise and opportunity. Member States have responded with the 2030 Agenda for Sustainable Development. The new agenda is a promise by leaders to all people everywhere. It is a universal, integrated, and transformative vision for a better world. It is an agenda for people, to end poverty in all its forms. An agenda for the planet, our common home. An agenda for shared prosperity, peace, and partnership. It conveys the urgency of climate action. It is rooted in gender equality and respect for the rights of all. Above all, it pledges to leave no one behind."

—Then UN Secretary-General Ban Ki-moon, speaking at the Summit for the Adoption of the Post-2015 Development Agenda[1]

In 2015, member states from 193 of the United Nations unanimously adopted a new global framework to end poverty, protect our planet, and promote peaceful and inclusive societies by 2030.[2] The agreement to develop the 2030 Agenda for Sustainable Development with 17 Sustainable Development Goals (SDGs) at its core aims to create a sustainable and inclusive future for everyone and leave no one behind.

Soon after the adoption of the SDGs, the global community took yet another significant step toward creating a better future, when the Paris Agreement was adopted in December 2015. The agreement represented a landmark for sustainable development efforts as the first legally binding multilateral agreement to take action toward curbing the climate crisis and significantly limiting the increase in global warming.[3]

A key element of the Paris Agreement requires each country (called *parties*) to "prepare, communicate, and maintain successive nationally determined contributions (NDCs) that it intends to achieve."[4] NDCs outline a country's plan to reduce national emissions and adapt to the impacts of climate change. The Paris Agreement is a critical component of the 2030 Agenda, especially for achieving SDG 13, which requires urgent action to combat climate change and its impacts and careful planning and monitoring from participating countries and strong multilateral cooperation.

This chapter delves into several of the SDGs to show how international organizations are using GIS to help to end hunger and poverty, war and political strife, and include all the peoples of the world in making the planet a sustainable place to live and thrive. This section offers snapshots of SDG 6, ensuring clean water and sanitation; SDG 15, protecting all terrestrial life; SDG 2, ending hunger; and SDG 16, promoting peace, justice, and inclusive societies.

The table of UN SDGs is an interconnected matrix of ambitious goals designed to deliver a better world to the planet in 2030.

# GIS SUPPORTING THE SUSTAINABLE DEVELOPMENT GOALS

**6 CLEAN WATER AND SANITATION**

## SDG 6

Educators at Earth Institute's Center for Sustainable Development and geochemists at Lamont-Doherty Earth Observatory at Columbia University in New York City collaborated on a citizen-science project in Alirajpur, India, to locate wells with safe drinking water. The resulting map, with more than 1,000 data points, helps governmental departments make water quality decisions and therefore brings the community closer to achieving SDG 6: Ensure the availability and sustainable management of water and sanitation for all.[5]

An all-woman team tests for fluoride in a village in Alirajpur District in Western India.[6]

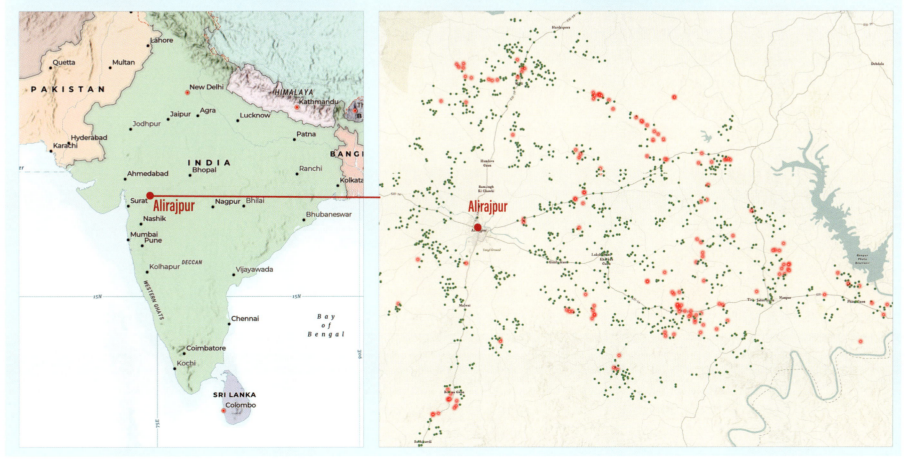

**15 LIFE ON LAND**

## SDG 15

The International Union for Conservation of Nature's (IUCN) Red List of Threatened Species uses spatial data for global assessments of more than 111,000 species.[7] Understanding where vulnerable species are located is critical for efforts to better protect them and achieve SDG 15: Protect, restore, and promote sustainable use of terrestrial ecosystems, sustainably manage forests, combat desertification, halt and reverse land degradation, and stop biodiversity loss.

*The SDGs aim to protect all life on Earth, including Madagascar's vulnerable fossa (Cryptoprocta ferox), a cat-like, carnivorous mammal.*

*The IUCN Red List objectively assesses species and their risk of extinction based on measurements of population size and geographic range, and their trends in the past, present, and future.[8] Established in 1964, the Red List has evolved to become the world's most comprehensive source on the global extinction risk status of animal, fungus, and plant species. The Red List's curated data was used to create this interactive story that highlights some of the 37,400 species (assessed in the database so far) that are threatened with extinction.*

*Illustration of fossa, circa 1927.*

## SDG 2

Geospatial satellite data provides useful information on environmental indicators that dictate crop production. Organizations such as Group on Earth Observations Global Agricultural Monitoring Initiative (GEOGLAM) use this data to forecast agricultural yields at national, regional, and global levels[9] and provide a public good of open, timely, science-driven information on crop conditions. Early warning of reduced agricultural production is a key component of achieving SDG 2: Zero hunger.[10]

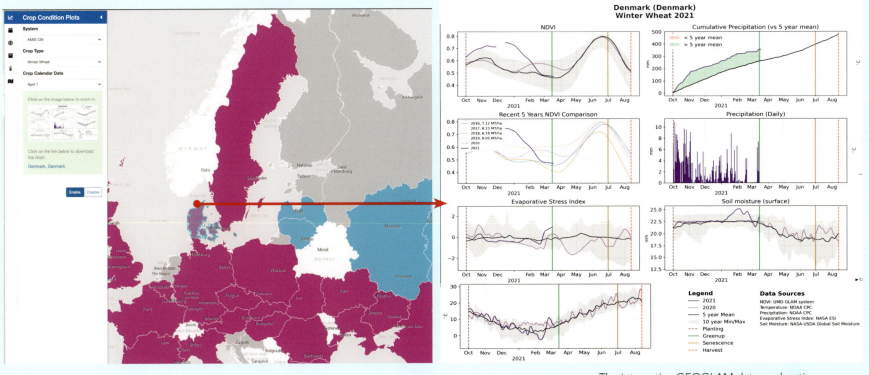

The interactive GEOGLAM data exploration tool delivers transparent, multisource consensus reports on global crop conditions when users click locations on the map.

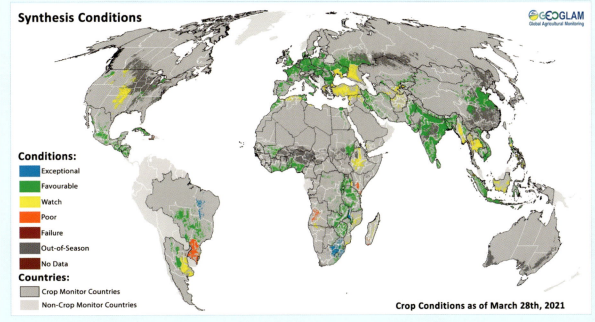

The synthesis maps provide a simplified overview of current global crop conditions. The cropped area displayed in the synthesis map is the total area of all crops, depending on the community of focus, depicted according to the observed crop condition categories.

SDG 16

Hindou Oumarou Ibrahim, SDG Advocate, environmental activist, and president of the Association for Indigenous Women and Peoples of Chad, leads 3D participatory mapping efforts in her own community to bridge indigenous peoples' traditional knowledge with sustainable management of ecosystems for nature-based resource conflict prevention at the local level. Mapping in 3D helps inform the design of agricultural and pastoral policies at the local, regional, and national level in the Sahel region, and is critical for achieving this mission-crucial goal.[11] SDG 16 promotes just, peaceful, and inclusive societies, and recently has focused on the world's response to the COVID-19 pandemic. "Responses that are shaped by and respect human rights result in better outcomes in beating the pandemic, ensuring health care for everyone and preserving human dignity," according to the UN.[12]

"3D participatory mapping brings together technology, science and indigenous peoples' traditional knowledge to help local communities adapt to climate change and mitigate its impact. It builds better understanding between farmers, pastoralists and local government authorities, resulting in fewer conflicts over limited resources. 3D participatory mapping has proven to be a valuable tool to contribute to the design of agricultural and pastoral policies at both the local, regional and national level in the Sahel region."[13]

—Hindou Oumarou Ibrahim, SDG Advocate

Hindou Oumarou Ibrahim leading 3D participatory mapping efforts in her local community.[14]

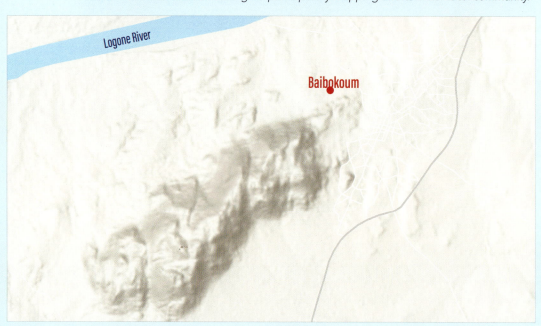

Participatory 3D modeling (P3DM) is a community-focused and executed mapping method that brings local spatial knowledge together with accurate elevation of the land to produce physical, scaled, and georeferenced relief models. Although this practice has been employed for decades, improved GIS tools are bridging the gap between the physical and the digital.

# MEASURING PROGRESS TOWARD THE GOALS

"Why is geospatial data important for the SDGs? When I was training in economics, we did not have modern GIS tools at hand. We may have looked at maps, but the data for our statistical models came mostly from national income accounts, or state and local data. Much public policy was therefore blind to geography. Policy decisions on poverty, schooling, health care, even infrastructure was taken with far too little regard for the spatial distribution of real needs and opportunities. Now we can do vastly better because of the powerful GIS tools and massive digital data now available.

Indeed, to meet the SDGs, we absolutely need reliable, quality, and timely geospatial data, for example, to monitor environmental changes in real time (climate change, deforestation, pollution) and to assess by region the most urgent economic needs (poverty, hunger, access to health care and education, access to safe water and sanitation, and so forth). Real-time geospatial data is crucial for implementing urgent policies such as controlling the COVID-19 epidemic, and for holding governments accountable for their commitments."[15]

—Jeffrey Sachs, president of the SDSN,
discussing the critical roles of GIS
and storytelling in solving global challenges

From community participatory mapping to global satellite data collection, GIS has quickly permeated the world of sustainable development as a critical tool for meeting environmental, social, and economic development targets. As a global community, our technological and financial capacity to protect the natural environment and ensure all basic human needs are met have never been greater, but our potential for actionable solutions, sustainable planning, and monitoring progress rely critically on knowing where people and resources are, and how things are changing in real time.

Beyond the 17 SDGs, The Inter-Agency and Expert Group on SDG Indicators (IAEG-SDGs)[16] defined 169 targets and 232 indicators to monitor the SDG framework and ensure that we meet the economic, environmental, and social objectives of the global goals.

To effectively track the status of the SDGs, we need quality, spatially disaggregated and timely data. Five years into the implementation of the SDGs, much of the data is either out-of-date or simply unavailable with far too many people left behind in the numbers. According to the 2021 Sustainable Development Report, only 59% of the data points needed to track progress on the SDGs are covered in the SDG Database (up from 41% in March 2019). Moreover, only 54% of environmental data points have observations for at least one year after 2015, and just 46% also have observations for multiple years (necessary for tracking changes over time).[17] There is an urgent need for quality, timely, and geographically disaggregated data for the SDGs.

The myriad of data and tools available cannot address the challenges of implementing the SDGs while the information is produced and used in silos. Often, data used for reporting or analysis at subnational, national, and global levels has not been effectively curated to convey or address the issues at hand. Key stakeholders, including policy makers, government officials, journalists, academics, students, and members of civil society, need a virtual space to access timely and spatially disaggregated data relevant to the SDGs and learn to use the data effectively to push the 2030 Agenda forward. As a leader in this effort, the UN Sustainable Development Solutions Network (SDSN) integrates research, policy analysis, the data revolution, education, and global cooperation to promote SDG implementation.

## UN Sustainable Development Solutions Network

The SDSN was established in 2012 under the auspices of the UN Secretary-General. The SDSN mobilizes global scientific and technological expertise to promote practical solutions for sustainable development. Its work includes implementing the SDGs and the Paris Agreement. The SDSN works closely with UN agencies, multilateral financing institutions, the private sector, and civil society.

"Timely data are the key to achieving the SDGs—if we don't know where we are and how we can progress, we certainly can't arrive at the future we want," Sachs said.

The SDSN organizes its work around three priorities: national and regional networks and membership, research and policy tools, and The SDG Academy. The SDSN's networks of universities, research centers, and other knowledge institutions span six continents and comprise more than 1,500 members. They promote the local implementation of the SDGs, develop long-term transformation pathways for sustainable development, provide education around the 2030 Agenda, and launch solution initiatives to solve particular challenges.

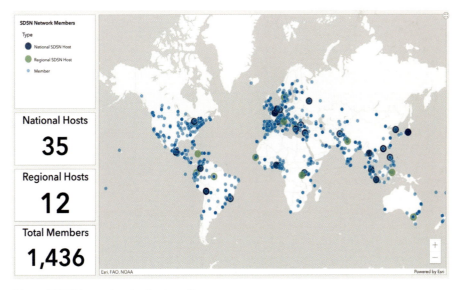

*Map of SDSN member institutions.*[18]

# SIX TRANSFORMATIONS

The SDGs require an integrated framework to support complementary actions across all sectors. The SDSN is committed to a holistic framework called the "Six Transformations to Achieve the SDGs".[19] Geospatial data is needed to understand how populations move, identify communities that lack access to essential services, track land use changes, and are essential to mapping the state of the SDGs at high temporal and spatial resolution. In recent years, the SDSN has improved efforts to determine how geospatial information can contribute to the six transformations.

To advance the global community toward a geospatial framework for sustainable development, the SDSN launched SDGs Today: The Global Hub for Real-Time SDG Data in partnership with Esri and the National Geographic Society. SDGs Today, an open access data platform, aims to advance the production and use of real-time and geospatial data for the SDGs. The initiative offers education and training resources to support countries, institutions, and civil society members to produce, share, and engage with the data to help meet the global goals by 2030.

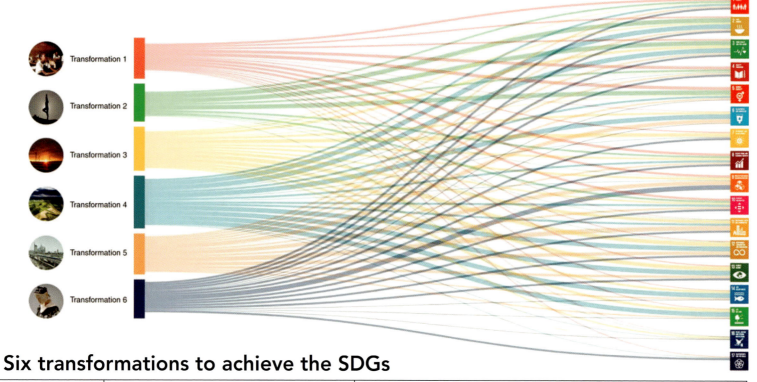

*Contribution of each SDG transformation toward the 17 SDGs.[20]*

## Six transformations to achieve the SDGs

| | Transformation | Objective | GIS-based contribution |
|---|---|---|---|
| 1 | Education, gender, and inequality | Aims to enhance education, which in turn improves economic growth, eliminates extreme poverty, supports decent work, and overcomes gender and other inequalities. | Geospatial data and technologies can map the links between spatial exclusion, access to education and other services, and socioeconomic and environmental correlates of education.[21,22] |
| 2 | Health, well-being, and demography | Promotes investments in health and well-being and health interventions in other sectors that can improve social determinants of health. | GIS-based analytics uses spatial, demographic, economic, and environmental data to study and analyze spatial relationships between communities, support health care planning, model disease-affected areas and accessibility to health care services, and enable other applications.[23,24] |
| 3 | Energy decarbonization and sustainable industry | Aims to improve universal access to modern energy sources, decarbonize the energy system in line with the Paris Agreement, and reduce industrial pollution of the soil, water, and air. | GIS-based energy system planning and modeling can optimize decisions on renewable installations, decarbonization, resource assessment, and economic needs.[25,26,27] |
| 4 | Sustainable food, land, water, and oceans | Encourages approaches to land use, ocean use, and water management to help manage competing claims on land and water for food production, urban development, industry and mining, ecosystem management, carbon sequestration, and biodiversity conservation. It has the same goal on the ocean for transport, food production, energy harvesting, and mining and tourism, per the Paris Agreement, and to reduce industrial pollution of the soil, water, and air. | GIS can be used for monitoring and risk quantification for drought, heat, cold, salinity, flooding, and pests; improving early warning systems; and tracking maritime traffic, biodiversity, coastal zones, pollution, etc.[28,29,30] |
| 5 | Sustainable cities and communities | Aims to enhance resilience against climate change and extreme weather events, ensuring access to the water supply, appropriate sewage and waste disposal, and efficient mobility. Also promotes more compact, safe, and healthy settlements. | GIS and analytical datasets such as the Global Human Settlement Layer[31] and Global Urban Footprint[32] can improve development planning, disaster monitoring, settlement mapping, measuring access to services, and strengthening infrastructure networks.[33,34] |
| 6 | Digital revolution for sustainable development | Aims to raise productivity, lower production costs, reduce emissions, expand access, reduce resource intensity of production processes, improve matching in markets, enable the use of big data, and make public services more readily available. | Geospatial data and technologies can enhance decision-making by using information with high temporal and spatial resolution. A digital transformation can change how we use existing and new sources of data and emerging technologies to measure and monitor progress across all geographies.[35,36,37,38] |

*GIS contributions related to the six transformations.*

# ESRI AND SDSN: GIS COLLABORATORS

"One of the most important things we can do as a society at this particular time is to help bring the nations of the world together to collaborate on and measure the progress of our collective work toward the Sustainable Development Goals. It is a crucial time for the nations of the world to work together to solve problems that transcend national borders, and which pose uniquely geographic challenges. GIS allows us to better understand these issues and measure the progress of our collective solutions."

—Jack Dangermond,
founder and president of Esri

SDGs Today endeavors to provide a snapshot of the state of sustainable development around the world in real time while enabling users to produce, access, and engage with timely data on sustainable development; obtain geographic information systems (GIS) training and education resources; and learn how to use the data effectively to drive action toward the 2030 Agenda for Sustainable Development. In early 2020, the SDSN partnered with Esri and the National Geographic Society to bring SDGs Today to life.

Esri possesses leading GIS technologies and training materials, and the SDSN provides a major propellant into the policy world at the highest levels and into a network of more than 1,000 universities. Working together, Esri and SDSN equip a wide audience of academics, policy makers, students, civil societies, and the public with the knowledge and tools required for evidence-based monitoring of the SDGs and for devising solutions across various communities globally.

The initiative strives to inspire and train the next generation of leaders, researchers, educators, and policy makers to cultivate the necessary knowledge and skills to carry out the implementation of the SDGs and future UN agendas. This initiative responds to the call of the UN secretary-general to accelerate sustainable solutions to the world's biggest challenges in the next decade.

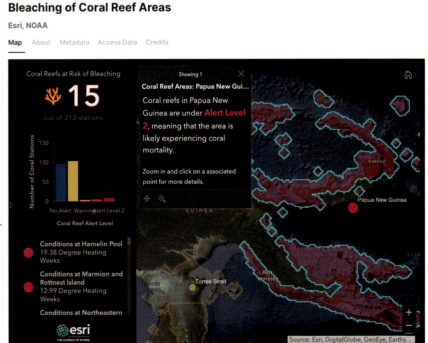

*The SDGs Today initiative aims to advance the production of real-time data and the integration of geospatial information within the SDG framework, explore new data sources and methods, and accelerate the production and use of timely and spatially disaggregated data for SDGs. More than just a data storage repository, SDGs Today acts as a hub to deliver a modern map-browsing interface, making it useful to everyone, not just spatial scientists. The hub is linked at GISforScience.com.*

# STORYTELLING WITH DATA

SDGs Today builds partnerships with many organizations to feature real-time and timely data on the SDGs after a comprehensive and systematic evaluation process that focuses on level of frequency, spatial coverage and disaggregation, validated methodology, reliability of sources, public accessibility, thematic relevance, ease of understanding, and sustainability of production.

SDGs Today also features stories that are created using ArcGIS StoryMaps and organized by specific SDGs throughout the data hub to help contextualize presented data and tell important stories that can effectively engage target audiences with the SDGs. These stories highlight a wide variety of SDG-related issues, demonstrate how various constituents can use key datasets, and help bridge the gap between data science and effective communication.

SDGs Today features free massive open online courses (MOOCs) and educational materials on sustainable development and the SDGs created by the SDSN's SDG Academy. It also includes guided Learn ArcGIS lessons on how to display and analyze data using maps designed by Esri.

The SDGs Today initiative provides data and analysis that informs policies, contributes to research, and enables national and global reporting on SDG-related efforts. The platform highlights a number of use cases on how timely data helps assess progress and the effectiveness of our work to meet the SDGs and leave no one behind in the process. One use case, for example, tells a story about the destructive force of Hurricane Dorian in 2019 in the context of SDG 13 and climate action.

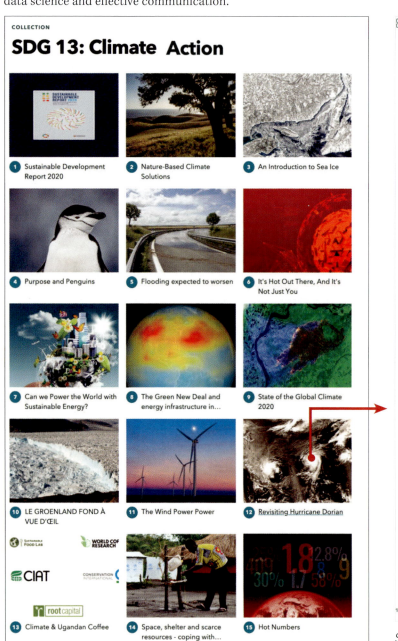

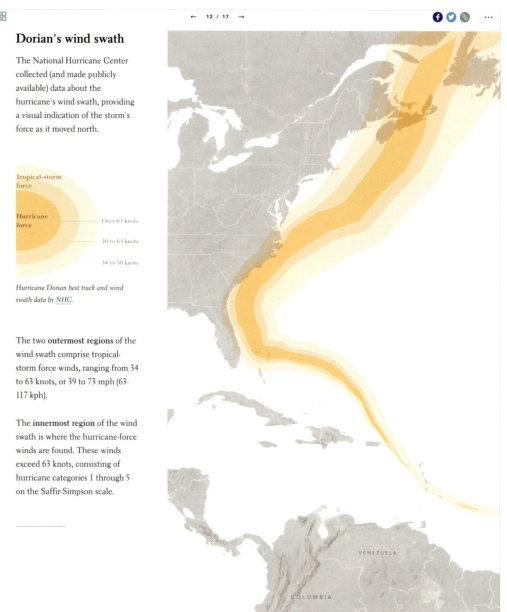

SDGs Today ArcGIS StoryMaps Collection for SDG 13.[39] Excerpt from the Revisiting Hurricane Dorian story shows the destructive course of the historic hurricane's wind path in 2019.

# GEOSPATIAL APPLICATIONS IN ACTION

> "Everything is interlinked—the global commons and global well-being. That means we must act more broadly, more holistically, across many fronts, to secure the health of our planet on which all life depends ... the world has not met any of the global biodiversity targets set for 2020. And so we need much more ambition and greater commitment to deliver on measurable targets and means of implementation, particularly finance and monitoring mechanisms."
>
> — UN Secretary-General António Guterres,
> *State of the Planet* address at Columbia University[40]

The SDGs Today hub includes real-time data layers from ArcGIS Living Atlas of the World. One of the layers features the World Database on Protected Areas (WDPA), which is crucial for the 2030 Agenda, particularly SDG 15: Life on land: the conservation, restoration, and sustainable use of terrestrial and freshwater ecosystems. Compiled and managed by the UN Environment Program World Conservation Monitoring Center as part of its Protected Planet project, the WDPA is the most comprehensive global spatial dataset measuring the coverage of marine and terrestrial protected areas. It is updated monthly and includes national parks, nature reserves, habitat management zones, natural monuments, protected landscapes, seascapes, and other areas. Governments submit data to the WDPA, which verifies and integrates the data into the dataset every month.[41]

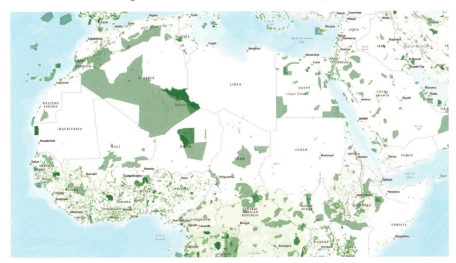

*Coverage of marine and terrestrial protected areas from the WDPA database.[42]*

The SDSN's Thematic Research Network on Data and Statistics (TReNDS) and Esri are also members of the POPGRID Data Collaborative,[43] which connects diverse data users, providers, and stakeholders from the public and private sectors working with georeferenced data on population, human settlements, and infrastructure. The size, demographics, and geographies of populations are changing globally. Fulfilling the SDG promise to leave no one behind requires frequent and reliable data on where people are and how they move.

Traditional population data sources, such as household surveys and population censuses, are not always updated in a timely manner. However, infrequent intervals of data collection, inaccessibility to certain geographic locations, and other social constraints (e.g., language barriers between enumerators and indigenous communities) often leave many people uncounted in the total population estimates. An increasing number of data providers are combining high-resolution satellite imagery with advanced GIS and remote sensing capabilities to produce gridded population datasets (using various modeling approaches to disaggregate population data using spatial data and imagery) that can complement census data and narrow timely data gaps.[44]

Gridded population data can be integrated in a wide range of application areas, such as in disaster response, health interventions, and sustainable planning.[45] The POPGRID Data Collaborative seeks to improve data access, timeliness, consistency, and utility; support data use and interpretation; identify and address pressing user needs; reduce duplication and user confusion; and encourage innovation and cross-disciplinary use. Many SDG indicators require total population count, and gridded population data is a valuable timely data resource for policy makers to measure population growth, monitor change, and plan interventions. The collaborative promotes cooperation in producing and harmonizing high-quality data products and services needed by a range of scientific and applied users with the aim to work toward the goal of leaving no one behind.

The collaborative uses gridded datasets for various applications. For instance, WorldPop[46] measures the availability and geographical accessibility of health care services at the national and subnational levels across sub-Saharan Africa. Improving women's access to health care, particularly reproductive services, is an important goal for SDG 3: Good health and well-being, and has been declared a priority by the United Nations.[47] However, monitoring of health care availability and access occurs primarily on the national level, which may obscure variations at the subnational level. In a 2020 study, WorldPop estimated the number of women of childbearing age, births, and pregnancies that are in and out-of-reach of essential health services and showed how data at varying spatial scales considerably impacts monitoring outcomes.[48]

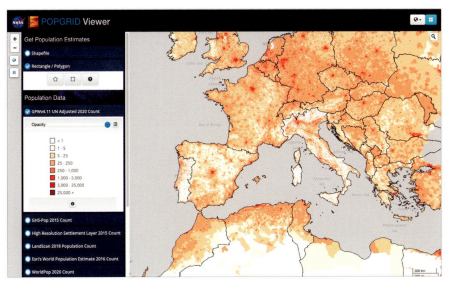

*Snapshot of the POPGRID Viewer.*

A different study, carried out by another one of the collaborative's members, the Center for International Earth Science Information Network (CIESIN), highlights the significance of gridded population data to measure SDG 9.1.1: The Rural Access Index (RAI). Open datasets are essential for producing globally comparable results in measuring progress toward the SDGs where high-quality local data is unavailable. CIESIN calculated RAIs for Nigeria, Colombia, and Spain using openly available roads data from OpenStreetMap and gridded population data from the High Resolution Settlement Layer and the Global Human Settlement Layer to measure the proportion of a country's rural population that lives within 2 kilometers walking distance of an all-season road[49]. Data on access to roads is critical to help ensure rural communities have access to health care facilities, local markets, schools, and other infrastructures and services necessary to enhance their quality of life.

The SDSN is a member of the Nature Map Consortium,[50] which developed an approach based on spatial priority that allows areas to be ranked based on their significance for biodiversity and carbon storage. This approach is already applied worldwide using globally consistent data on species distribution and carbon stocks. More recently, the Nature Map Consortium has been requested to adapt the data and methods developed by the project to provide scientifically robust estimates of the potential contribution that achieving global biodiversity targets can make to climate change mitigation at global and national scales. This work aims to quantify the emissions reductions and removals that could result from achieving a range of global biodiversity targets and make the results available in appropriate forms to climate policy leaders, national decision-makers, and the scientific community. Global progress requires national action, so it is crucial to apply these analytical approaches to inform national-scale decision-making, including the revision and delivery of NDCs.

The SDSN partners with many organizations to assess progress toward SDG achievement at national and local levels. The methodology to assess progress at national and subnational levels was audited by the European Commission Joint Research Center in 2019[51] and measures distance to targets for each of the SDGs. The SDG Indices aim to complement the official UN monitoring efforts and identify priorities for action, understand key implementation challenges, track progress, ensure accountability, and identify gaps that must be narrowed by 2030. SDGs Today and the SDGs Index and Monitoring will integrate geospatial data and analysis to provide SDSN with real-time assessment of SDGs progress and inform more timely decision-making processes.

## Building a better future with GIS

The potential of GIS applications for the SDGs is boundless, especially when these applications draw from a combination of policy and technical expertise. Looking ahead, we have only just scratched the surface on using GIS to address many critical challenges. SDSN's Science Panel for the Amazon, composed of more than 40 scientists and researchers from the eight Amazonian countries and one territory, combines scientific research on the Amazon and proposes solutions that will secure the future of the region and its inhabitants.[52] GIS and real-time data have and will continue to play a vital role in helping these scientists understand and guide solutions for the region's most pressing socioecological challenges.

In early 2021, the SDSN launched a new initiative to convene leading experts in urban planning and geodesign to support the green and digital transformation of cities globally. The SDSN's Six Transformations framework emphasizes the role of "Sustainable cities and communities" (Transformation 5) and "Harnessing the digital revolution for sustainable development" (Transformation 6) to achieve the SDGs and requires robust geospatial, disaggregated, and real-time data to achieve them. From the remote depths of our most precious ecosystems to the center of our rapidly changing cities, GIS will play a profound role in the global community's effort to achieve the SDGs.

SDGs Today: The Global Hub for Real-Time SDG Data will continue to grow and adapt to the needs of policy makers, researchers, students, and the public to ensure they can effectively use the best data and tools to track progress and implement solutions for the SDGs. The SDSN looks forward to continued collaboration and partnership with leaders in GIS technologies to accelerate progress toward the 2030 Agenda and beyond.

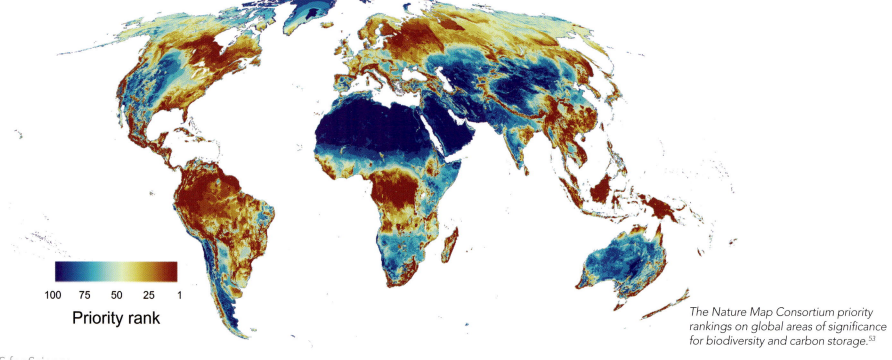

*The Nature Map Consortium priority rankings on global areas of significance for biodiversity and carbon storage.*[53]

Priority rank

100  75  50  25  1

# NOTES

1. Ban Ki-moon, "Secretary-General's Remarks at Summit for the Adoption of the Post-2015 Development Agenda," (September 25, 2015).

2. "The 17 Goals," *The 2030 Agenda for Sustainable Development*, The United Nations.

3. UNFCC, "Paris Agreement to the United Nations Framework Convention on Climate Change," (Dec. 12, 2015).

4. UNFCC, "Paris Agreement to the United Nations Framework Convention on Climate Change."

5. The Earth Institute, Columbia University, "It Takes a Village," last modified June 22, 2020, https://www.earth.columbia.edu/videos/view/it-takes-a-village-citizen-science-in-rural-india-grades-9-12.

6. The Earth Institute, Columbia University, "It Takes a Village."

7. IUCN, "The IUCN Red List of Threatened species," accessed June, 2021.

8. IUCN, "The IUCN Red List of Threatened Species."

9. Inbal Becker-Reshef et al., "Strengthening Agricultural Decisions in Countries at Risk of Food Insecurity: The GEOGLAM Crop Monitor for Early Warning," *Remote Sensing of Environment*, 237 (2020).

10. Eleonora Bonaccorsi, "GIS for a Sustainable World: Bringing the Power of Maps to the SDGs," International Institute for Sustainable Development (April 24, 2018).

11. Hindou Ibrahim, "Advocating Indigenous Peoples' Knowledge in Fighting Climate Change," The Global Sustainable Technology and Innovation Community (2021).

12. United Nations, "COVID-19 and Human Rights" (April 2020): https://doi.org/10.18356/514718a2-en.

13. Ibrahim, "Advocating" (2021).

14. Ibrahim, "Advocating" (2021).

15. StoryMaps Community, "Jeffrey Sachs Discusses the Critical Roles of GIS and Storytelling in Solving Global Challenges," *ArcGIS Blog* (August 7, 2020).

16. "Inter-agency and Expert Group on SDG Indicators," Sustainable Development Goal Indicators, https://unstats.un.org/sdgs/iaeg-sdgs.

17. J. Sachs, C. Kroll, G. Lafortune, G. Fuller, and F. Woelm, *Sustainable Development Report 2021: The Decade of Action for the Sustainable Development Goals* (Cambridge: Cambridge University Press: 2021).

18. Networks SDSN, "SDSN's Networks Program Advances Partnerships Globally," https://www.unsdsn.org/networks-overview.

19. Jeffrey Sachs et al., "Six Transformations to Achieve the Sustainable Development Goals," *Nature Sustainability*, 2 (2019): 805–814.

20. Jeffrey Sachs, "Six Transformations" (2019).

21. Local Burden of Disease Educational Attainment Collaborators, "Mapping Disparities in Education Across Low- and Middle-income Countries," *Nature* no. 577,7789 (2020): 235–238, doi:10.1038/s41586-019-1872-1.

22. S. Agrawal and R.D. Gupta, "School Mapping and Geospatial Analysis of the Schools in Jasra Development Block of India," *The International Archives of the Photogrammetry, Remote Sensing and Spatial Information Sciences*, XLI-B2, XXIII ISPRS Congress, 12–19 (Prague, Czech Republic: July 2016).

23. Bandar F. Khashoggi and Abdulkader Murad, "Issues of Health Care Planning and GIS: A Review," ISPRS *International Journal of Geographic Information* 9, no. 6: 352 (2020): https://doi.org/10.3390/ijgi9060352.

24. Center for International Earth Science Information Network (CIESIN), Columbia University, "SEDAC Global COVID-19 Viewer" (Palisades, New York: NASA Socioeconomic Data and Applications Center, 2020): https://sedac.ciesin.columbia.edu/mapping/popest/covid-19.

25. United Nations, "Towards Sustainable Renewable Energy Investment and Deployment: Trade-offs and Opportunities with Water Resources and the Environment," *ECE Energy Series* No. 63 (2020).

26. SDSN, Zero Carbon Action Plan, (New York: Sustainable Development Solutions Network, 2020).

27. W. Horan, S. Byrne, R. Shawe, R. Moles, and B. O'Regan, "A Geospatial Assessment of the Rooftop Decarbonisation Potential of Industrial and Commercial Zoned Buildings: An Example of Irish Cities and Regions," *Sustainable Energy Technologies and Assessment* no. 38 (April 2020): https://doi.org/10.1016/j.seta.2020.100651.

28. C. Mbow et al., "Food Security," in *Climate Change and Land: An IPCC Special Report on Climate Change, Desertification, Land Degradation, Sustainable Land Management, Food Security, and Greenhouse Gas Fluxes in Terrestrial Ecosystems*, eds. P.R. Shukla et al. (Intergovernmental Panel on Climate Change, 2019).

29. FABLE Consortium, *Pathways to Sustainable Land-Use and Food Systems* (Laxenburg and Paris: International Institute for Applied Systems Analysis and Sustainable Development Solutions Network, 2020).

30. UN Environment, "Regional Seas Follow-up and Review of the Sustainable Development Goals (SDGS)," *UN Environment Regional Seas Reports and Studies* no. 208 (2018).

31. European Commission, Joint Research Centre (JRC), "Global Human Settlement Layer," accessed June 2021, https://ghsl.jrc.ec.europa.eu/datasets.php.

32. German Aerospace Center, "Global Urban Footprint," accessed June 2021, https://www.dlr.de/eoc/en/desktopdefault.aspx/tabid-9628/16557_read-40454/.

33. German Aerospace Center, "Global Urban Footprint" (2021).

34. UN-HABITAT, *Tracking Progress towards Inclusive, Safe, Resilient and Sustainable Cities and Human Settlements: SDG 11 Synthesis Report 2018* (New York: United Nations Publications, 2018).

35. Group on Earth Observations, *Earth Observations in Support of the 2030 Agenda for Sustainable Development*, Japan Aerospace Exploration Agency (JAXA) on behalf of GEO under the EO4SDG Initiative (March 2017).

36. G. Scott and A. Rajabifard, "Sustainable Development and Geospatial Information: A Strategic Framework for Integrating a Global Policy Agenda into National Geospatial Capabilities," *Geo-spatial Information Science*, 20:2, 59–76, (2017), doi.org/10.1080/10095020.2017.1325594.

37. Ram Avtar et al., "Utilizing Geospatial Information to Implement SDGs and Monitor Their Progress," *Environmental Monitoring and Assessment* 192, no. 1 (2020).

38. Cameron Allen et al., "A Review of Scientific Advancements in Datasets Derived from Big Data for Monitoring the Sustainable Development Goals," *Sustain Sci* (2021): https://doi.org/10.1007/s11625-021-00982-3.

39. SDGs Today, "SDG 13: Climate Action," ArcGIS StoryMaps Collection, UN Sustainable Development Solutions Network, accessed June 2021, https://storymaps.arcgis.com/collections/3baa103e40a5469bbed7503fc203b76d.

40. António Guterres, "Secretary-General's Address at Columbia University: The State of the Planet" (December 2, 2020).

41. Heather C. Bingham et al., "User Manual for the World Database on Protected Areas and World Database on Other Effective Area Based Conservation Measures 1:6," *UNEP-WCMC* (2019).

42. SDGs Today, "SDG 13: Conservation," UN Sustainable Development Solutions Network, accessed June 2021, https://sdgstoday.org/dataset/conservation.

43. POPGRID Data Collaborative, accessed June 2021, https://www.popgrid.org.

44. SDSN TReNDS, "Leaving No One Off the Map: A Guide for Gridded Population Data for Sustainable Development," *Thematic Research Network on Data and Statistics* (May 14, 2020).

45. SDSN TRENDS, "Leaving No One Off the Map."

46. A.S. Wigley et al., "Measuring the Availability and Geographical Accessibility of Maternal Health Services across sub-Saharan Africa," *BMC Medicine* 18, no. 237 (2020), https://doi.org/10.1186/s12916-020- 01707-6.49.

47. UN Women, "Gender Equality: Women's Rights in Review 25 years After Beijing," (2020).

48. SDGs Today, "Access to Healthcare Services for Women of Childbearing Age," WorldPop, UN Sustainable Development Solutions Network, accessed June 2021.

49. CIESIN and Esri, "Rural Access to Roads," ArcGIS StoryMap (2020).

50. SDSN, Nature Map, https://www.unsdsn.org/nature-map, accessed June 2021.

51. Eleni Papadimitriou et al., *JRC Statistical Audit of the Sustainable Development Goals Index and Dashboards* (Luxembourg: Publications Office of the European Union, May 2019).

52. SDSN, "Science Panel for the Amazon," accessed June 2021.

53. SDSN, Nature Map, https://www.unsdsn.org/nature-map.

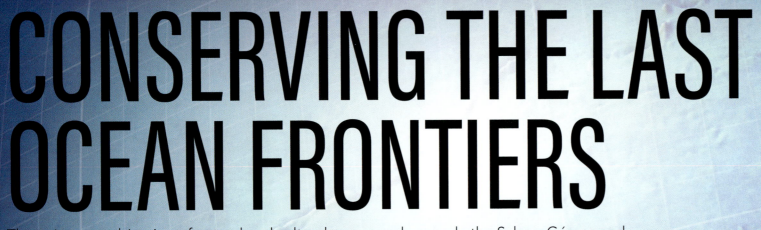

# CONSERVING THE LAST OCEAN FRONTIERS

The unique combination of natural and cultural resources has made the Salas y Gómez and Nazca Ridges a top priority for protection on the high seas. With support from global datasets and GIS analysis, researchers have identified this area off the west coast of South America as a key location to conserve and protect marine biodiversity without impacting industries.

By Daniel Wagner and T. 'Aulani Wilhelm, **Conservation International**; Alan M. Friedlander, **National Geographic Society, University of Hawai'i**; Richard L. Pyle, **Bernice P. Bishop Museum**; Kristina M. Gjerde, **International Union for Conservation of Nature**; Erin E. Easton, **University of Texas, Rio Grande Valley**; Carlos F. Gaymer and Javier Sellanes, **Universidad Católica del Norte**; Cassandra M. Brooks, **University of Colorado, Boulder**; Liesbeth van der Meer, **Oceana Chile**; and Lance E. Morgan and Samuel E. Georgian, **Marine Conservation Institute**

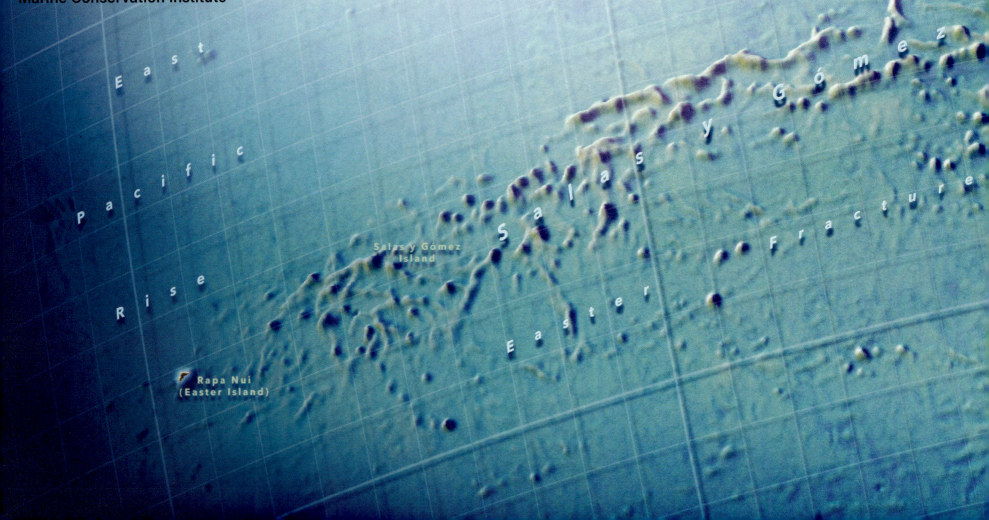

Peru-Chile

Peru-Chile Trench

Ridge

Ridge

Nazca

Desventuradas
Islands

Ridge

Zone

Chile
Rise

The Salas y Gómez and Nazca Ridges include more than 100 unique submarine mountains—
seamounts—of volcanic origin, and collectively stretch west from South America across nearly
2,900 kilometers to Rapa Nui (Easter Island) in the southeastern Pacific.

# INTRODUCTION

Covering more than 61% of the ocean's surface and 95% of its volume, marine areas known as the high seas represent one of the last science, conservation, and resource management frontiers on Earth. The laws of any one country are insufficient to protect the high seas because they lie beyond all national jurisdictions. These vast ocean expanses include the majority of the total inhabitable space for life on Earth. As such, they provide critical habitat to millions of species, the vast majority of which remain to be discovered. The high seas are crucial for sustaining life on Earth, because they produce nearly half of the oxygen we breathe, capture more than 1.5 billion tons of carbon dioxide each year, and contain nearly 90% of all life in the ocean by mass.

Although most people give little if any thought to these remote areas, the high seas have played a pivotal role in many seafaring cultures, which for millennia have navigated these areas to sustain themselves physically and spiritually. Even today, close to 90% of the world trade is carried out through international shipping on the high seas, and nearly 10 million tons of fish are caught there each year. Unfortunately, many of these activities are not well regulated. For example, some estimate that close to 90% of global fish stocks are either depleted or fully exploited, with much of this poorly regulated fishing taking place on the high seas. Various intergovernmental organizations, each with their own legal mandate, are responsible for regulating fishing, shipping, seabed mining, and other activities in these areas. The lack of a coordinated approach to conservation on the high seas makes resource management less effective.

Recognizing this critical gap, the United Nations in 2015 committed to develop a legally binding treaty to conserve and sustainably use marine biodiversity on the high seas. Plans for further negotiations offer hope that a legal mechanism will soon establish marine protected areas on the high seas. Currently, less than 8% of our ocean has any protection measures, far less than the 30% minimum that many scientific assessments conclude is necessary to limit the widespread impacts of climate change and arrest global declines in biodiversity. The high seas are by far the largest and least protected portion of our planet, and therefore our best opportunity for reaching the global target of protecting 30% of our oceans by 2030. However, scientific information is disproportionately scarce on the high seas, thereby complicating resource management and conservation planning.

This chapter presents a case study to show how various global datasets and GIS analyses supported the identification of one of the most promising places to establish a marine protected area on the high seas—the Salas y Gómez and Nazca Ridges. Numerous studies identified this region as one of the top protection priorities on the high seas because of its unique collection of natural and cultural resources. Fishing and other commercial activities are still minimal in this region, thus providing a narrow window of opportunity to proactively protect its natural and cultural resources without impacting industries. This chapter also is designed to serve as a foundation for guiding large-scale ocean conservation efforts elsewhere.

ISLA SALAS Y GÓMEZ

SOUTH PACIFIC OCEAN

25°S

Isla de Pascua
(Easter Island,
Rapa Nui)
(Chile)

Isla Salas y Gómez
(Chile)

MOTU MOTIRO
HIVA MARINE PARK

SALAS Y GÓMEZ RIDGE

EAST PACIFIC RISE

200 nautical mile limit

110°W

105°

30°

100 miles
100 kilometers

*The Salas y Gómez and Nazca Ridges are anchored on the west side by the islands of Salas y Gómez and Rapa Nui, which have extraordinary significance in Rapa Nui culture and include one of the most renowned archaeological sites on Earth.*

*Situated 250 nautical miles from its nearest neighbor, the incredibly remote Salas y Gómez Island, Chile, is one of the few spots along the Salas y Gómez and Nazca Ridges that rise above sea level.*

# JURISDICTIONAL BOUNDARIES

Unlike land, where the borders of many countries have been established and refined for many centuries, the delineation of countries' maritime borders is a relatively recent development. In 1982, the UN Convention on the Law of the Sea established the general rules that govern the uses of the ocean and its resources. Among many other things, the Convention provides the framework for how maritime borders are drawn, including those between adjacent coastal states, as well as those between coastal states and areas beyond national jurisdiction. Specifically, the Convention defines exclusive economic zones, an area in which countries have exclusive rights to explore, exploit, conserve, and manage all resources located on the seafloor and in the water column, including fishery resources, oil, gas, offshore energy, and seabed minerals. The Convention further gives nation states exclusive rights to establish artificial islands, installations, structures, conduct scientific research, and protect the marine environment within their exclusive economic zones. The boundaries of countries' exclusive economic zones generally extend to 200 nautical miles (370 kilometers) from their coastal baselines, with the exception of when that limit would overlap with another country.

The Convention also defines continental shelf boundaries of countries, which extend at least 200 nautical miles from coastal baselines, but potentially up to 350 nautical miles, if scientific evidence shows that the continental margin lies farther offshore. Countries that want to delimit their outer continental shelf beyond the 200 nautical miles limit can submit scientific justifications to the UN Commission on the Limits of the Continental Shelf. If the commission grants the claim, the country obtains sovereign rights over the seafloor and its resources in these additional areas. However, the country does not obtain rights over those resources found in the water column above it, which are still considered high seas. Consequently, areas claimed as extended continental shelf are governed in a mixed way, with activities on the seafloor regulated by the claimed country and activities in the water column regulated by the respective intergovernmental organization.

The Convention thus provides the framework for delimiting the maritime borders of countries and for those with areas beyond national jurisdiction. Seventy-three percent of the Salas y Gómez and Nazca Ridges are located in areas beyond national jurisdiction, with smaller portions located in the national waters of Chile and Peru. Specifically, the northeastern section of the Nazca Ridge is located in the national waters of Peru, whereas both ends of the Salas y Gómez Ridge are located within the Chilean exclusive economic zone. Additionally, in December 2020, the Chilean government submitted an extended continental shelf claim for 550,000 km² east of Salas y Gómez Island to gain sovereignty over the seafloor resources in this area, although the water column above it would still be considered high seas.

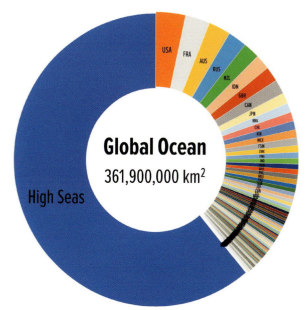

*Marine areas beyond national jurisdiction, commonly known as the high seas, cover more than 61% of the surface of the global ocean. Jurisdiction over the other 39% of the global ocean surface is divided among 157 countries.[1]*

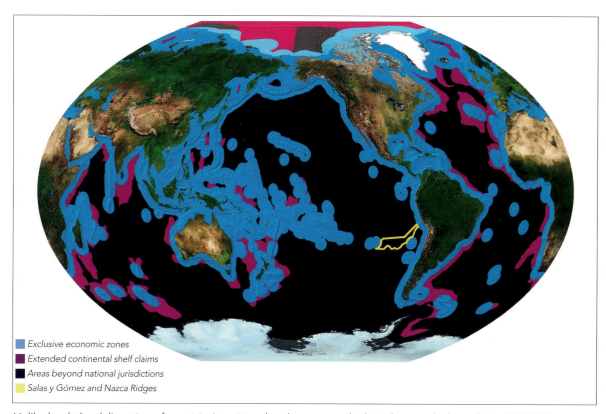

Exclusive economic zones
Extended continental shelf claims
Areas beyond national jurisdictions
Salas y Gómez and Nazca Ridges

*Unlike land, the delineation of countries' maritime borders occurred relatively recently through the 1982 UN Convention on the Law of Sea. Coastal states jurisdictions extend over all seafloor and water column resources within their exclusive economic zones,[1] which are generally located within 200 nautical miles of their coastlines. States can claim additional jurisdiction over seafloor resources if they can provide scientific evidence that their continental shelf extends beyond 200 nautical miles from shore.[2] The UN Convention of the Law of Sea therefore delimits both the maritime borders of countries, as well as those of areas beyond national jurisdiction, which cover the majority of the Salas y Gómez Ridges.*

# THE SALAS Y GÓMEZ AND NAZCA RIDGES

Extending more than 2,900 kilometers of seafloor off the west coast of South America, the Salas y Gómez and Nazca Ridges are two adjacent underwater mountain chains of volcanic origin. The more adjacent ridge to the South American continent, the Nazca Ridge, stretches across roughly 1,100 kilometers between the coast of Peru and the Desventuradas Islands. The Salas y Gómez Ridge spans approximately 1,600 kilometers between the Desventuradas Islands and Rapa Nui, also known as Easter Island. The Desventuradas Islands, Rapa Nui, and its close neighbor, Salas y Gómez Island, are the only places where the Salas y Gómez and Nazca Ridges rise above sea level. All of the other 110 peaks of these seamounts lie underneath the sea surface, where they create important habitats and migration corridors for many unique species.

The islands and seamounts that make up the ridges are thought to have been produced by a common geological hot spot located close to the present location of Salas y Gómez Island. Magma that first erupted here more than 27 million years ago grew to form seamounts that were carried eastward with the tectonic movement of the Nazca Plate. New seamounts formed with new eruptions, which then followed the journey of their predecessors eastward on the Nazca Plate. These seamounts provide a detailed chronological record of the geological formation of this region that tracks the movement of the Nazca Plate eastward before it gets subducted under South America. In addition to becoming older from west to east, these seamounts generally become progressively deeper moving eastward and range between just a few meters below the surface on the western portion of the ridges to more than 3,000 meters toward the northeastern end. Drowned fringing and barrier reefs are still evident on many of these deep seamounts, reminding us that these features were all near the sea surface at some point in their past.

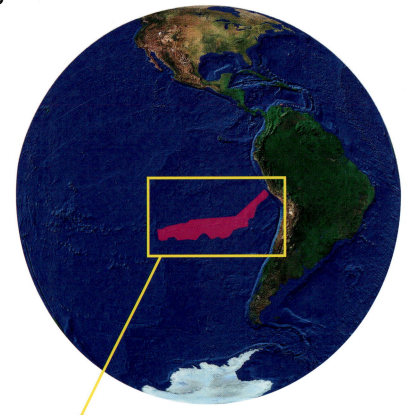

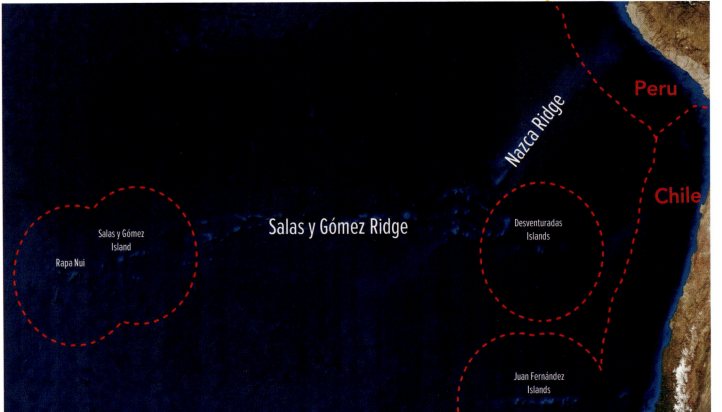

Located off the west coast of South America, the Salas y Gómez and Nazca Ridges[3] are two underwater mountain chains that collectively stretch across 2,900 kilometers in the South Pacific. The Salas y Gómez Ridge spans 1,600 kilometers between Rapa Nui and the Desventuradas Islands, where it connects to the Nazca Ridge, which covers another 1,100 kilometers of seafloor before meeting the coast of Peru.

# HISTORIC SCIENTIFIC SURVEYS

The general lack of exploration of the most remote high seas—compared with waters closer to shore—is evident in that one-third of all recorded species on the high seas is represented by a *single* record in the most comprehensive repository of global ocean biodiversity, the Ocean Biogeographic Information System. For the Salas y Gómez and Nazca Ridges, however, the Ocean Biogeographic Information System contains a relatively dense collection of records, including more than 14,000 occurrences of 930 species.[4] This number includes numerous records of ecologically important species such as whales, sea turtles, sharks, and reef-building corals, which is quite unique for an area on the high seas.

These species records result from various scientific expeditions, most notably a series of Russian expeditions that surveyed the ridges from 1973 to 1987. The Chilean National Oceanographic Committee surveyed the high seas portions of the ridges in 1999 and 2016. And in 2019, the Japan Agency for Marine-Earth Science and Technology led an expedition that included Chilean and other partners.

These scientific explorations noted that the marine biodiversity of this region is composed of a high proportion of endemic species, or species that are not known to occur anywhere else on Earth. For many groups of organisms, nearly half of the species are endemic to the region, the highest level of marine endemism known anywhere on Earth. Furthermore, previous surveys noted that every seamount of these ridges appears to have a unique community composition, with few species shared between opposite ends of the ridges. This finding highlights the need to protect all the ridge features to safeguard representative biodiversity.

*Russian survey vessel Ikhtiandr.*

Professor Mesyatzev  Professor Shtokman  Mirai  Cabo de Hornos

*Compared to many still virtually unexplored regions of the world's high seas, the international waters of the Salas y Gómez and Nazca Ridges have been surveyed by a series of Russian, Chilean, and Japanese expeditions beginning in the early 1970s. These vessels carried the latest technologies of the day. Data from these expeditions have since made it into GIS databases and applications such as Esri's Ocean Basemap, which can be accessed at GISforScience.com.*

# SEAMOUNT AND SEAFLOOR MAPPING DATA

Seamounts are defined as underwater mountains that rise at least 1,000 meters above the seafloor. Seamounts and other steep topographical features significantly impact the deeps-sea environment, because they accelerate currents, increase upwelling, and increase food supply, and thereby provide favorable habitat for a wide variety of organisms. Many conservation initiatives have prioritized seamounts for protection because they are regarded as some of the most diverse and productive habitats in the ocean.

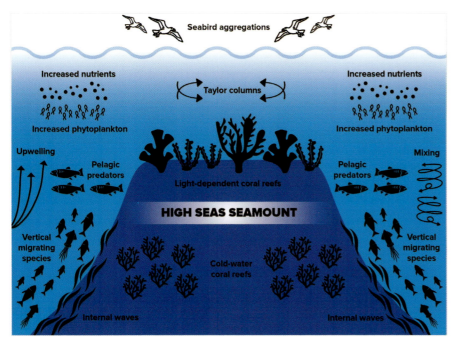

*Seamounts are widely regarded as one of the top ocean conservation priorities globally, because these features create favorable conditions and habitats for a wide variety of marine life.[5]*

The most accurate way to identify seamounts is to map the seafloor using echosounders mounted on ships or on submersible vehicles. Using such modern mapping technologies, researchers have mapped about 20% of the seafloor to date at a resolution high enough to identify fine topographic details (~100 meters).[6] Most of the areas lacking detailed mapping are located on the high seas, where there have been almost no dedicated mapping surveys, and only few surveys during opportunistic ship transits. This pattern is also true for high seas waters surrounding the Salas y Gómez and Nazca Ridges, the majority of which has not been mapped in detail using modern mapping systems.

In areas that lack high-resolution mapping, satellite altimetry data can help generate coarser maps of the seafloor that can resolve large topographical features to a resolution of about 1 kilometer. Satellite-derived bathymetry data have been used to generate various databases on seamounts and other steep topographical features. These datasets indicate that the Salas y Gómez and Nazca Ridges contain a large aggregation of seamounts and other steep topographical features, such as ridges and escarpments, which are known to provide suitable habitat for diverse marine life. The Salas y Gómez and Nazca Ridges contain an estimated 110 or more seamounts, or approximately 41% of all seamounts found in the southeastern Pacific Ocean.

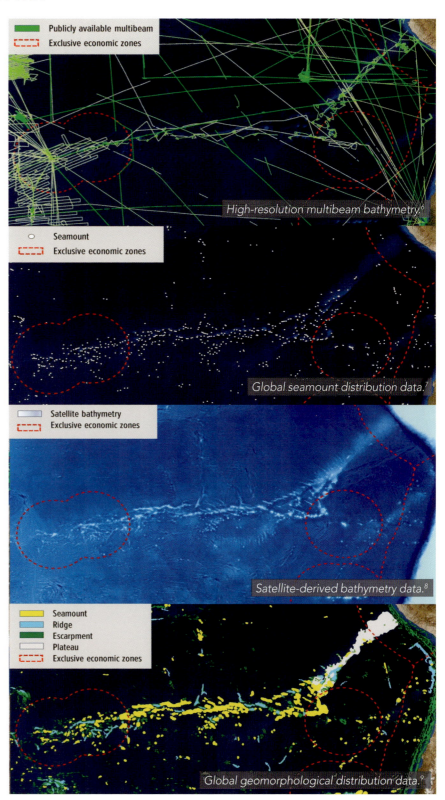

*Maps showing seafloor mapping data surrounding the Salas y Gómez and Nazca Ridges from four source technologies.*

# INTERNATIONAL DISTINCTIONS

In 2014, the Conference of the Parties to the Convention of Biological Diversity[3] recognized the Salas y Gómez and Nazca Ridges as an ecologically or biologically significant marine area. This distinction recognizes globally significant marine areas that need protection. These areas are evaluated based on seven criteria, including uniqueness, special importance for life stages of species, importance for threatened or endangered species, vulnerability, biological productivity, biological diversity, and naturalness

*Seamounts provide critical habitats and ecological stepping-stones for whales, sea turtles, corals, and many other ecologically important species, including 82 threatened or endangered species that are known to inhabit the Salas y Gómez and Nazca Ridges.[11]*

The Global Ocean Biodiversity Initiative and the Global Census of Marine Life on Seamounts also recognized the ridges as an important area, and Mission Blue identified the ridges as a Hope Spot, which are classified as special places critical to the health of the ocean.[12] Based on spatial distribution data of the International Union for Conservation of Nature Red List of Threatened Species,[13] the ridges provide important habitats to 82 threatened or endangered species, including 25 species of sharks and rays, 21 species of birds, 16 species of corals, 7 species of marine mammals, 7 species of bony fishes, 5 species of marine turtles, and 1 species of sea cucumber. Salas y Gómez Island and the Desventuradas Islands are considered Important Bird Areas by BirdLife International,[14] because they host important colonies of Christmas Island shearwater, masked booby, white-throated storm petrel, de Filippi's petrel, and Chatham petrel.

The region has also been recognized internationally based on its rich cultural heritage. The island of Rapa Nui includes one of the most renowned archaeological sites on Earth, which has been distinguished globally as a World Heritage Site by the United Nations Educational, Scientific and Cultural Organization (UNESCO).[15] The broader region that contains the Salas y Gómez and Nazca Ridges represents the easternmost corner of the Polynesian Triangle, a region with an exceptionally rich and long history of seafaring cultures. Polynesian and other seafarers have sailed across these ridges for centuries, and archival research indicates several shipwrecks likely occurred in this region.

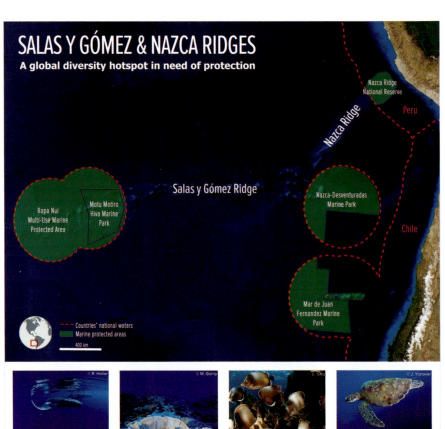

## SALAS Y GÓMEZ & NAZCA RIDGES
### A global diversity hotspot in need of protection

Nazca Ridge National Reserve

Peru

Nazca Ridge

Salas y Gómez Ridge

Nazca-Desventuradas Marine Park

Chile

Rapa Nui Multi-Use Marine Protected Area

Motu Motiro Hiva Marine Park

Mar de Juan Fernandez Marine Park

- - - Countries' national waters
Marine protected areas
400 km

The Salas y Gómez and Nazca Ridges are one of the most unique diversity hotspots on Earth. Located off the west coast of South America, they include over 110 seamounts that stretch across 2,900 kilometers.

Over 73% of these ridges lie in areas beyond national jurisdiction, where they are unprotected and under threat from overfishing, plastic pollution, climate change, and potential deep-sea mining.

The region harbors one of the most unique collections of ecosystems and species on Earth. For many groups of organisms, nearly half of the species are endemic to the region and found nowhere else on our planet.

The ridges provide critical habitats and migration corridors for whales, sea turtles, and numerous other ecologically important species, including 82 threatened or endangered species.

The Salas y Gómez and Nazca Ridges represent the easternmost corner of the Polynesian Triangle, a region with an exceptionally rich and long history of seafaring cultures.

Due to its high productivity, the region provides important feeding grounds for a wide variety of seabirds.

Recent explorations in this region have documented one of the deepest light-dependent coral reefs on Earth, as well as numerous species that are new to science.

Commercial activity is still low in this region. We must act now to protect its unique resources before they are lost forever.

*Excerpt from a Salas y Gómez and Nazca Ridges fact sheet.*

The international distinctions and supporting scientific data prompted Chile and Peru to establish several marine protected areas in the portions of this region that fall under their jurisdiction.[16] These areas include the Rapa Nui Multiple-Use Coastal Marine Protected Area, the Motu Motiro Hiva Marine Park, the Nazca-Desventuradas Marine Park, and the Juan Fernández Multiple-Use Coastal Marine Protected Area, all established by Chile in the last decade. In June 2021, Peru created the Nazca Ridge National Reserve,[17] which protects the seafloor portion of the Nazca Ridge that falls in Peruvian national waters. Despite these important advances, more than 73% of the Salas y Gómez and Nazca Ridges lack any protected status.

# HUMAN ACTIVITIES AND MANAGING AUTHORITIES

Within areas beyond national jurisdiction, the authority to regulate human activities belongs to various intergovernmental organizations, including the International Maritime Organization for shipping activities, the International Seabed Authority for seabed mining activities, and regional fishery management bodies, which regulate fishing activities for specific fishery species and geographic regions.

## CONSERVATION DISTINCTIONS

**1966** ▪ Chile establishes the Rapa Nui National Park to protect ~40% of the terrestrial areas of Easter Island for their extraordinary cultural significance.

**1976** ▪ Chile establishes the Salas y Gómez Nature Sanctuary to protect the entire land of Salas y Gómez Island.

**1995** ▪ The United Nations Educational, Scientific and Cultural Organization recognizes the Rapa Nui National Park as a World Heritage Site for its exceptional cultural significance.

**2010** ▪ Chile establishes the Motu Motiro Hiva Marine Park to protect 150,000 km² around Salas y Gómez Island.
▪ BirdLife International recognizes the islands of Salas y Gómez, San Felix, and San Ambrosio as Important Bird Areas.

**2011** ▪ The Salas y Gómez and Nazca Ridges are recognized as an important area by the Global Ocean Biodiversity Initiative and the Census of Marine Life on Seamounts.

**2014** ▪ The Salas y Gómez and Nazca Ridges are recognized as an ecologically or biologically significant area at the 12th Meeting of the Conference of the Parties to the Convention on Biological Diversity.
▪ Chile passes the vulnerable ecosystem law, thereby protecting all Chilean waters surrounding the Salas y Gómez and Nazca Ridges from bottom trawling.

**2015** ▪ Chile establishes the Nazca-Desventuradas Marine Park to protect 300,035 km² around the islands of San Félix and San Ambrosio, also known as the Desventuradas Islands.
▪ Mission Blue recognizes the Salas y Gómez and Nazca Ridges as a Hope Spot, which are special places that are critical to the health of our global ocean.

**2016** ▪ Chile designates the Mar de Juan Fernández Multiple-Use Coastal Marine Protected Area to protect 12,000 km² around the Juan Fernández Archipelago.

**2018** ▪ Chile establishes the Rapa Nui Multi-Use Marine Coastal Protected Area designated, which protects 579,368 km² around Easter Island, thereby making it the largest marine protected area in the Americas.
▪ Chile establishes the Mar de Juan Fernández Marine Park to protect 286,000 km² around the Juan Fernández Archipelago.
▪ Chile expands the Mar de Juan Fernández Multiple-Use Coastal Marine Protected Area to protect 24,000 km² around the islands of Juan Fernández.

**2021** ▪ Peru creates Nazca Ridge National Reserve to protect 62,392 sq. km. of seafloor in Peruvian waters around the Nazca Ridge.

*As a result of its unique collection of natural and cultural resources, the region has been acknowledged for its exceptional significance by numerous organizations.*

The International Maritime Organization is responsible for regulating international shipping activities on the oceans, which includes having the authority to implement conservation measures to prevent potential environmental impacts on fragile ecosystems. Specifically, the International Maritime Organization has the authority to designate particular sensitive sea areas, which may be protected by ship routing measures, such as areas to be avoided by all ships, or by certain classes of ships. There have been no particular sensitive sea area designations anywhere on the high seas, nor are there any shipping route limitations around the Salas y Gómez and Nazca Ridges. However, with the exception of the northern section of the Nazca Ridge, this region does not contain major commercial shipping routes.

While deep-sea mining has not yet occurred, regulations for this developing industry are currently being created. The International Seabed Authority, an intergovernmental institution established under the 1982 UN Convention on the Law of the Sea, is mandated with developing deep-sea mining regulations in areas beyond national jurisdiction, commonly known as the mining code. The mining code is currently only applicable to the prospecting or exploration for deep-sea minerals, but not yet to the exploitation or collection of minerals for commercial purposes, regulations for the latter of which are still under development. To date, the International Seabed Authority has approved 31 licenses to explore for seabed minerals, which are located in the Clarion-Clipperton Zone, the North West Pacific, the Mid-Atlantic Ridge, the Rio Grande Rise (South Atlantic), and several areas in the Indian Ocean. No contracts have been issued for the exploration of deep-sea minerals in the South Pacific. While commercially valuable seabed minerals occur on seamounts of the Salas y Gómez and Nazca Ridges, the International Seabed Authority has not yet developed exploration regulations in this region. Thus, this area could be proactively closed to seabed mining without having any impact on this developing industry.

Fishing activities on the high seas surrounding the ridges are managed by two regional fishery management organizations: the Inter-American Tropical Tuna Commission, which manages highly migratory fishery species such as tuna, billfish, and sharks; and the South Pacific Regional Fisheries Management Organisation, which manages non-highly migratory fishery species, such as jack mackerel, giant squid, and orange roughy. Some historical fishing has targeted Chilean jack mackerel, giant squid, tuna, striped bonito, marlin, and swordfish on the ridges. However, today most of the fishing in this region targets pelagic species and is primarily focused on high seas waters outside Peru. For fisheries managed by the South Pacific Regional Fisheries Management Organisation, fishing is virtually nonexistent. The orange roughy fishery has been closed in this region since 2006, and total effort for giant squid and Chilean jack mackerel have collectively accounted for less than one day in eight years. Similarly, catch data for most species of billfishes and sharks have been zero in this region in the last 10 years, with the exception of black marlin, blue marlin, striped marlin, and swordfish. However, low catch rates for these species have cumulatively accounted for only 40 metric tons in the last 10 years. In contrast, tuna fisheries targeting skipjack, bigeye, and yellowfin tuna, are active in the region; however, these activities are strongly localized in the area just outside Peruvian national waters. These activities are all conducted by distant water fishing fleets, as more than 96% of the fishing effort in this region is conducted by vessels flagged by China, Spain, Japan, Taiwan, and the Republic of Korea. Importantly, these tuna fishing fleets are much more successful when operating in areas outside the ridges. Thus, closing the Salas y Gómez and Nazca Ridges to commercial fishing activities would have little or no impact on fishing industries.

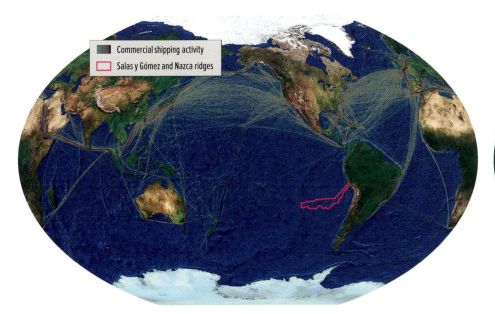

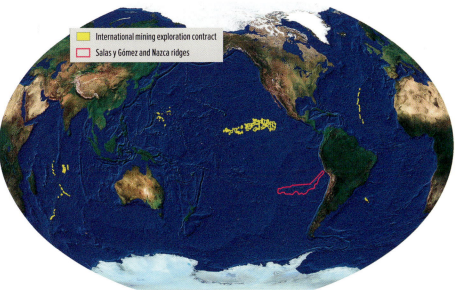

Close to 90% of the world trade is carried out through international shipping[18] on the high seas, activities that are regulated by the International Maritime Organization. International shipping currently does not have a major footprint on most of the Salas y Gómez and Nazca Ridges. With the exception of the northern section of the Nazca Ridge, this region does not intersect major shipping routes.

Mining activities in the international seabed are regulated by the International Seabed Authority, which by 2021 has issued 31 contracts allowing for the exploration of seabed minerals[19] and is in the process of developing regulations that would allow exploitation of these resources. Deep-sea mineral exploration has not yet occurred on the ridges, despite having commercially valuable minerals, thus providing a window of opportunity to proactively protect this area without significantly impacting the developing mining industry.

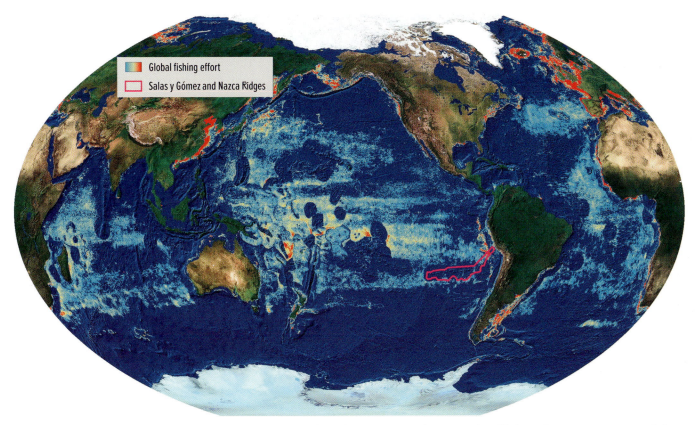

Commercial fishing activities are minimal to nonexistent on the Salas y Gómez and Nazca Ridges.[20] The only exception are tuna fisheries concentrated just outside Peruvian national waters. However, these fisheries targeting skipjack, bigeye, and yellowfin tuna are much more successful when operating outside this area. Consequently, closing the Salas y Gómez and Nazca Ridges to commercial fishing activities would have minimal to nonexistent impacts on fishing industries.

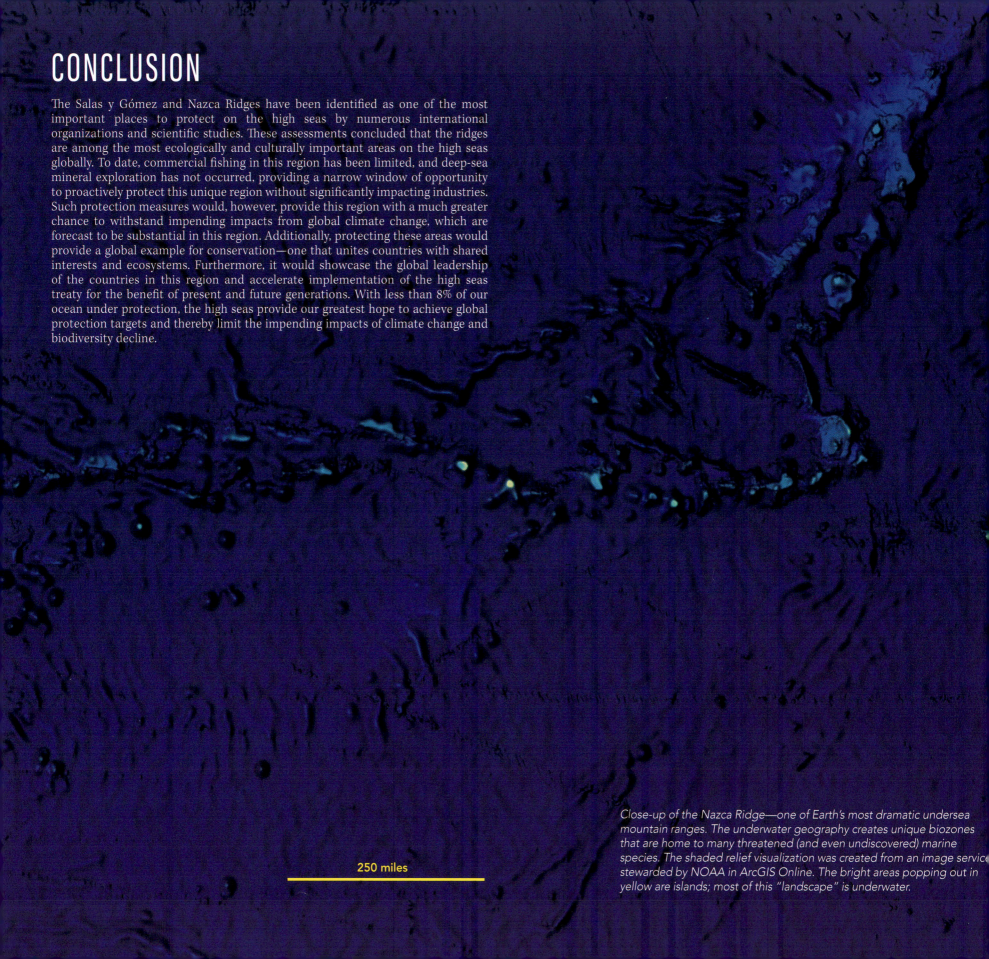

# CONCLUSION

The Salas y Gómez and Nazca Ridges have been identified as one of the most important places to protect on the high seas by numerous international organizations and scientific studies. These assessments concluded that the ridges are among the most ecologically and culturally important areas on the high seas globally. To date, commercial fishing in this region has been limited, and deep-sea mineral exploration has not occurred, providing a narrow window of opportunity to proactively protect this unique region without significantly impacting industries. Such protection measures would, however, provide this region with a much greater chance to withstand impending impacts from global climate change, which are forecast to be substantial in this region. Additionally, protecting these areas would provide a global example for conservation—one that unites countries with shared interests and ecosystems. Furthermore, it would showcase the global leadership of the countries in this region and accelerate implementation of the high seas treaty for the benefit of present and future generations. With less than 8% of our ocean under protection, the high seas provide our greatest hope to achieve global protection targets and thereby limit the impending impacts of climate change and biodiversity decline.

250 miles

*Close-up of the Nazca Ridge—one of Earth's most dramatic undersea mountain ranges. The underwater geography creates unique biozones that are home to many threatened (and even undiscovered) marine species. The shaded relief visualization was created from an image service stewarded by NOAA in ArcGIS Online. The bright areas popping out in yellow are islands; most of this "landscape" is underwater.*

# NOTES AND DATA SOURCES

Full author affiliations: Daniel Wagner and T. 'Aulani Wilhelm, Conservation International, Center for Oceans; Alan M. Friedlander, National Geographic Society, University of Hawai'i; Richard L. Pyle, Bernice P. Bishop Museum (Hawaii); Kristina M. Gjerde, International Union for Conservation of Nature; Erin E. Easton, University of Texas, Rio Grande Valley, School of Earth, Environmental, and Marine Sciences; Carlos F. Gaymer and Javier Sellenas, Millennium Nucleus for Ecology and Sustainable Management of Oceanic Islands, Universidad Católica del Norte, Coquimbo, Chile; Cassandra M. Brooks, University of Colorado, Boulder, Environmental Studies Program; Liesbeth van der Meer, Oceana Chile, Santiago; and Lance E. Morgan and Samuel E. Georgian, Marine Conservation Institute.

1. Flanders Marine Institute, *Maritime Boundaries Geodatabase: Maritime Boundaries and Exclusive Economic Zones* (200NM), version 11 (2019).

2. United Nations Environmental Program, Extended continental shelf shapefiles (2021).

3. *Convention on Biological Diversity*, Salas y Gómez and Nazca Ridges ecologically or biologically significant areas (2021).

4. *Ocean Biogeographic Information System*, Nazca and Salas y Gómez, https://obis.org/area/10172 (2021).

5. D. Wagner, A.F. Friedlander, R.L. Pyle, C.M. Brooks., K.M. Gjerde, and A.M. Wilhelm, "Coral Reefs of the High Seas: Hidden Biodiversity Hotspots in Need of Protection," *Frontiers in Marine Science* 7 (2020): 567428.

6. NOAA National Centers of Environmental Information, *Multibeam Bathymetry Database (MBBDB)* (2021).

7. C. Yesson, M.R. Clark, M. Taylor, and A.D. Rogers, "The Global Distribution of Seamounts Based on 30-second Bathymetry Data," DeepSea Research Part I: *Oceanographic Research Papers* 58 (2011): 442–453.

8. GEBCO Compilation Group, with 12× vertical exaggeration. Land imagery: NASA (2020).

9. P.T. Harris, M. Macmillan-Lawler, J. Rupp, and E.K. Baker, "Geomorphology of the Oceans," *Marine Geology* 352, 1 (June 2014): 4–24, https://doi.org/10.1016/j.margeo.2014.01.011.

10. D. Wagner, L. van der Meer, M. Gorny, J. Sellanes et al., "The Salas y Gómez and Nazca Ridges: A Review of the Importance, Opportunities, and Challenges for Protecting a Global Diversity Hotspot on the High Seas," *Marine Policy* 126 (2021): https://doi.org/10.1016/j.marpol.2020.104377.

11. A.M. Friedlander, W. Goodell, J. Giddens, E.E. Easton, and D. Wagner, "Deep-Sea Biodiversity at the Extremes of the Salas y Gómez and Nazca Ridges with Implications for Conservation," *PLOS ONE* 16, no. 6 (2021): e0253213, https://doi.org/10.1371/journal.pone.0253213.

12. Mission Blue: Sylvia Earle Alliance, "Hope Spots," *Mission Blue*, https://mission-blue.org/hope-spots.

13. International Union for Conservation of Nature, *IUCN Red List of Threatened Species*, IUCN, Spatial Data and Mapping Resources, https://www.iucnredlist.org/resources/ spatial-data-download.

14. BirdLife International, *Important Bird and Biodiversity Area Digital Boundaries: September 2019 version* (Cambridge: BirdLife International, 2019).

15. UNESCO, United Nations Educational, Scientific, and Cultural Organization World Heritage Sites data, https://whc. unesco.org/en/syndication.

16. *World Database of Protected Places*, Marine Protected Areas, UN Environment World Conservation Monitoring Centre, (2021).

17. MINAM, Supreme Decree that Established the Nazca Ridge National Reserve (June 2021).

18. B.S. Halpern, M. Frazier, J. Potapenko, K.S. Casey et al., "Spatial and Temporal Changes in Cumulative Human Impacts on the World's Ocean," *Nature Communications* 6 (2015): 7615.

19. International Seabed Authority, International Seabed Authority Maps (2020).

20. Global Fishing Watch, *Global Fishing Watch* map and data (2020).

John Nelson, opening map of Salas y Gómez and Nazca Ridges, Esri; Bathymetry, GEBCO Compilation Group with 12× vertical exaggeration; and Land Imagery, NASA (2020).

Photos on page 95 courtesy of Javier Sellanes, IORAS and Tsusima.ru.

Photos on page 97 courtesy of Rodolphe Holler, Matthias Gorny, Enric Sala, Jeff Yonover, Daniel Wagner, and Maël Imirizaldu.

# PART 3
# HOW WE LOOK AT EARTH

Successfully understanding how Earth works and looks requires integrative and innovative approaches to observation and measurement. These approaches include earth observation in varying forms, such as from sensors on satellites, aircraft, drones, and ships. They also include the important data science issues of conducting analysis; modeling, developing, and documenting useful datasets for science; and interoperating between these datasets and between various approaches.

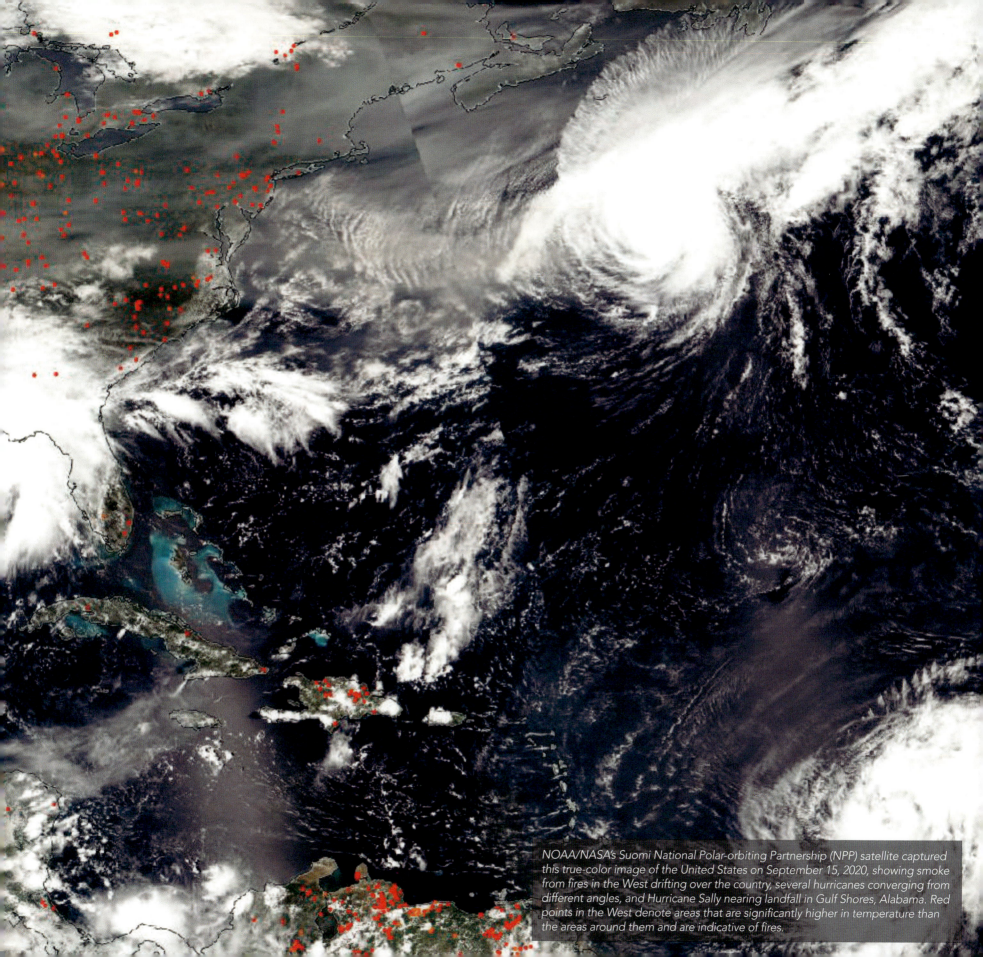

NOAA/NASA's Suomi National Polar-orbiting Partnership (NPP) satellite captured this true-color image of the United States on September 15, 2020, showing smoke from fires in the West drifting over the country, several hurricanes converging from different angles, and Hurricane Sally nearing landfall in Gulf Shores, Alabama. Red points in the West denote areas that are significantly higher in temperature than the areas around them and are indicative of fires.

# AI FOR GEOSPATIAL ANALYSIS

Marine and earth scientists are using treasure troves of emerging geospatial data to apply artificial intelligence to their work as never before. Pervasive cloud computing and rapidly improving machine and deep learning algorithm capabilities have combined to create a new GIS neural network.

By Bonnie Lei, **Microsoft;** Kate Longley-Wood and Zach Ferdana, **The Nature Conservancy;** Susanna De Beauville-Scott, **Organisation of Eastern Caribbean States;** and Julian Engel, **OceanMind**

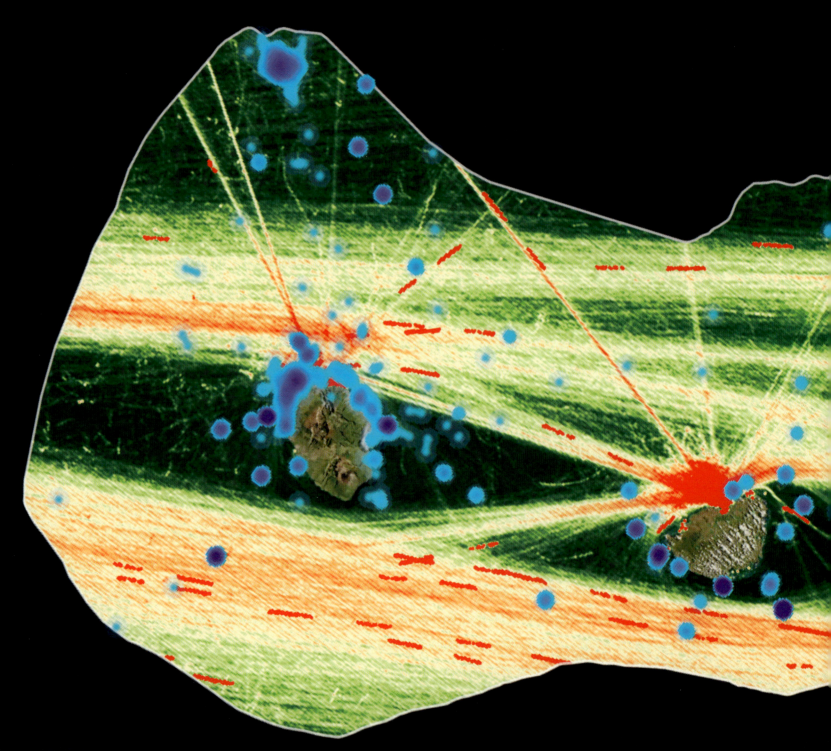

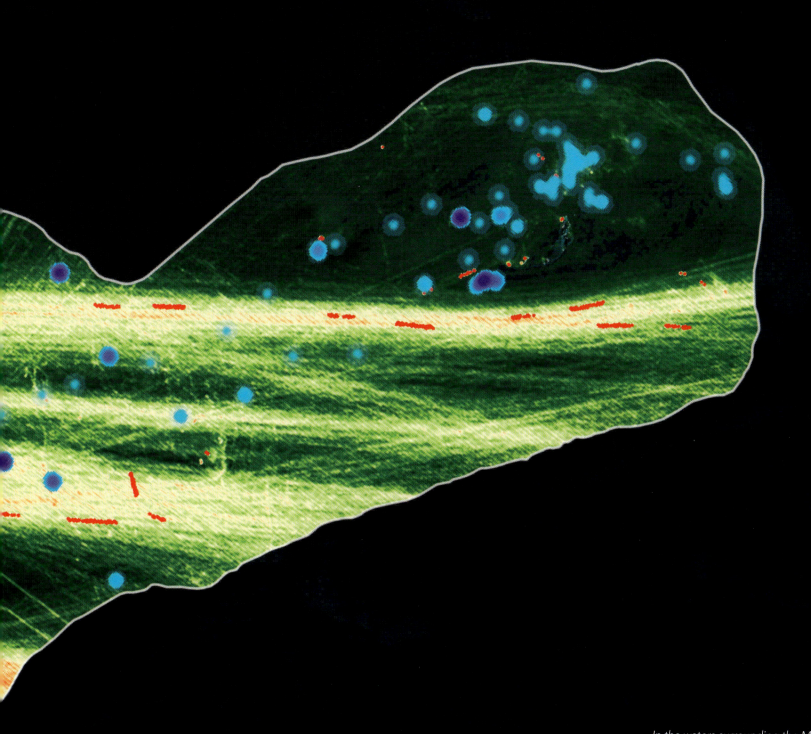

*In the waters surrounding the Mascarene Islands, marine mammal aggregation points (shown in blue) and possible pathways between those in comparison to vessel activity.*

# GIS ON THE NEURAL NETWORK

Facing unprecedented challenges, the world needs the sciences to help to mitigate climate change, ensure abundant water supplies, feed a growing population, and confront the catastrophic loss of biodiversity. Finding solutions can be daunting when it comes to understanding the natural world because we face an information drought.

On the forefront of the global effort to bring better information to bear on these pressing challenges, scientists are seeing the emergence of artificial intelligence (AI) as a path toward the application of information technology to design better ways of doing things, and hopefully accelerating humanity down the path toward a more sustainable future. AI allows us to automatically collect and process huge amounts of data that would otherwise require vast amounts of human capital. Recent machine learning advancements allow these systems to match and even surpass human accuracy for tasks such as image classification, population mapping, and habitat monitoring. The intersection of AI and geographic information systems (GIS) presents an opportunity to solve some of the world's most pressing environmental challenges.

Increasing numbers of sensors, satellites, and drones provide more available data and meaningful analysis and at a greater rate than previously possible. The growing capability of cloud computing, machine learning algorithms, and deep neural networks all support GIS, which in turn becomes more useful when it capitalizes on powerful cloud computing capabilities, ever-improving machine learning algorithms, and deep neural networks.

Microsoft AI for Earth empowers organizations to innovate at the intersection of environmental science and data and computer science. The program, according to Microsoft, focuses its AI and research technology on four areas of environmental sustainability: agriculture, water, biodiversity, and climate change. Recognizing the importance of applying AI approaches at cloud scale to geospatial data, the AI for Earth grants program was launched in 2017. By partnering with Esri from the beginning, grantees received access to Microsoft Azure and Esri ArcGIS software. Esri's data analytics and mapping technologies paired with Microsoft's cloud and AI solutions deliver crucial insights to conservationists worldwide. To date, the program has supported more than 700 grantees working in more than 100 countries.

This chapter will present two projects that have used geospatial AI approaches to work on important issues in conserving marine biodiversity—one from the Caribbean and the other featuring the remote Mascarene Islands east of Madagascar.

The importance of the ocean to global environmental health cannot be overstated. According to the United Nations Educational, Scientific and Cultural Organization (UNESCO), the ocean is the source of more than half of the oxygen we breathe. It profoundly affects climate and weather and absorbs nearly a third of the carbon dioxide emitted into the atmosphere because of human activities. The ocean has socioeconomic impacts on many nations, with the livelihoods of many impacted by the ocean's health. Because the ocean is so vast, pairing AI with remote sensing can help us better understand and monitor the breadth and depth of marine ecosystems.

*Saint Kitts and Nevis were among the first islands in the Caribbean to be colonized by Europeans. Today, the islands' rich but fragile marine ecosystem benefits the region economically, in part through tourism.*

# MAPPING OCEAN WEALTH

The Caribbean is more dependent on the travel and tourism sector than any other region worldwide, accounting for more than 10% of the gross domestic product, and 15.2% of jobs in the region. This sector is almost entirely focused on coastal areas, notably through beach-based activities, cruise tourism, and in-water activities including sailing and diving.

With support from the Global Environment Facility (GEF), the Organisation of Eastern Caribbean States (OECS), and in partnership with the World Bank, The Nature Conservancy (TNC) used an approach called Mapping Ocean Wealth to develop ecosystem service models and maps at the scale of the Eastern Caribbean. This work in Dominica, Grenada, Saint Lucia, St. Kitts and Nevis, and St. Vincent and the Grenadines supports the Caribbean Regional Oceanscape Project (CROP), an effort to foster this region's blue economy and promote greater consideration of the region's ecosystem functions and services through ocean policy development and coastal and marine spatial planning.

## Mapping ecosystem services

Coral reefs that encircle most islands draw tourists, either directly through scuba and snorkeling or indirectly through beach-related activities and access to seafood.

Even so, the connections between these ecosystems and the personal, direct benefits they generate across these societies can be difficult to quantify. Without such information, opportunities to enrich society through improved management, reduced user or use conflict, and restoration, or enhancement of ecosystem functions may be lost. Mapping Ocean Wealth quantifies and maps the linkages between vulnerable natural habitats such as coral reefs and tourism revenues associated with these habitats, asserting that protecting these habitats is a win-win for economies, livelihoods, and ecosystems alike.

One persistent challenge when mapping the spatial footprint of tourism activities is the lack of authoritative, spatially explicit data describing what tourists do and where. Previous studies addressed this gap by using crowdsourced data to evaluate patterns of tourism associated with a specific activity. In 2017, the Mapping Ocean Wealth team, including partners at the World Resources Institute, the Natural Capital Project, University of Cambridge, and University of Edinburgh, published a study and global map describing worldwide patterns of coral reef tourism and related expenditures using keyword searches of tagged photos uploaded to the photo-sharing website Flickr. Using the coordinate embedded in photo metadata, the researchers plotted locations of scuba and snorkeling activities which, combined with other data inputs, allowed them to map activity and tourism values to the locations of coral reefs.

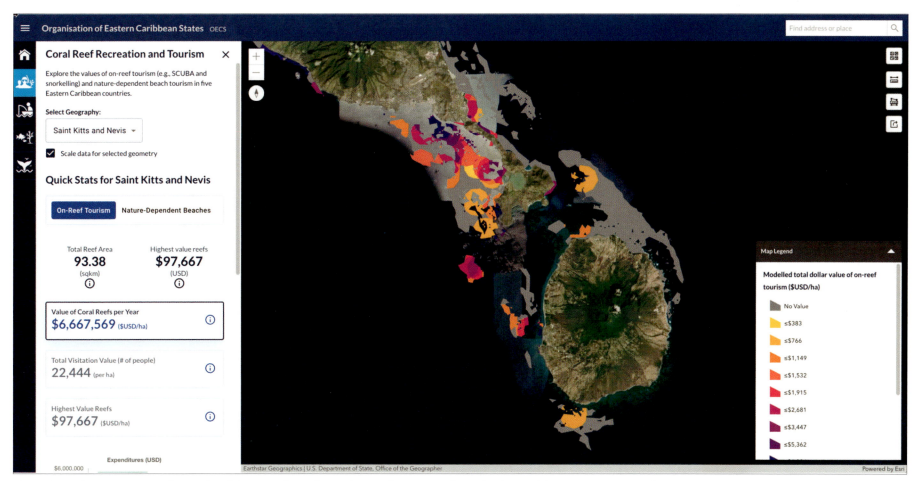

*This web application presents the results of the AI analysis of crowdsourced imagery collected in the Caribbean as a way to catalog ecosystem services provided by coral reefs and their surrounding environments.*

The initial work showed that coral reefs generate $36 billion in tourism spending annually and drive almost 70 million visits per year worldwide. Developing economies, including Small Island Developing States (SIDS), relied heavily on coral reefs for tourism. For many of these countries, tourism is a lifeline, generating livelihoods, wealth, and foreign exchange. The research provided a much clearer understanding of the dependence of many of these nations on coral reefs, and the resolution of the work was already sufficient to support the management of these fragile ecosystems in some countries.

## Innovations in AI in Mapping Ocean Wealth

The Mapping Ocean Wealth project under CROP allowed for several key advancements on the global model, notably, the application of emergent artificial intelligence/machine learning (AI/ML) technologies and methodologies to crowdsourced data. These methods yield datasets that can be used as data inputs into the ecosystem service models, improving traditional keyword searches used in previous studies and allowing us to map these and other distinct subcomponents of reef-related tourism, such as:

- **On-reef tourism**: Map of use intensity and values of activities that take place on or near a reef, specifically, scuba and snorkeling.

- **Nature-dependent beaches**: Map of use intensity and values of beaches that tourists perceive as exceptionally natural or pristine, perhaps because of clear and calm turquoise water, white sands, and foliage, which nearby reefs can influence.

- **Paddle sports**: Heat map of nonmotorized watersports activities (e.g., kayaking and paddleboarding). These activities are often reef-associated and clearly depend on healthy natural coastal waters.

- **Seafood restaurants**: Heat map of seafood restaurants in the region. These restaurants are often largely dependent on the ability to serve fresh fish caught on or near reef habitats.

Each of the models incorporated the results of AI computer vision techniques and methodologies in image classification applied to Flickr photos and Tripadvisor photos. Logistic regression and natural language processing methods were applied to Tripadvisor reviews. We developed training data for our image classification models by selecting a significant sample of images from Flickr and Tripadvisor, while our text analytics models used selected text labels and terms explicitly from Tripadvisor reviews that best represented the elements to capture in the models. For example, for on-reef recreation and tourism, we selected photographs depicting underwater scenes or reviews that describe diving or snorkeling experiences; for beaches, images of beaches with dominant natural elements (e.g., white sands, turquoise waters, vegetation); a review describing a visit to a resort where the visitor ate delicious red snapper and rented a kayak would be tagged as "seafood" and "reef-adjacent activity."

The team used Microsoft's Azure Cognitive Services to classify the remainder of the photos from Flickr and Tripadvisor, iteratively refining the training layers until high levels of accuracy were obtained. To analyze text, another Microsoft team then applied a random-forest regression model to automatically classify the remainder of the reviews and return a list of reviews that matched each set of criteria. Our models were evaluated based on metrics of precision and recall, as frequently found in the machine learning literature.

Data output could then be plotted on a map using either known locations of Tripadvisor locations to which photos and reviews were associated, or, in the case of Flickr, using the coordinates embedded in photo metadata.

These points were then combined with other data inputs, including names and locations of dive sites, dive shops, hotels, other attractions, and statistics and reports on tourism arrivals, expenditures, and reef-related activities. For on-reef activities and nature-dependent beaches, data were considered sufficiently reliable to build a direct model of value, developing a national-level estimate of importance to be expressed as tourism numbers and expenditure. These quantitative values were then linked to the maps of use intensity to spread actual values to coastal areas or ecosystems.

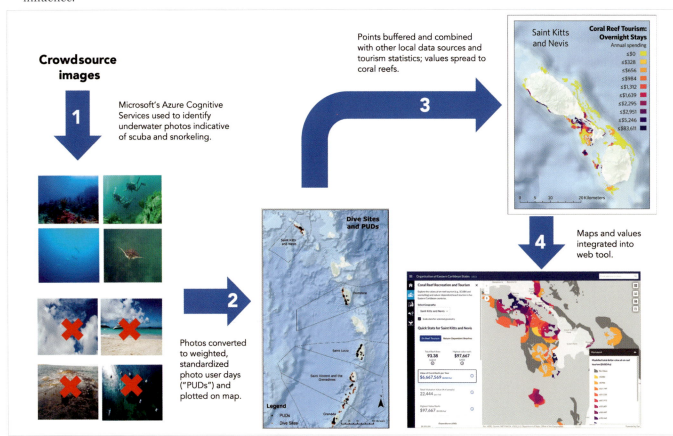

*The high-level workflow for a crowdsourced image analysis and reporting system that uses AI and cloud computing.*

For paddle sports and seafood restaurants, simple heat maps were generated, but no quantitative values were applied.

## Results

Across the combined CROP countries, tourism expenditure directly linked to on-reef activities is estimated at $118 million annually. This expenditure can also be expressed in terms of visitor numbers, with 83,000 overnight visitors and 60,000 cruise visitors choosing these islands for their on-reef activities. Natural values of the beaches in the CROP countries are estimated to be generating some $318 million of tourism expenditure annually with 143,000 overnight visitors and 565,000 cruise visitors attracted specifically to the pristine, natural aspects of the region's beaches.

The on-reef values explored here are considerably higher than those projected from the 2017 global assessment, which estimated a total value of $73 million per year. This increase is largely driven by increases in tourism arrivals and expenditure between these two studies.

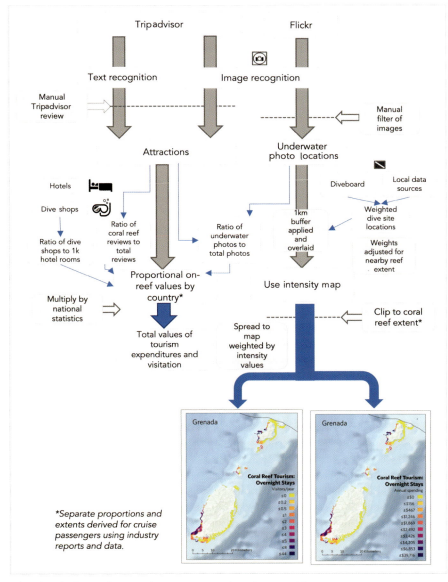

*The detailed crowdsource data capture and analysis workflow.*

## Conclusion

This is the first time that these components of nature-based tourism associated with coral reefs have been so extensively mapped and analyzed at these resolutions. The results are of considerable use for understanding the value of coral reefs and coastal ecosystems at local scales applicable to management. They should support a broad range of users from the public to industry to government to better plan and manage the tourism industry and other sectors whose actions could either support or jeopardize these values. The maps are also valuable for future planning. In this context, looking at places within countries and between countries might provide models for future natural resource management, for example, in restoring natural values as a way to expand natural benefits to new locations.

User-generated content from large crowdsourced datasets such as Flickr and Tripadvisor serves as a powerful tool for understanding relatively fine-scale patterns in tourism. Concerns have been raised about accuracy bias and the reality that any public-sourced datasets have a high ratio of errors. Nevertheless, it is the high volume of data that makes these datasets valuable, enabling us to smooth over the occasional errors. Other platforms, notably social media platforms, would represent another rich source of data. However, such platforms do not allow large-scale data extraction and cannot at present be used for data mining in this way.

One key element of this work is the high degree of local engagement that enabled us to enhance the data from these more international sources, and proof, corroborate, or correct the final models and output maps. While AI/ML methodologies are an improvement on previous studies, the benefit from extensive regional engagement cannot be overstated. Local data collectors obtained country-level data from government agencies, scientists, and other sources. These datasets serve to ground-truth and enhance data derived from global sources, including AI/ML output. Data were further verified through stakeholder engagement activities, bringing together experts who have deep familiarity with data inputs, and the locations being characterized. This level of participation is not feasible for modeling activities at the global or even regional scale, but it allowed the team to refine our methodologies. These data products are directly responsive to the needs of people who could ideally use this information to support their work. This process also validates the AI/ML output that can then be used in CROP work related to marine spatial planning.

Given the impact of COVID-19 on tourism in the Caribbean and the expected changes in demands from a recovering tourism sector, it is highly likely that future tourism will depend even more on natural values and lower-density locations. Because sites of high natural value will likely show an increasing proportional relevance for the recovering sector, the need for conservation in these areas is more important than ever.

## Acknowledgments

This work would not have been possible without the support of our TNC colleagues, including Dr. Mark Spalding, for leading the development of the spatial models and co-authoring the report from which this text is derived; Valerie McNulty, for creating the maps; Giselle Hall, for her assistance in data cleaning and development; Dr. Sherry Constantine, for her overall project guidance and regional expertise; and Dr. Francesco Tonini, for additional AI/ML support. Barry Nickel and Aaron Cole at the University of California, Santa Cruz, led the development of the Cognitive Services models. Darren Tanner at Microsoft developed the text recognition and classification models. Charlie Ballard, formerly of Tripadvisor, provided data access and advice on its use. We are grateful for additional modeling and data advice provided by Dr. Peter Schuhmann, Nikoyan Roberts, and Lauretta Burke.

# PREVENTING ILLEGAL, UNREPORTED, AND UNREGULATED FISHING

Fish are a crucial global resource, providing the primary source of protein in the diets of as many as 3 billion people worldwide. As much as 12% of the world's population derives its livelihood from fishing and the seafood industry. But this resource is seriously threatened. As reported by the United Nations Food and Agriculture Organization, human consumption of fish has grown on average about 1.5% per year since 1961. A third of fish stocks are now overfished and no longer biologically sustainable, and about 60% are maximally sustainably fished, with as little as 7% remaining underfished.

OceanMind was founded with the mission to help governments more effectively prevent illegal, unreported, and unregulated (IUU) fishing and to help the seafood industry more responsibly comply with regulations, thereby improving the sustainability of fishing. The nonprofit began as a partnership between the Pew Charitable Trusts and the Satellite Applications Catapult, a UK government-funded research and development organization designed to help grow the UK economy through the application of satellite technologies.

Worldwide, fishing fleets collectively include more than four million vessels—a relatively huge number in the even more vast environment of the oceans, where physical borders are often amorphous. These fleets operate in a variety of jurisdictions. National governments patrol their own coastal waters and usually form treaties to jointly manage nearby areas in between. Regional fisheries management organizations also govern some areas of international waters, while other large areas of the oceans are unregulated. With this complicated situation, it's challenging to prevent IUU fishing. OceanMind and its partners invest in developing a system to gather satellite data for analyzing the behavior of fishing vessels to help curb illegal fishing.

OceanMind also helps governments and conservation organizations map the intersection between human activity and marine wildlife to inform policy decisions and marine spatial planning.

## Mascarene Islands Important Marine Mammal Areas

OceanMind analyzed 115 Important Marine Mammal Areas (IMMAs), studying fishing vessel activity to assess the risk of *bycatch*, the unintended capture and death of a marine mammal during fishing, and shipping lanes for *ship strikes*, in which vessels hit a mammal resulting in a fatality. OceanMind provided case studies to highlight risk areas and suggest ways to preserve endangered and charismatic animals.

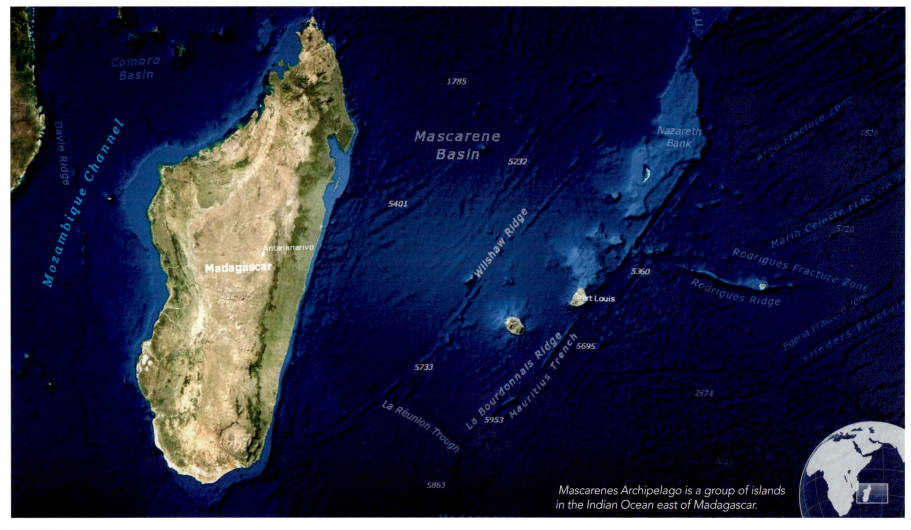

Mascarenes Archipelago is a group of islands in the Indian Ocean east of Madagascar.

OceanMind chose the Mascarene Islands IMMA as one of its focus areas, partly because the three islands represent one of the most important breeding and feeding habitats for sperm whales in the southwestern Indian Ocean. The islands also are habitat for calving, nursing, and mating humpback whales and for three dolphin species. The area has high vessel transit activities, creating a higher probability of ship strikes. While some data indicate the presence of marine mammals, the region is still understudied, which was another reason for the focus. With the support of the World Wildlife Fund and Globice (Groupe Local d'OBservations et d'Identification des CEtacés), OceanMind identified risk areas where ship strikes and bycatch by fishing vessels were more likely to occur.

## Methodology

1.  **Data**: Mammal tracking data, vessel and satellite data (from photo images to radar), automated tracking signals, and even ship lights can determine a vessel's movement behavior. The Mascarene Islands IMMA is a data-poor environment for marine mammal occurrences, so the work focused on what evidence the limited data can provide. The Mascarene Islands IMMA was limited to data from a handful of studies and the global assessment of marine mammals by the Ocean Biodiversity Information System (OBIS), which presented numerous gaps compared to other regions such as the North Pacific, North Atlantic, and Southern Oceans, where extensive studies on marine mammal movements have been undertaken. Data on marine mammal movements were derived from visual and acoustic surveys and from a limited number of satellite tracking tags.

2.  **Geospatial visualization of vessel tracking data:** OceanMind highlighted marine mammal aggregation points (shown on the heat map of marine mammal sightings) and possible pathways of mammals compared to vessel activity. Reported sightings of marine mammals were mostly anecdotal because satellite tags registered only a few data points of their positions. So OceanMind created a grid to compile available information and therefore highlight where the most marine mammals were previously seen. While not free of bias, the grid showed where ship strikes most likely occur, and given the importance of this IMMA to breeding and nursing, where marine mammals are at most risk from fishing activities.

3.  **AI algorithm:** OceanMind uses an AI algorithm to identify where fishing activities occur and what type of fishing gear may be in use. Fisheries experts trained the algorithm using vessel tracking data representing thousands of vessel-years of fishing of different kinds to make this possible. Where vessel tracking data such as Automatic Identification System (AIS) is available, OceanMind can accurately detect fishing to the point of recognizing different phases of fishing, such as the deploying of gear, soaking gear to catch fish, and the recovery of gear. Once fishing activities have been identified, the data is cross-referenced with vessel registries and licensing information to establish compliance. Suspected noncompliance is highlighted for an expert user to investigate. Once identified, fishing activities can be compared with marine mammal movements to determine the risk of bycatch and entanglement. Where no vessel tracking data are available (so-called *dark targets*), OceanMind turns to satellite observations to detect the location of vessels in sensitive locations such as IMMAs. Satellite radar can cover a wide area of ocean and detect commercial fishing vessels through clouds, at night, and in bad weather. In good weather, satellite imagery can help detect the smallest fishing vessels and identify features on larger vessels.

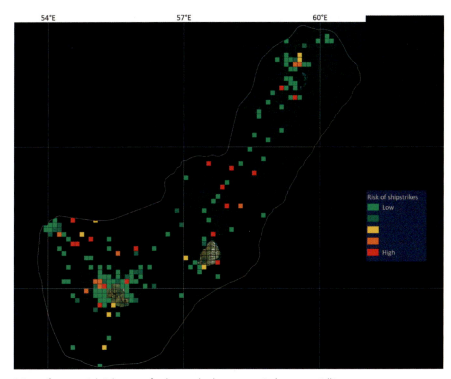

Map of potential risk areas for bycatch shown on .1 degree gridlines.

Arrangement of the Mascarene Islands and the local bathymetry layer.

4. **Decision-making:** Esri tools were used to create visualizations for stakeholder presentations. Solutions for reducing instances of ship strikes and bycatch could include stopping traffic through the islands or limiting the allowed transit speeds in this area. These solutions could help improve shipping lanes to reduce ship strikes and allow fishing vessels to operate without the risk of catching smaller marine mammals, such as dolphins.

## Conclusion

OceanMind's work would be impossible without modern computing resources, including cloud computing, AI, and, machine learning. At any one time, hundreds of vessels operate across wide areas involving many jurisdictions and thousands of regulations. Basing its work on temporal data, OceanMind depends on GIS to manage, analyze, and present information. Using the capabilities of ArcGIS software, OceanMind creates detailed analysis of multiple datasets, and uses the outputs to help inform partners about human activity on the ocean. This use of technology enables OceanMind to analyze more data in real time and do it more quickly, accurately, and cost effectively. The goal is to deliver insights to any government, in real time, telling them of suspected noncompliance in their waters. This capability can make a big difference in the work to reduce bycatch, ship strikes, and illegal fishing.

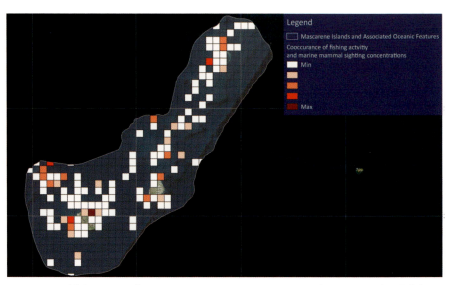

*Heat map of fishing vessel AIS positions transmitting on speeds associated with fishing activity and marine mammal sightings, normalized around the highest value. The outline indicates the Mascarene Islands and associated oceanic features with a .2 degree grid.*

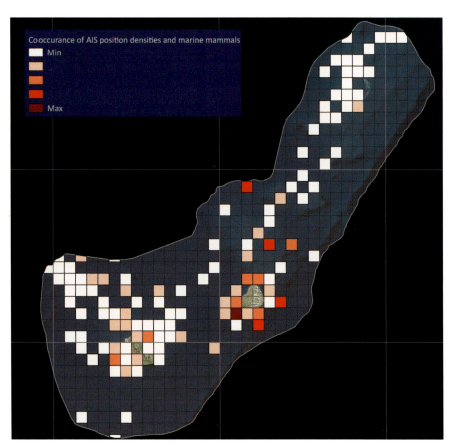

*Heat map of fishing vessel AIS positions transmitted and marine mammal sightings, normalized around the highest value. The outline indicates the Mascarene Islands and associated oceanic features with a .2 degree grid.*

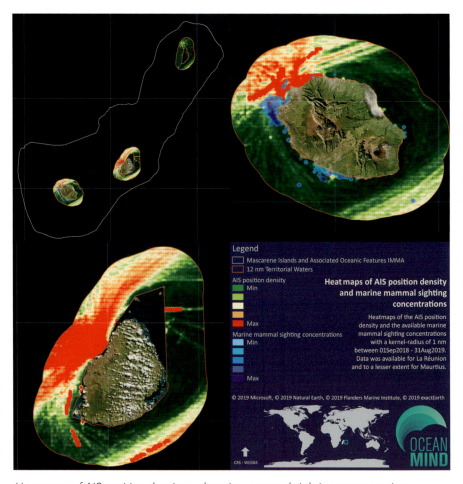

*Heat maps of AIS position density and marine mammal sighting concentrations.*

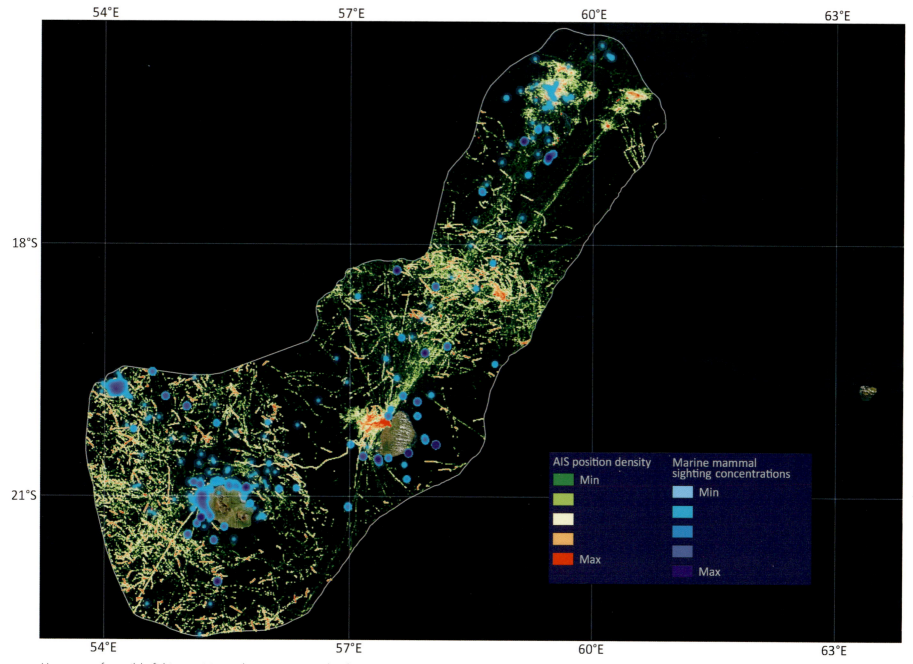

Heat map of possible fishing activity and marine mammal sighting concentrations.

# TERRESTRIAL SPOTLIGHTS

## Trust for Public Land

Interest has grown in recent years to understand the value of city parks and open spaces, not just as social and recreational areas, but also as sources of economic and health benefits. The American Planning Association (APA) convened its City Parks Forum in the early 2000s to create a nationwide conversation about the role and importance of city parks. Subsequent research found that in addition to recreational, social, and aesthetic benefits, parks and open spaces provide health and economic benefits.

With this knowledge, the Trust for Public Land (TPL) has long worked with urban communities to create parks and protect public land to benefit everyone. Its ParkScore Index evaluates the effectiveness of parks in the largest U.S. cities, and TPL plans to expand that work to many more communities nationwide. This index is designed to help local communities improve their park systems and identify where new parks are needed most. TPL built on the concept of ParkScore by launching ParkServe, an interactive platform to track urban park access nationwide. The central concept of ParkServe (and ParkScore) is to determine and map the number of city residents who have access to a park, playground, or publicly accessible

protected area within a 10-minute walk. By mapping the existing parks, population density, and other demographics, ParkServe also reveals what areas of a city are underserved by parks.

TPL aims to eventually scale up to nationwide coverage, serving nearly 14,000 municipalities and more than 80% of the U.S. population. Such national-level analysis requires more computing power and infrastructure than TPL's existing in-house systems and would make it difficult for TPL to perform other mission-critical work. With an AI for Earth grant, TPL performs cloud computing through the Microsoft Azure platform. TPL also uses ArcGIS Pro housed in a Microsoft Azure Geo AI data science virtual machine (Geo-DSVM). The Azure Geo-DSVM was specifically designed in collaboration with Esri to support geospatial analytics capabilities through interoperation with ArcGIS Pro. The resulting analysis will help TPL explore trends and impacts around parks and communities, such as how much the parks are being used or how well each park provides health and environmental benefits.

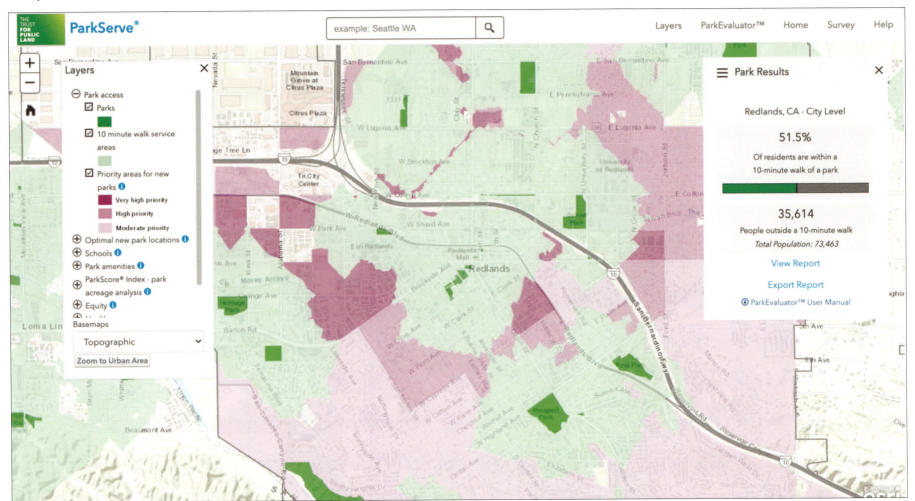

*ParkServe, an interactive platform to track urban park access nationwide. The central concept is to map how many people living in cities have access to a park, playground, or publicly accessible protected area within a 10-minute walk.*

# Optimizing streamflow prediction

Students at Brigham Young University's (BYU) Hydroinformatics Lab saw the potential to optimize streamflow prediction using cloud computing and GIS modeling systems. This strategy allows researchers to use a web-based platform, eliminating the need for expensive, high-performance local computer hardware. It also ensures evergreen data and reduces the need for local storage capacity. Using Esri's ArcGIS platform plus custom software, students produced streamflow forecasting, flood mapping, and data sharing tools that stakeholders can access. With advanced modeling, mapping, and visualization, water resource managers can make more informed decisions. This resulted in a set of global streamflow forecast maps, web applications and an API, which produces a 15-day view of streamflow and flood risk. This capability allows affected regions to better prepare for flooding and become more resilient after floods and other disasters.

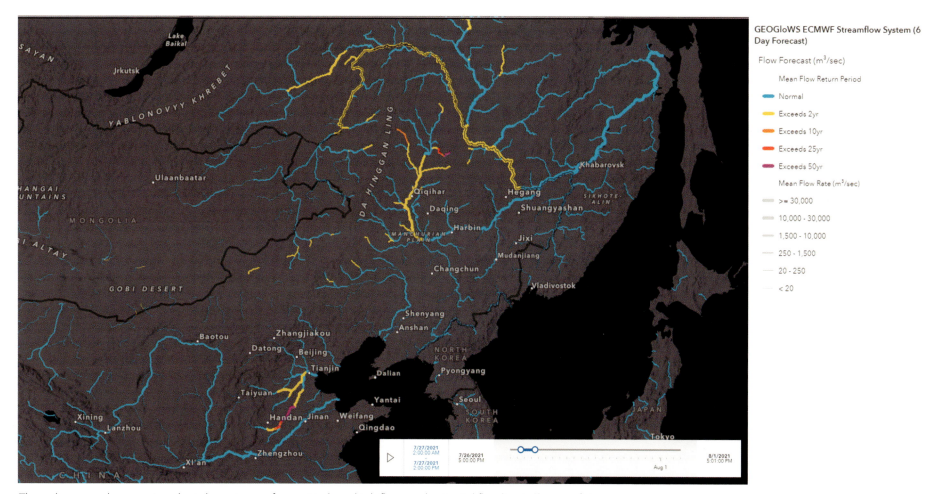

*The web map and companion chart show streams forecast to have high flows and potential flooding in the near future.*

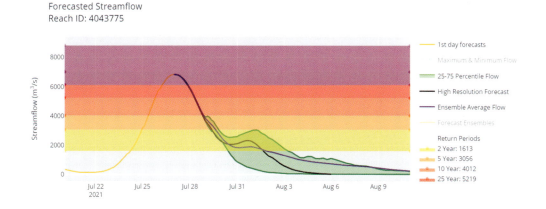

# Sustainable land management

Forests cover one-third of the United States, totaling about 749 million acres that can be managed to improve wildlife habitat and purify drinking water. More than half of these woods are privately owned, mostly by families and individuals with 100 or fewer acres. These properties are often vulnerable to development and fragmentation because owners lack the information and resources to practice modern forest management and participate in ecosystem services markets. By paying landowners for the environmental services their land provides, ecosystem services markets can help protect forests and reward sustainable management.

Sustainable management requires a deep understanding of the forest. But most forest owners have little quantitative information about their land because forests are large and extremely complex. Foresters typically gather data by conducting a *timber cruise*, which involves surveying sample plots by foot to inventory timber and assess other ecological values, such as habitat and carbon. Timber cruises are labor-intensive and expensive, and because only a small portion of the area is surveyed, the results are often imprecise. SilviaTerra was founded with the idea that widely available forest inventory data could help people make better land management decisions. SilviaTerra migrated its datasets to the Azure cloud and scaled its computing process from thousands of acres to hundreds of millions of acres, creating a .05-acre pixel resolution map of U.S. forestland via state-by-state deployment. The process combines field measurements from the U.S. Forest Service Forest Inventory and Analysis dataset with a stack of remotely sensed imagery. SilviaTerra's detailed inventory system monitors the status of a forest on a small scale, significantly reducing the costs for assessing the carbon value of the forest.

SilviaTerra is also creating an application programming interface (API) on Azure that will allow landowners, foresters, and researchers to efficiently access insights from the forest inventory layer. Many forest managers today use ArcGIS as their geographic information system. ArcGIS holds forest inventory estimates for each management unit of the forest; however, these estimates are often years out of date. Normally, foresters must collect new stats in the field, which could take months. But because the Microsoft Azure API interoperates with ArcGIS, foresters can simply click a forest management unit and instantly update it with the latest information from the SilviaTerra Basemap national forest inventory.

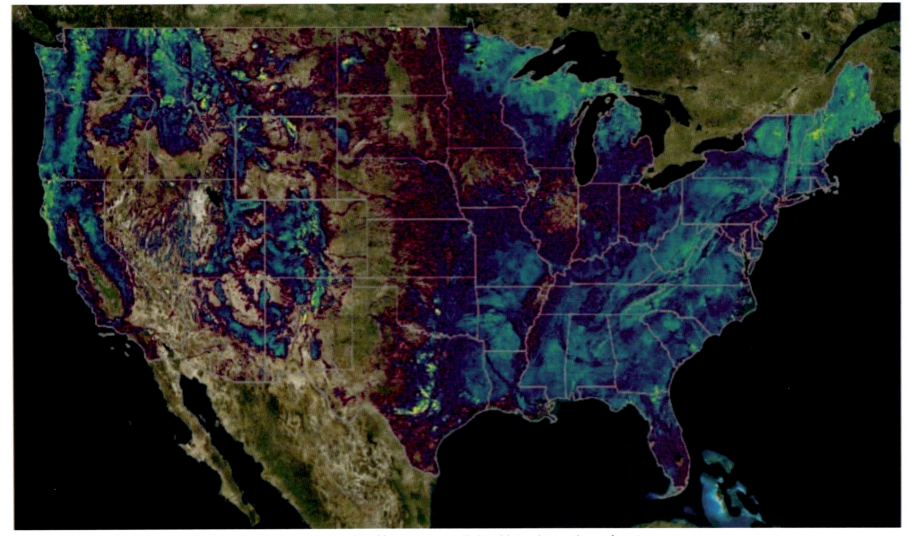

*SilviaTerra Basemap forest inventory. The gradient indicates the density of forest coverage (lighter blue indicates denser forest).*

## Conserving biodiversity

Biodiversity is under threat globally from anthropogenic pressures such as habitat loss, fragmentation, and degradation. The World Wildlife Fund for Nature's *Living Planet Report 2016* found that populations of vertebrate animals have declined by 58% overall from 1970 to 2012, with freshwater species dropping as much as 81%. The National Audubon Society's 2014 *Birds and Climate Change Report* found that habitat loss caused by global warming threatened more than half of the 588 North American bird species in the study. Disturbances such as hurricanes cause significant erosion and extensive land cover change, resulting in loss of critical habitats important to breeding, migratory, and wintering birds.

To conserve coastal birds and other biodiversity, it's essential to assess disturbance effects quickly and accurately and identify areas in critical need of habitat restoration. Bird watchers are among the most enthusiastic citizen scientists. For years, the National Audubon Society has relied on this volunteer community to help count birds. However, volunteer bird monitoring is often extremely difficult to do well, particularly for sites that have limited accessibility or can only be surveyed during narrow time windows (such as islands with tide-limited beach access). Given the challenges of traditional bird counting, Audubon is augmenting in-person monitoring with drones. Drones, however, produce a large amount of data and require extensive computer resources and human labor to process using traditional analytical methods.

Audubon's new project to advance the rapid evaluation of bird monitoring and habitat assessment is twofold: documenting changes in bird habitats because of disturbances such as major storms, and more accurately counting the various species of birds living in those habitats. Developing machine learning and cognitive algorithms that can rapidly process remotely sensed data to census birds and wildlife and classify land cover will change how Audubon accomplishes its conservation goals. To this end, high-resolution imagery from drones and aerial surveys, and lidar elevation data for land cover classification was loaded into Azure data science virtual machines for high-throughput processing with ArcGIS Pro. Machine learning algorithms were implemented in Machine Learning Studio using the MicrosoftML package, and additional tools, such as the ArcGIS Pro Spatial Analyst Image Classification tool, were used to help train the algorithms. Once trained, these algorithms enabled Audubon to quickly obtain accurate count and colony size data from remote sites and across large spatial scales that were impractical or impossible to attempt using traditional ground-based surveys. The algorithms developed from this project will vastly increase the speed of image processing, identification, and count estimation for future surveys, empowering Audubon to meet the conservation challenges posed by accelerating global change.

*Nesting bird locations can be safely photographed with drones and then mapped and processed with machine learning to arrive at accurate counts.*

## Planetary Computer

Humans still do not know enough about species, biodiversity, and ecosystems that are vital to our health and prosperity. Simply understanding the locations of the world's forests, fields, and waterways remains a daunting task of environmental accounting. Understanding what species call those ecosystems home and why they thrive or decline is largely unknown. We can't solve a problem we don't fully understand.

The UN launched the first worldwide assessment of natural systems in 2000, and it took nearly five years and more than 1,300 experts to complete. A more recent assessment by the UN Intergovernmental Platform on Biodiversity and Ecosystem Services (IPBES) intended to close the gap between simple scientific insight and more effective policy implementation, and was published in 2019. As our environmental challenges intensify, the world needs greater access to better and better environmental data to assess, diagnose, and treat the natural systems that society depends on. Data powered by machine learning will make that possible.

Assessing the planet's health must become a more sustained, integrated practice that allows us to understand what is happening in time to enable smart decision-making. Fortunately, technology potentially can revolutionize our environmental assessment practices, so they are faster and cost-efficient on a global scale. It should be as easy to learn about the environmental state of the planet as it is to search the internet for driving directions. We must use the architecture of the Information Age—data, compute, algorithms, application programming interfaces, and end-user applications—to accelerate a more environmentally sustainable future.

Several years ago, we took our first step in this direction by launching Microsoft's AI for Earth program to put AI technology into the hands of the world's leading ecologists, conservation technologists, and organizations that are working to protect our planet. Yet for all the great work of our AI for Earth community, we have also learned it still needs much greater access to data, more intuitive access to machine learning tools, and a greater ability to share work and build on the work of others than our program currently provides.

Our community needs a new kind of computing platform—a Planetary Computer platform that can provide access to trillions of data points collected by people and by machines in space, in the sky, in and on the ground, and in the water. One that would allow users to search by geographic location instead of keyword. Where users seamlessly go from asking about what environments are in their area of interest to asking where a particular environment exists globally. A platform that allows users to provide new kinds of answers to new kinds of questions by providing access to state-of-the-art machine learning tools and the ability to publish new results and predictions as services available to the global community.

*The Planetary Computer connects myriad data sources and AI analysis of Earth's condition and what can be done to accelerate meaningful changes in how humankind manages the planet's finite resources.*

Microsoft envisions its Planetary Computer as a way to combine a "multi-petabyte catalog of global environmental data with intuitive APIs, a flexible scientific environment that allows users to answer global questions about that data, and applications that put those answers in the hands of conservation stakeholders." The Planetary Computer will provide insights into critical issues that scientists and conservation organizations address regularly in their work. For example:

- Understanding tree density, land use, and forest size has implications for biodiversity conservation and climate change mitigation. Organizations often conduct expensive on-the-ground surveys or build customized solutions to understand local forests. The Planetary Computer will provide satellite imagery, state-of-the-art machine learning tools, and user-contributed data about forest boundaries from which forest managers will have an integrated view of forest health.

- Needing to make more than educated guesses about land management, urban planners and farmers depend on forecasts of water availability and flood risks. The Planetary Computer will provide satellite data, local measurements of streams and groundwater, and predictive algorithms that will empower land planners and farmers to make data-driven decisions about water resources.

- Combining information about terrain types and ecosystems with accurate and available data about where species live, the Planetary Computer will enable a global community of wildlife biologists to benefit from each other's data. This capability supports wildlife conservation organizations that have relied on their own local surveys, global views of wildlife populations, and suitable habitats for wildlife.

- Combating climate changes requires organizations to measure and manage natural resources that sequester carbon, such as trees, grasslands, and soil. The Planetary Computer will combine satellite imagery with AI to provide up-to-date information about ecosystems and a platform for using predictive models to estimate global carbon stocks and inform decisions about land use that impact our ability to address climate change.

The AI for Earth program is dedicated to building this Planetary Computer platform through dedicated investments in infrastructure development. The system provides access to the world's critical environmental datasets and a computing platform to analyze those datasets. We will also further invest in specific environmental solution areas such as species identification, land cover mapping, and land use optimization.

This Planetary Computer, which is emerging from beta testing in 2021, is incredibly complex, and we cannot build it alone. We must continue to learn from the work and demands of our grantees while partnering with the organizations best suited to advance global environmental goals. That is why we are deepening our partnership with Esri, a company with years of experience building environmental monitoring solutions.

Our partnership with Esri began at the launch of AI for Earth through shared technology granting programs. Microsoft and Esri share the goals of making geospatial data and analysis—meaning the gathering, display, and manipulation of information about Earth systems—available to every sustainability researcher and practitioner, and ensuring that every conservation organization can contribute its local data back to that global repository. Through hands-on collaboration and grants, Esri has helped conservation organizations transform their operations to use digital spatial information in areas of endangered species conservation, land protection, and the basic science that allows us to understand the natural world. From mapping forest loss to combating elephant poaching, organizations depend on Esri's tools and expertise to understand and protect the ecosystems in which they operate.

We are deepening our partnership around the development of the machine learning-based geospatial solutions that are the foundation of the Planetary Computer. The partnership will build on the work started with the launch of AI for Earth. Key geospatial datasets will be available on Azure and accessible through Esri tools in 2021. The partnership also will continue providing grants that ensure conservation organizations have access to the datasets, compute, and other resources.

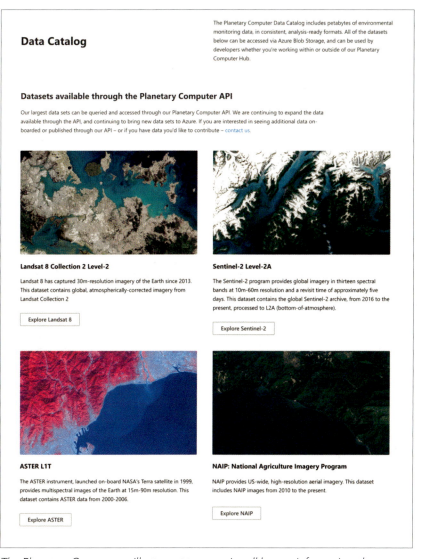

*The Planetary Computer will attempt to organize all known information about Earth and, more importantly, make the information meaningful and actionable.*

# MAPPING EXTREME EVENTS FROM SPACE

Observations from land, ships, balloons, aircraft, satellites, and the International Space Station all help NASA understand our planet's atmosphere, land cover, ice fields, and oceans. These datasets, collected from even the most remote areas of Earth, are freely and openly available to anyone.

By the **NASA Earth Science Division**[†]

*Orbiting 250 miles above North America, the International Space Station captures wildfires burning in the Mendocino National Forest north of the San Francisco Bay Area in Northern California.*

# A UNIQUE VANTAGE POINT

Wildfires, hurricanes, floods, drought, record-breaking heat, and scarce food and water supplies have affected vast areas of Earth in recent years. As these extreme events cause billions of dollars in damage and untold human suffering, how do the planet's natural landscapes and interconnected ecosystems respond? To answer this question, NASA's Earth Science Division (ESD) collects petaybytes' worth of observations to help scientists learn more about how these events affect all aspects of life on our planet.

A fleet of NASA satellites and other observation sources give researchers around the world the opportunity to combine and use vast amounts of data once separated by specific research questions or hazard types. By integrating data sources from multiple spatial scales, NASA's missions give researchers unprecedented insights into Earth's systems.

Instruments on land, satellites, the International Space Station, airplanes, balloons, and ships enable data collection about Earth's atmospheric motion and composition; land cover, land use, and vegetation; ocean currents, temperatures, and upper-ocean life; and ice on land and sea. Collected from the most remote areas of the planet, these datasets are freely and openly available to anyone.

NASA advocates a collaborative culture using technology that empowers the open sharing of data, information, and knowledge within the scientific community and the wider public to accelerate scientific research and understanding to benefit society.

Earth's ecosystems constantly respond to changing climates, extreme weather conditions, and human activities. The NASA Ecological Forecasting Program promotes the use of NASA earth observations to monitor, analyze, and forecast these changes and develop resource management strategies. In addition, the NASA Biological Diversity Program supports basic research to increasingly understand why and how biological diversity is changing.

NASA explores patterns of biological diversity on land and in water using observations from satellites, airborne and seaborne platforms, and on-site surveys. These observations help NASA understand how biological diversity affects and interacts with Earth's dynamic system—its hydrosphere, atmosphere, biosphere, and exosphere.

These observations are well-suited for detecting such patterns, especially at the ecosystem level, but also at finer community and species levels. This research includes documenting and identifying factors that determine the distribution, abundance, movement, demographics, physical or genetic characteristics, behavior, and physiology of organisms on Earth.

NASA supports open access for open science and the development of remote sensing tools, techniques, and associated models that enable this understanding. These observations can be studied based on location: mapping, visualizing, and analyzing where these events take place and how species in that area respond. Geographic information systems (GIS) technology allows NASA data to be visualized, analyzed, and shared to a wide community of researchers, scientists, educators, government partners, NGOs, nonprofit organizations, and the public.

NASA's global repository of current and historical data can help us better understand extreme climate events and apply that knowledge to a range of biodiversity variables. This chapter explores three major themes: wildfires, flooding and drought, and extreme heat.

*Typhoon Soudelor photographed from the International Space Station on August 4, 2015, in the western Pacific. The storm produced sustained winds of more than 160 mph at the time of this photograph.*

# WILDFIRES

Each year, fires ravage millions of acres of land, destroying thousands of homes and properties around the world. In 2020, California experienced its single greatest wildfire season in its recorded history. The LNU Lightning Complex fires burned across a wide area of Northern California—Lake, Napa, Sonoma, Solano, and Yolo Counties—in the late summer and fall of 2020. That year, wildfires across California destroyed more than 10,000 structures and caused more than $12 billion in damage and firefighting costs.

The magnitude of physical devastation and financial reverberation requires careful study of the event and its aftermath. NASA is uniquely positioned to supply emergency management authorities near real-time situational and long-term environmental impact data to the legions of land planners, land management agencies, and private sector professionals dealing with the fallout.

Studying the relationships between environmental factors and fires can minimize risk. Many of NASA's Earth-observing instruments contribute to our understanding of fire in the Earth system. NASA's satellite instruments, which are often the first to detect remote wildfires, automatically send their locations to land managers worldwide within hours.

NASA instruments detect actively burning fires, track the movement of smoke, provide current information to fire managers, and map changes post-event to ecosystems, capturing the extent and severity of burn scars.

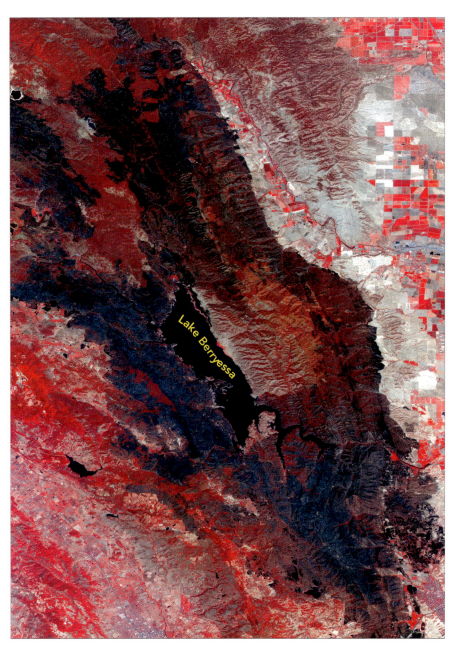

This false color image captured by the Advanced Spaceborne Thermal Emission and Reflection Radiometer (ASTER) instrument shows the burn scar from the LNU Lightning Complex fires in Northern California. Shades of red depict healthy vegetation whereas shades of black show burned areas. Little healthy vegetation remained within the burn scar, which can negatively impact forest connectivity.

The Hennessey and Spanish Fires burn toward Lake Berryessa in Napa County on August 18, 2020.

# Analysis of forest connectivity after the 2020 Oregon fires

While fire plays critical roles in sustaining ecosystems, including many that benefit biodiversity, shifts in vegetation conditions post-fire can influence the connectivity of landscapes for species that avoid recently burned areas. These reductions in connectivity can hinder wildlife access to important habitat areas and limit gene flow among neighboring populations.

Similarly, the mosaic of vegetation conditions that arise after burning prevents seeds from surviving trees from reaching the interior of the burn scar, a condition that impedes forest regeneration. Connectivity modeling tools based on electrical circuit theory can provide insight into these potential changes in "ecological flows" across landscapes as fires change the distributions of various land cover types.

By any measure, the 2020 western U.S. wildfire season was one of the worst in recorded history. Even Portland, Oregon, normally associated with cool weather and summer rains, experienced unusually high winds and continued dry weather, which in turn caused many fires to spread quickly throughout the state. In Oregon, more than one million acres burned, leading to more than 40,000 evacuations and putting a half million people within evacuation warning areas. With so much land altered in one series of events, forest researchers relied on NASA data to evaluate the immediate impact on forest connectivity.

NASA's Applied Sciences Program worked in conjunction with The Nature Conservancy to illustrate the potential influence of fires observed in 2020 on the connectivity of forests in the western United States. To begin, Visible Infrared Imaging Radiometer Suite (VIIRS) thermal hot spot data for August through October 2020 was combined with percentage of forest cover and land cover data from the National Land Cover Database (2016), both rescaled to 180 m pixels.

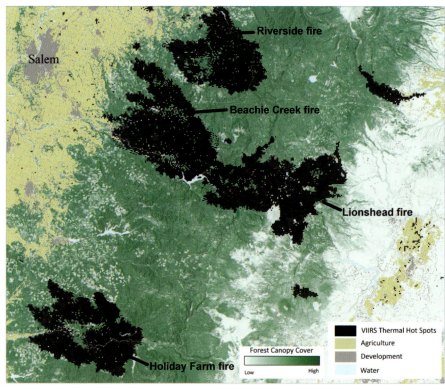

*Land cover in western Oregon overlaid with VIIRS thermal hot spot imagery. The black areas delineate the most recent burn scars from the 2020 fires.*

*Smoke-filled skies over downtown Portland, Oregon, on September 9, 2020.*

Using the connectivity modeling software Omniscape, the team compared the current flow across forested pixels before and after a modeled reduction of forest cover from these fires. In this simple illustration, they developed an input resistance grid with five land cover classes. Higher resistance values indicate reduced movement potential.

In the first model run (map A), researchers divided the forest cover values into four categories (low: <20%, medium: 20–40%, high: 40–60%, and very high: >60%) and scored those as 9, 6, 3, and 1 (no resistance). Areas in a non-forested but natural land cover type received a resistance weight of 10; agriculture pixels were scored 15; and urban areas, roads, and water were assigned the highest resistance value of 20. In the second connectivity model run (map B), the team integrated the VIIRS hot spot data and weighted burned areas in the low forest cover (resistance of 9) category. For both input maps, the Omniscape tool used algorithms based on electrical circuit theory (as implemented in Circuitscape software) to calculate current flow from all forested pixels in a 50 km radius moving window to a central pixel, with the amount of current emerging from each forested pixel assigned by reversing the resistance weight values (current source strength is 9 for pixels with very high forest cover and 1 for low forest cover), so the most current comes from the most highly forested pixels. The current flow outputs are summed for each moving window, and the final maps represent the cumulative movement potential for generalized forest-dependent species as a function of the structural pattern of forest cover on the landscape. The maps show areas in purple and red indicating high current flow, or movement potential, across large extents of forest. Orange and yellow areas indicate narrow bands of forest where current accumulates due to higher resistance in other neighboring pixels. Post-fire (map B), the modeled loss of forest cover, suggests further concentration of current (bright yellow) as the burned areas are avoided due to their high resistance and highlights the importance of narrow bands of forest along the edges of fires and between fire scars as potential corridors for forest-dependent species while these forests recover.

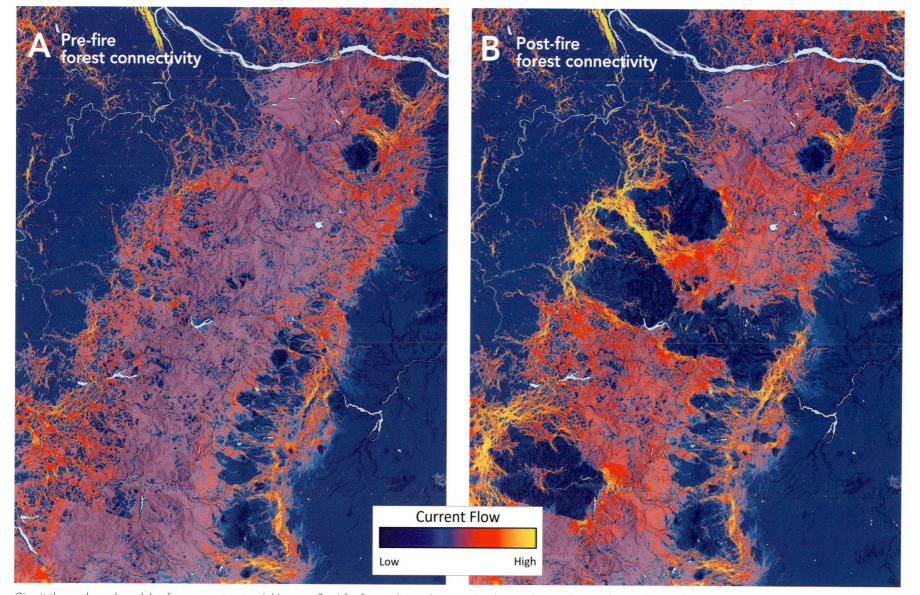

*Circuit theory–based models of movement potential (current flow) for forest-dependent species that might avoid recent burns depicted in the VIIRS thermal hot spot imagery.*

## Comprehensive look at the Australian bushfire disasters

Around the world, the 2019–2020 Australian fire season was also one of the worst on record. Australia experienced unprecedented heat waves, including temperatures reaching 120° F (49.1° C) in January across central and eastern Australia. NASA's satellites tracked the event in real time and collected large volumes of data that scientists and researchers can use to study the regional and global effects of the disaster. Using NASA data to visualize the footprint of the fires, the height of the smoke plumes, and the transport of particulate matter allows patterns to be revealed.

With ArcGIS StoryMaps, users can interact with NASA's data and perform time-series analysis, zoom to areas of interest, and toggle data layers to reveal patterns and trends. For example, a web app focused on the 2019–2020 Australian Bushfires using NASA data allows anyone to explore the causes and effects of the fires. The NASA app can be found linked at GISforScience.com.

## Fire Weather Index

By 2019, a combination of long-term warming, rainfall deficiency, and oceanic circulation anomalies left ground conditions in Australia extremely susceptible to fires. NASA's Goddard Space Flight Center has developed the Global Fire WEather Database (GFWED) to better understand the Fire Weather Index, a parameter that reveals how different weather factors are influencing the likelihood of a vegetation fire starting and spreading. The Fire Weather Index is similar to fire danger indices used operationally in Australia to account for current and antecedent rainfall, humidity, temperature, and wind speed.

During Australia's unprecedented 2019 fire season, brushfires engulfed many parts of the continent, including coastal areas in the more populous south and southeast. Wildfires burned more than 10 million hectares, an area roughly the size of Iceland. Flames destroyed 5,900 buildings, including 2,800 homes, according to the United Nations, and the dire impact on the ecosystem included the deaths of millions of animals.

Additionally, initial studies estimated that fires destroyed roughly 20% of the continent's forests and put much more flora at risk, including the extremely rare wild Wollemi pines. Fewer than 200 of these prehistoric trees remain in a secluded area about 125 miles northwest of Sydney, and firefighters were deployed on a special mission to save them.

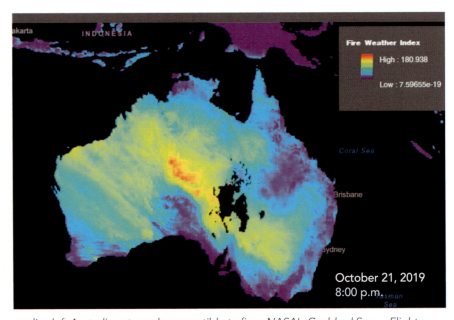

*A cane fire in New South Wales, Australia, 2019.*

*By 2019, a combination of long-term warming, rainfall deficiency, and oceanic circulation anomalies left Australia extremely susceptible to fires. NASA's Goddard Space Flight Center developed GFWED, integrating different weather factors influencing the likelihood of a vegetation fire starting and spreading (see the map of October 21, 2019). It is based on the Fire Weather Index (FWI), the most widely used fire weather system in the world.*

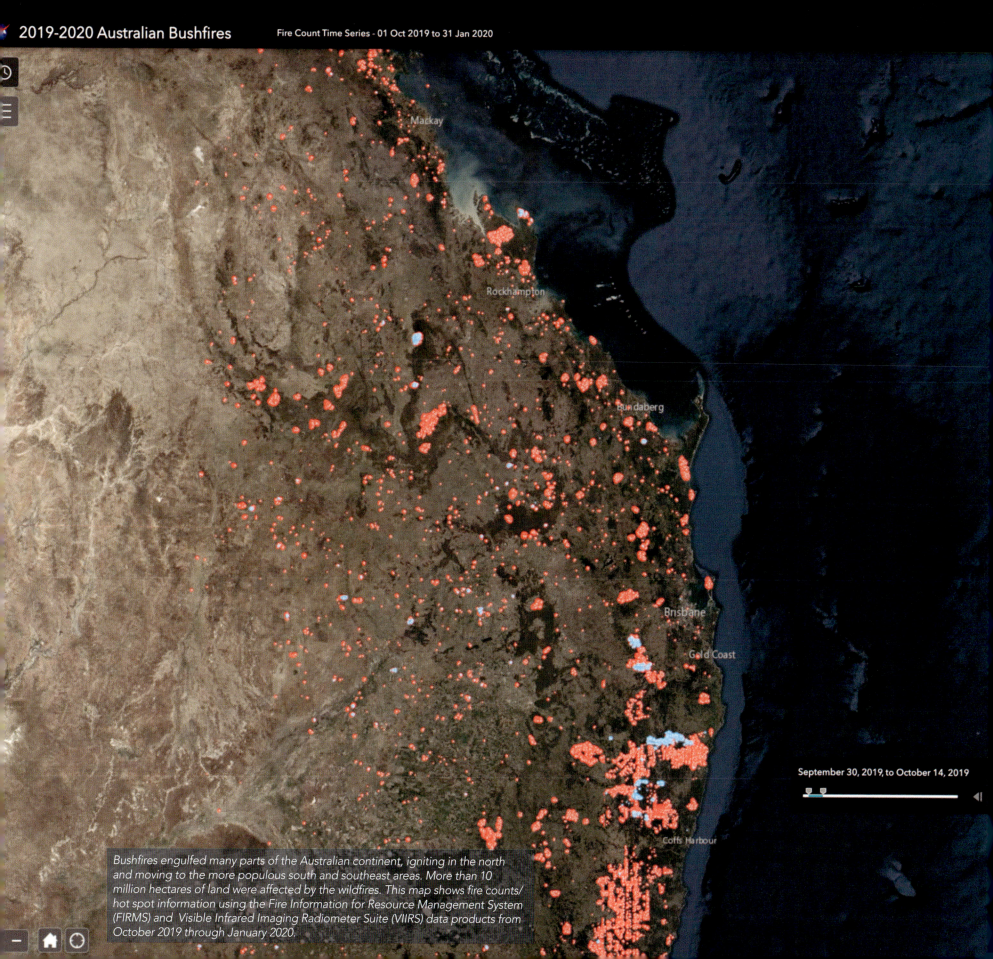

Mackay

Rockhampton

Bundaberg

Brisbane

Gold Coast

Coffs Harbour

September 30, 2019, to October 14, 2019

Bushfires engulfed many parts of the Australian continent, igniting in the north and moving to the more populous south and southeast areas. More than 10 million hectares of land were affected by the wildfires. This map shows fire counts/hot spot information using the Fire Information for Resource Management System (FIRMS) and  Visible Infrared Imaging Radiometer Suite (VIIRS) data products from October 2019 through January 2020.

# FLOODING AND DROUGHT

Water is an essential requirement for life. It is in the atmosphere above us; the oceans, lakes, and rivers around us; and the soil and rocks beneath us. It is also locked in the snow, ice caps, and glaciers that cover a good part of the planet. NASA's remote sensing data help us understand the availability and distribution of Earth's water and the impacts of climate change on the processes of the water cycle—evaporation, advection, convection, and precipitation. Using this data and other information, scientific studies have found changes in the spatial and temporal variability of precipitation: wet regions are becoming wetter, dry regions are becoming drier, and extreme events such as floods and droughts are becoming more frequent and intense. Researchers track the sensitivity of precipitation to climate change in terms of intensity, spatial patterns, and frequency of consecutive wet and dry days. Extreme precipitation events exert stresses on soil moisture and on stream and river flow, whereas the lack of rain stresses vegetation and water reservoirs, and when the drought frequency increases, water reservoirs are less likely to recover before the next dry spell.

NASA's Earth Observing System (EOS) is a coordinated series of polar-orbiting and low-inclination satellites for long-term global observations of the land surface, biosphere, atmosphere, and oceans. NASA's Goddard Earth Sciences Data and Information Services Center archives EOS data related to atmospheric composition, the carbon cycle and ecosystems, weather and climate variability, and the water and energy cycles. NASA will design a new set of Earth-focused missions called the Earth System Observatory (ESO) to provide key information to guide efforts related to climate change, disaster mitigation, fighting forest fires, and improving real-time agricultural processes. Each satellite in the ESO will be uniquely designed to complement the others, working in tandem to create a 3D, holistic view of Earth, from bedrock to atmosphere.

## The Integrated Multi-satellitE Retrievals for Global Precipitation Measurement

The Global Precipitation Measurement (GPM) mission is an international project of NASA and JAXA—the Japan Aerospace Exploration Agency. The program estimates surface precipitation over most of the globe. Its key components are the Orbiting Core Observatory carrying the GPM Microwave Imager (GMI) and a special instrument called the Dual-frequency Precipitation Radar (DPR). The GMI and DPR also serve as precipitation and radiometric standards for a virtual constellation of partner satellites. Both capture precipitation intensities, horizontal patterns, and the vertical structure (profiles) of hydrometeors.

The complex algorithm called the Integrated Multi-satellitE Retrievals for GPM (IMERG, Huffman, et al. 2019) optimally fuses precipitation estimates from this partner constellation with the Core Observatory, geostationary infrared sensors, and rain gauges to produce half-hourly near-global estimates at 10 km resolution.

The *early* and *late* products are available approximately four hours (early) and 14 hours (late) after a satellite observation. The IMERG *final* satellite-gauge product combines the satellite observations with monthly gauge analyses data and is available approximately 3.5 months after the observations. Each successive IMERG run provides a better estimate by including more data in exchange for a longer latency.

IMERG precipitation and quality index of precipitation variables are valuable mainly for their fine spatial and temporal resolution. Researchers use them to study precipitation microphysics, storm structures, and large-scale atmospheric processes. The long temporal span of GPM IMERG, dating to the year 2000, allows

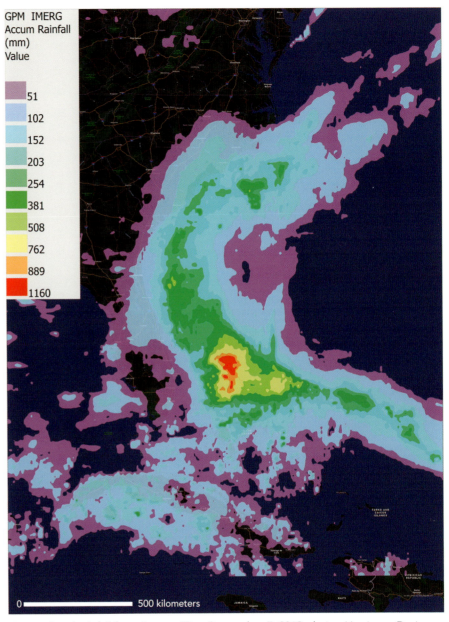

GPM IMERG Accum Rainfall (mm) Value

51
102
152
203
254
381
508
762
889
1160

*Accumulated rainfall from August 30 to September 5, 2019, during Hurricane Dorian, a Category 5 hurricane on the Saffir-Simpson scale which battered the Commonwealth of the Bahamas and the U.S. East Coast. The accumulated rainfall map is derived from a GPM IMERG final run.*

scientists to study Earth's water cycle, climate variability, climate sensitivity, and feedback processes.

This information allows scientists to issue expedited estimates of accumulated precipitation during an extreme event such as a hurricane, including its spatial extent. It also allows for timely updates regarding landslide potential and early warnings for potential flooding downstream. The use of rainfall data can help forecast crop yields, monitor freshwater resources, and assess hazards related to glacial lakes.

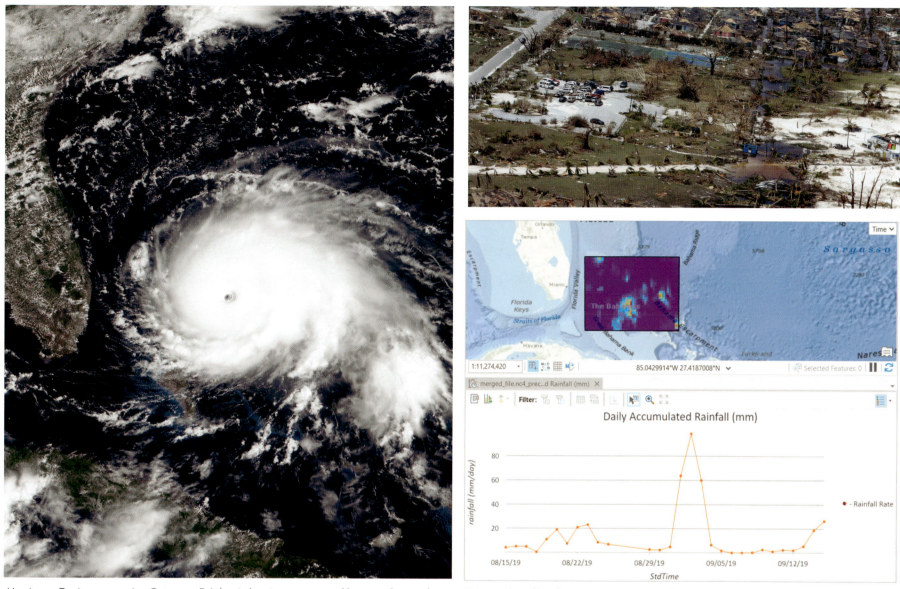

*Hurricane Dorian, a massive Category 5 Atlantic hurricane captured here on September 1, 2019, became the most intense tropical cyclone to strike the Bahamas in recorded history. The event was recognized as the country's worst natural disaster.*

*Temporal profile of Hurricane Dorian rainfall in the Bahamas region. Six and a half feet of water flooded the island nation in a 24-hour period.*

*A workflow to derive a time-averaged map using NASA's Giovanni tool and publish the output as a web map and web application. Giovanni is a web-based application developed by NASA's Goddard Earth Sciences Data and Information Services Center that provides a simple and intuitive way to visualize, analyze, and access vast amounts of earth science remote sensing data.*

## NASA's Land Data Assimilation System

NASA's work to identify devastating food shortages has led to the Famine Early Warning Systems Network Land Data Assimilation System (FLDAS). This network provides early warning and analysis to identify famine and monitor acute food insecurity globally (McNally et al. 2017). FLDAS, which includes a crop water balance model, is a custom instance of NASA's Land Information System (LIS), a software framework for high-performance terrestrial hydrology modeling.

FLDAS has been applied to study hydrometeorological conditions such as drought and their impact on agriculture and food security. FLDAS output variables, such as soil moisture, evapotranspiration, and groundwater, are used to study water availability in data-sparse regions of the world. Similarly, FLDAS helps to study runoff and rainfall flux anomalies during and after major rainfall events. For example, map A shows the monthly anomaly of rainfall flux from October 2019 to June 2020 obtained from FLDAS for Lake Victoria in East Africa. Map B shows the same data globally.

NASA's earth observations are critical in understanding extreme events and the effects of climate change. One key variable often associated with this science is atmospheric temperature. However, the impact of global warming is far greater than just increasing temperatures. Warming modifies rainfall patterns, amplifies coastal erosion, lengthens the growing season in some regions, melts ice caps and glaciers, and alters the ranges of some species and infectious diseases. Most of these changes are already distinctly observable in data-variable trends.

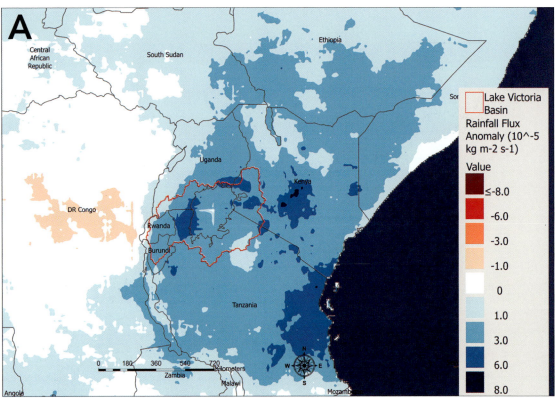

*Global monthly anomaly of rainfall flux obtained from FLDAS Noah Land Surface Model from October 2019 to June 2020, during which water levels in Lake Victoria, the largest lake in East Africa, rose above historic water records.*

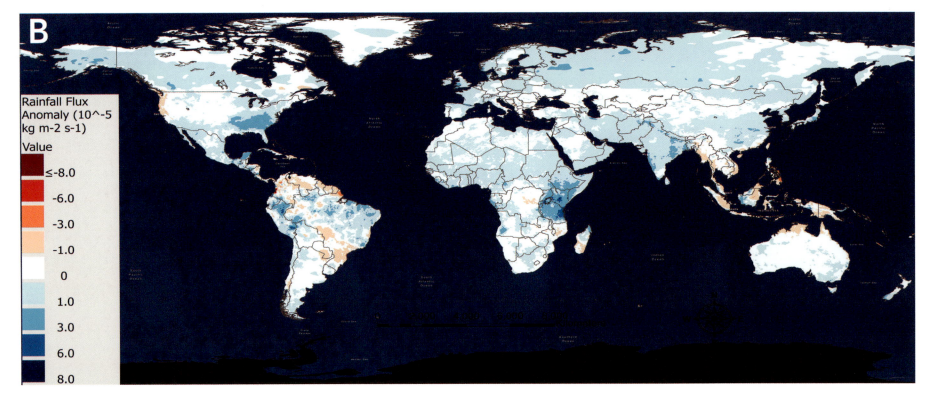

# RISING HEAT

Thermometer readings around the world show that temperatures have been rising since the Industrial Revolution. Scientists have high confidence that global temperatures will continue to rise for decades, largely due to greenhouse gases produced by human activities. Because the warming is superimposed on a naturally varying climate, the temperature rise has not been, and will not be, uniform or smooth across a country or over time. Scientists benefit from a high-resolution daily dataset that provides a long temporal range of weather data. The Daymet dataset provides gridded, continuous measurements of near-surface meteorological conditions often where no instrumentation exists. Weather parameters generated include daily surfaces of minimum and maximum temperature, precipitation, vapor pressure, shortwave radiation, snow water equivalent, and day length produced on a 1 km × 1 km gridded surface.

Daymet is a data product founded on ground-based observational data. It is derived from a collection of algorithms and computer software designed to interpolate and extrapolate from daily discrete meteorological observations to produce continuous gridded estimates of daily weather parameters. Having estimates of these surfaces is critical to understanding many processes in the terrestrial biogeochemical system.

Additional model inputs include a digital elevation model derived from the NASA Shuttle Radar Topography Mission (SRTM), derived horizon files, and a land water mask derived from the Moderate Resolution Imaging Spectroradiometer (MODIS) Land Water Mask. Observations of daily maximum temperature, minimum temperature, and precipitation from ground-based meteorological stations are provided by the National Centers for Environmental Information.

In the Daymet algorithm, researchers perform spatially and temporally explicit empirical analyses of the relationships between temperature and precipitation to elevation. A daily precipitation occurrence algorithm is introduced as a precursor to the prediction of daily precipitation amount. Water vapor pressure is generated as a function of the predicted daily minimum temperature and the predicted daily average daylight temperature. Daily incident solar radiation is estimated as a function of sun-slope geometry and interpolated diurnal temperature range. Snowpack, quantified as snow water equivalent, is estimated as part of the Daymet processing to reduce biases in shortwave radiation estimates.

Studying the relationship between living organisms and climate is an important component in understanding biodiversity, ecological forecasting, and biogeochemical cycles. It also helps to understand ecosystems in terms of structure and function. Further analysis of these data allows researchers to detect trends and patterns over space and time.

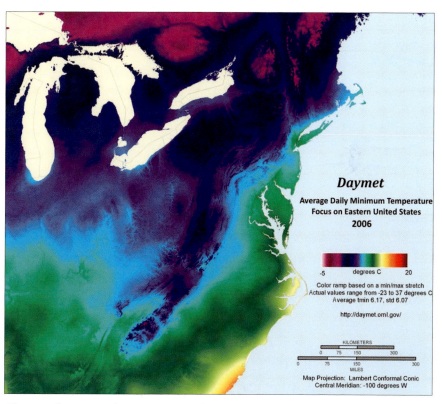

*This map focuses on the eastern United States, displaying Average Daily Minimum Temperature in 2006. Tracking temperature change can help us understand its impact on forest biomass and tree species distribution. Datasets such as Daymet allow us to analyze the relationship of terrestrial ecosystems, and living organisms, responses to extreme weather by calculating anomalies, and other higher-level bioclimatic variables.*

*Daymet produces gridded estimates of multiple weather parameters for North America. This group of maps showcases a sampling of parameters: Annual Total of Daily Precipitation, and Annual Average of Daily Vapor Pressure, Daily Maximum Temperature, and Daily Minimum Temperature.*

## Plight of the Carolina chickadee

Extreme weather events such as heat waves and drought affect the presence of certain bird species. The small, round Carolina chickadee (*Poecile carolinensis*) is frequently spotted in backyards in the Southeastern United States. When it's not at the bird feeder, the chickadee resides in wooded areas. But as temperatures change, these birds may be less-commonly sighted in the Southeast. For example, research by Cohen et al. (2020), finds that several bird species, including the Carolina chickadee, have shifted their behavior in response to extreme weather. Datasets such as Daymet can support research that monitors species distributions based on climate anomalies and extreme events.

Scientists working with a Daymet dataset using ArcGIS Pro can derive bioclimatologies using multidimensional tools and functions. It provides a platform to work with many types of scientific data. The multidimensional raster data mode and the analysis tools work with Network Common Data Form (netCDF), Hierarchical Data Format (HDF), a file format called GRIB, and many time-series satellite data produced from NASA missions. This example workflow uses ArcGIS Pro to process Daymet daily data, perform multidimensional raster analysis, and produce useful analytical products including climatologies, zonal analysis, spatial variation, anomalies, and trend analysis.

To start, Daymet daily data are distributed in netCDF file format. Performing time-series analysis in ArcGIS Pro using Daymet's daily data involves first creating a multidimensional Cloud Raster Format (CRF, an optimized raster format), then using geoprocesses in the multidimensional toolbox.

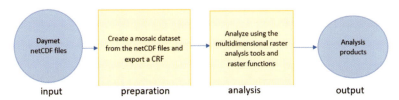

*Using the methods in Cohen et al. (2020) as direction, this ArcGIS Pro workflow demonstrates how to derive similar evaluations of extreme temperature. This graphic shows where the mean of weekly June temperatures between 2000 and 2018 exceeded 100° F. Data are clipped to the continental United States.*

Daymet data can be used to generate long-term temperature and precipitation profiles and determine the spatial distribution of areas that have experienced long-term extreme events. One such analysis product—using Daymet Daily Maximum Temperature—reflects the long-term average weekly temperature for June. Further analysis products can derive anomalies and show variance from the mean. These analyses show the spatial distribution of extreme temperature and extreme precipitation, which can affect species habitat suitability. A high-level explanation of this analysis is described later in the chapter.

Having determined where the anomaly is greater than 2 times the standard deviation, the ArcGIS Pro Find Argument Statistics tool can be used to find the number of months in each year that the anomaly consecutively exceeds 2, or 2 times the standard deviation. When the anomalies of the temperature and precipitation consecutively exceed their mean, we can see spatial patterns where weather can affect the habitat of birds and other species.

*Here, the ArcGIS Pro workflow is demonstrated to calculate the monthly anomaly for the precipitation variable, and then an anomaly z-score is determined. The z-score shows standard deviation from the mean instead of difference from the mean, so that we can use the same measurement across variables such as temperature and precipitation.*

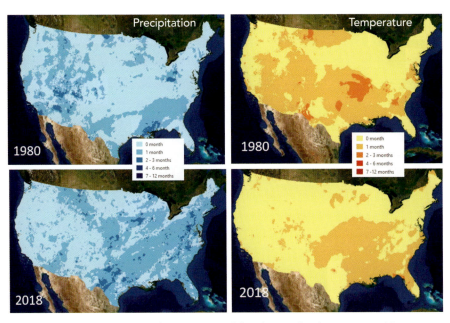

*Analyzing habitat suitability: This ArcGIS workflow begins by deriving monthly precipitation and temperature anomalies. Monthly anomalies are determined for each variable, and then an anomaly z-score is determined. The z-score shows how many standard deviations the anomaly is from the mean. Using the z-score, instead of difference from the mean, allows the same measurement on the precipitation and temperature variables. The images are determined by calculating where the anomaly is greater than 2 for each month, and then finding the number of months in each year where the anomaly consecutively exceeds 2 (i.e., 2 times the standard deviation). A researcher can determine spatially where the anomalies consecutively exceed their mean and record that data use in habitat suitability studies.*

# Urban heat islands

Urbanization transforms more than just the ecosystem. It also changes fundamental variables that influence weather and climate, such as land surface temperature, surface roughness, and evaporation. For many years, scientists have documented the changes in land surface temperature that result when natural or agricultural vegetation is replaced with parking lots, streets, and buildings. NASA describes land surface temperature as how hot Earth's surface would feel to the touch in any particular location, ranging from the roof on a building or grass on a lawn to the snow on a highway or ice on a frozen pond. In urban areas, temperatures can be up to 8 degrees warmer than surrounding suburban or natural landscapes, increasing the amount of energy a city needs to keep its residents cool and comfortable.

The urban heat island effect occurs primarily during the day, when impervious surfaces such as roads and parking lots absorb more sunlight than the surrounding vegetated areas. Trees, grasses, and other vegetation naturally cool the air as a by-product of photosynthesis. They release water back into the atmosphere in a process called evapotranspiration, which cools the local surface temperature in much the same way that sweat cools a person's skin as it evaporates. Trees with broad leaves, such as those found in many deciduous forests on the U.S. East Coast, have more pores to exchange water than trees with needles, so they have more of a cooling effect.

In the United States, the presence or scarcity of vegetation is an essential factor in urban heating. Using data from multiple satellites, NASA researchers modeled urban areas and their surroundings and found that areas covered partly by impervious surfaces—such as city centers, suburbs, or interstate highways—had an average summer temperature 1.9°C (3.4° F) higher than surrounding rural areas. In winter, the temperature difference was 1.5 °C (2.7° F) higher.

The highest urban temperatures relative to surrounding rural areas appeared along the Interstate-95 corridor from Boston to Washington, around Atlanta and the I-85 corridor in the Southeast, and near the major cities and roads of the Midwest and West Coast. Smaller cities had less pronounced increases in temperature compared to the surrounding areas. In desert cities, such as Phoenix, the urban area was actually cooler because irrigated lawns and trees provide cooling that dry, rocky areas do not.

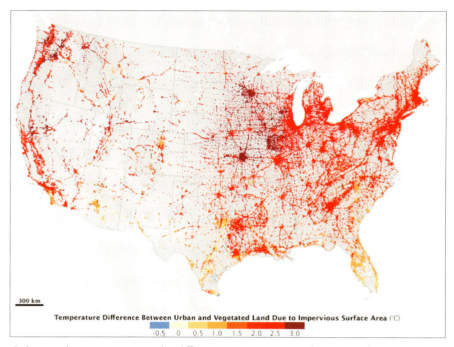

**Temperature Difference Between Urban and Vegetated Land Due to Impervious Surface Area** (°C)

-0.5  0  0.5  1.0  1.5  2.0  2.5  3.0

*Colors on the map represent the difference in temperatures between urban areas and the land that directly surrounds them. The urban heat island effect can raise the temperature of cities by several degrees Fahrenheit compared to neighboring rural and semirural areas due to the presence of impervious surfaces that absorb heat and disrupt the natural cooling effect provided by vegetation. The presence or scarcity of vegetation is an essential factor in urban heating.*

*Scatterplot depicting the relationship between impervious surface cover values and surface temperature.*

## Analysis of the urban heat island effect in ArcGIS Pro

These images provide views of Atlanta, Georgia, on May 6, 2020. Located in the center of the images is the urban core. The first Landsat image in the set displays a photo-like view of the area, where trees and other vegetation are bright green, roads and dense development appear cement gray, and bare ground appears tan or pink. The second image is a Landsat Provisional Surface Temperature map in which cooler temperatures are blue and hotter temperatures are red and yellow. Because vegetation cools the surface through evaporation of water, the most densely vegetated areas (bright green in the first image) are the coolest areas (dark blue in the second image). Where development is densest, the land surface temperature is near 38° C (100° F). The last image is a map derived from Landsat of fractional artificial impervious surfaces in which darker areas have more impervious cover.

An initial analysis of the images indicates a relationship between impervious surfaces and urban heat. To quantify this assumption, surface temperature and impervious surface cover values can be extracted and plotted using the ArcGIS Pro scatterplot charting tool.

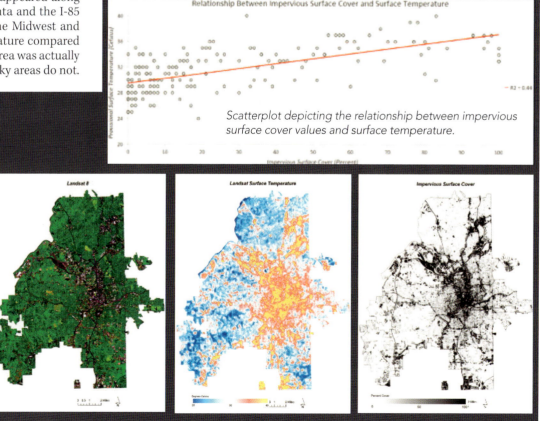

# CONCLUSION

Ecosystems are dynamic communities, but despite their vigor, these entities are sensitive to change. Understanding potential changes and how they might affect ecosystems provides the ability to detect the negative and often detrimental effects of extreme events such as the record-shattering heat wave in late June 2021 in the U.S. Pacific Northwest and western Canada, where temperatures reached 121° F in British Columbia, Canada, the highest temperature ever recorded in that country. Increasing knowledge of how ecosystems change under current conditions also helps us to model ecosystem and habitat changes under different climate scenarios. NASA's new ESO will guide efforts related to climate change, disaster mitigation, fighting forest fires, and improving real-time agricultural processes. NASA remote sensing data, coupled with ground-based data, aids in our understanding impact to biological diversity.

NASA strives to advance open science data systems for the next generation of missions, data sources, and user needs. NASA promotes the full and open sharing of all data, metadata, documentation, models, images and research results, and the source code used to generate, manipulate, and analyze them. Open science is a collaborative culture enabled by technology that empowers the open sharing of data, information, and knowledge within the scientific community and the wider public to accelerate scientific research and understanding. Merely providing users with access to the data is insufficient. Shifts in technology have unlocked the potential to use NASA's earth science data in new ways. As stewards of these data, NASA is committed to guiding users in using these new paradigms.

† Chapter contributors:
Leah Schwizer, NASA Earth Science Data Systems GIS Team (EGIST); Cyndi Hall, NASA Earth Science Data Systems Community Coordinator; Madison Broddle, Joseph Koch, and Sanjana Paul, NASA Langley Research Center's Atmospheric Science Data Center (ASDC); Binita KC, Jennifer Wei, and Allison Alcott, NASA Goddard Earth Science Data and Information Services Center (GES DISC); Desiray Wilson, NASA Langley Research Center's My NASA Data Team; Jeremy Kirkendall, and Garrett Layne, NASA Applied Sciences Program, Disasters Program; Viral Shah, Julia Computing; Ranjan Anantharaman, Massachusetts Institute of Technology (MIT); Vincent Landau, Conservation Science Partners; Kimberly Hall, Melissa Clark, Aaron Jones, and Jim Platt, The Nature Conservancy; and Michele Thornton, NASA Oak Ridge National Laboratory (ORNL) Distributed Active Archive Center (DAAC).

## General
NASA Earth Science Data Systems Program. "Data Pathfinders," updated April 28, 2021.

NASA Earth Science Data Systems Program. "Data Toolkits," updated October 21, 2020.

## Section: Wildfire

NASA Atmospheric Science Data Center (ASDC) and NASA's Applied Sciences Program. "Studying the 2019–2020 Australian Bushfires Using NASA Data," last modified May 2020.

NASA Applied Sciences Disasters Program. "Exaggerated Plume Height Application," created January 16, 2020.

NASA Johnson Space Center Earth Science & Remote Sensing Unit and Applied Sciences Disasters Program. "ISS Imagery: Australia Fires," created September 17, 2020.

NASA Earth Science Data and Information System (ESDIS) project. "Fire Information for Resource Management System," accessed March 15, 2021.

NASA Goddard Institute for Space Studies. "Global Fire WEather Database (GFWED)," last modified April 30, 2020.

NASA Missions: Fire and Smoke. "NASA Covers Wildfires Using Many Sources," updated October 2, 2020.

NASA Global Climate Change: Vital Signs of the Planet. "A Drier Future Sets the Stage for More Wildfires," published July 9, 2019.

NASA Earth. "NASA Study Finds a Connection Between Wildfires and Drought," published January 9, 2017.

## Section: Flooding and drought

NASA Goddard Earth Sciences Data and Information Services Center (GES DISC) and ArcGIS DAAC collaboration. "Lake Victoria Rising Water Levels," published December 9, 2020.

NASA Applied Sciences Disasters Program. "NASA Products for Hurricane Dorian 2019," accessed March 15, 2021.

NASA GES DISC and ADC. "Cyclone Amphan," last updated June 11, 2020.

My NASA Data. "Hurricanes as Heat Engines," last modified November 10, 2020. Huffman, G.J., E.F. Stocker, D.T. Bolvin, E.J. Nelkin, and J. Tan. 2019. "GPM IMERG Final Precipitation L3 Half Hourly 0.1 degree x 0.1 degree V06" (Greenbelt, Maryland: Goddard Earth Sciences Data and Information Services Center (GES DISC). https://doi.org/10.5067/GPM/IMERG/3B-HH/06.

McNally, A., K. Arsenault, S. Kumar, S. Shukla, P. Peterson, S. Wang, C. Funk, C.D. Peters-Lidard, and J.P. Verdin. 2017. "A Land Data Assimilation System for Sub-Saharan Africa Food and Water Security A Applications." Sci Data 4, no. 1: 170012. https://doi.org/10.1038/sdata.2017.12. ISSN: 2052–4463.

McNally, A., and NASA/GSFC/HSL. 2018. FLDAS "Noah Land Surface Model L4 Global Monthly Anomaly 0.1 x 0.1 degree (MERRA-2 and CHIRPS)." (Greenbelt, Maryland: Goddard Earth Sciences Data and Information Services Center (GES DISC). Accessed November 25, 2020, 10.5067/GNKZZBAYDF4W.

NASA Global Climate Change: Vital Signs of the Planet. "Earth's Freshwater Future: Extremes of Flood and Drought," published June 13, 2019.

## Section: Rising Heat

My NASA Data. "Creation of Urban Heat Islands Story Map," created July 10, 2019. "Vegetation Limits City Warming Effects." NASA Earth Observatory, August 26, 2015.

Thornton, M.M., R. Shrestha, Y. Wei, P.E. Thornton, S. Kao, and B.E. Wilson. 2020. "Daymet: Daily Surface Weather Data on a 1-km Grid for North America, Version 4." ORNL DAAC, Oak Ridge, Tennessee, USA. https://doi.org/10.3334/ORNLDAAC/1840.

NASA Earth Observatory. "World of Change: Global Temperatures," published January 29, 2020.

NASA Earth Observatory. "Global Warming," published June 3, 2010.

NASA Global Climate Change: Vital Signs of the Planet. "The Effects of Climate Change," Site last updated: April 5, 2021.

Xu, Hong. "Mapping Extreme Weather Zones for Birds," published May 17, 2021.

## Image credits
Carolina chickadee image credit Jeremy Cohen via Flickr.com.
Hennessey fire photo by Dripwoods via Flickr.com.

*The wildfires in California north of the San Francisco Bay Area were photographed by an Expedition 61 crew member as the International Space Station orbited 254 miles above the Golden State.*

# THE SCIENCE OF OCEAN ACOUSTICS

Understanding the soundscape of the world involves using an array of global tracking data, acoustic propagation models, and GIS. Together, these data and technologies help scientists identify Earth's loud and quiet places and learn about the implications of those results.

By Chris Verlinden, Sarah Rosenthal, Jennifer Brandon, Kevin Heaney, and James Murray, **Applied Ocean Sciences**

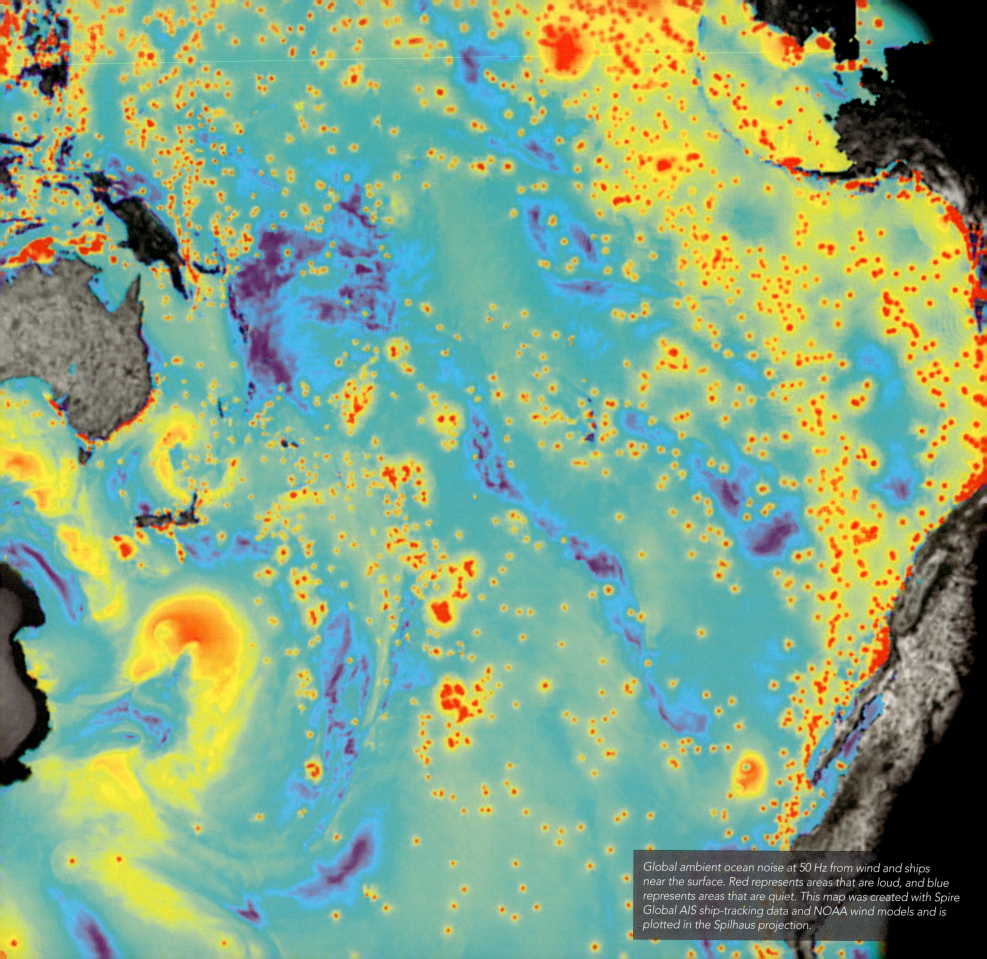

Global ambient ocean noise at 50 Hz from wind and ships near the surface. Red represents areas that are loud, and blue represents areas that are quiet. This map was created with Spire Global AIS ship-tracking data and NOAA wind models and is plotted in the Spilhaus projection.

# OCEAN SOUNDSCAPE

The Monterey Bay Aquarium Research Institute (MBARI) defines the ocean soundscape as "a continuously changing mosaic of sounds that originate from living organisms (communication and foraging), natural processes (breaking waves, wind, rain, earthquakes, etc.), and human activities (shipping, construction, resource extraction, etc.)." The ocean soundscape is 4D, varying in x, y, z, and t.

The soundscape is important because sound is one of the primary ways that we interact with and sense the ocean, which means the soundscape has implications for oceanographers who use acoustic sensors to measure the ocean, marine mammals that use sound to communicate, and the military that uses sound to find vessels such as submarines. The soundscape is a challenge to model, measure, represent, and understand because it is fundamentally 4D. The ocean is volumetric, and the soundscape along with the ocean properties it depends on (such as temperature and salinity) have values in every voxel (3D grid cell), and those values change over time. This 4D workflow can be difficult to manage throughout the process, from gathering and curating the datasets, managing high-bandwidth computational modeling mathematics that are applied to those datasets, making those results available to derivative tools and processes to conduct analysis, and representing and communicating the results of this analysis. That is where effective GIS is paramount.

In this chapter, we discuss the GIS workflow used to model global ambient ocean noise (the soundscape) derived from ship movement and weather phenomena such as wind and waves. This chapter presents changes in the global soundscape that occurred during the coronavirus disease 2019 (COVID-19) pandemic. Our work focuses on the Arctic as a critical example of this soundscape analysis and uses the impact of anthropogenic sound—such as shipping noise on bowhead whales—as an example of the impact of this type of analysis.

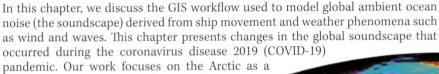

*The ocean soundscape is composed of all the sources of noise that contribute sound to the environment. They include sounds made by humans (anthropogenic; orange sound waves), the environment (natural sounds; green sound waves), and biological sources (animals: marine mammals, fish, and invertebrates; blue sound waves).*

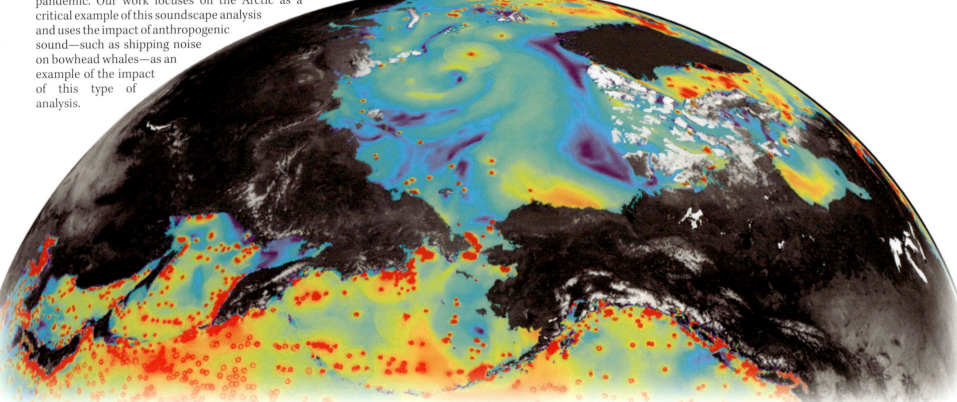

*Plot of global integrated ship and wind noise at a single moment in time—April 15, 2020.*

# OCEAN ACOUSTICS

In air, we use electromagnetic radiation to interact with our environment. We use visible light, detected by our eyes, to see, radio waves to communicate, and lasers to determine the range to an object (lidar). Underwater, this electromagnetic radiation does not travel very far because it attenuates rapidly. For this reason, researchers use sound as a tool to study the ocean and acoustics as a way to "see" it. For example, the military uses sound (sonar) to find submarines, and oceanographers use it to measure temperature, salinity, and other properties of the ocean.

## Acoustic propagation modeling

Scientists rely on acoustic propagation modeling to study ocean soundscapes, which is underpinned by an understanding of the fundamental physics of sound. Sound is a pressure wave. A vibrating body, such as the membrane of a speaker, presses on the fluid surrounding it (water or air), and those fluid molecules in turn push on the molecules next to them. The resulting disturbance spreads, or propagates, out in all directions as a pressure wave. The speed of sound is the speed at which this pressure disturbance propagates. The speed of sound in a media depends on the compressibility of the material. Fluids of different properties, such as air versus water, or water at different temperatures, have different sound speeds. For example, air is more compressible than water, so the speed of sound in air is approximately 350 meters per second (m/s); while in seawater the speed of sound is approximately 1,500 m/s.

Sound propagates faster in warm water than cold water because warm water is less compressible. Similarly, sound propagates faster in water under pressure. This gives the ocean a variable refractive index, which causes sound to curve away from warmer water or water under pressure and curve toward colder water. This movement can cause sound to follow a deep sound channel called the SOFAR (sound fixing and ranging) channel, at a depth where the sound speed is at a minimum.

The physics of how sound travels through seawater enables researchers to use acoustics to sense the ocean, but these same physics also require researchers to measure the ocean everywhere to successfully model acoustic propagation. The fact that the temperature, salinity, and pressure of seawater cause the sound to travel along different paths at different speeds allows us to use the travel time of underwater sound to measure these seawater properties using a technique called tomography. Conversely, if the seawater properties are well known, researchers can accurately simulate sound propagation and use these simulations to communicate underwater, navigate submersibles, or find whales.

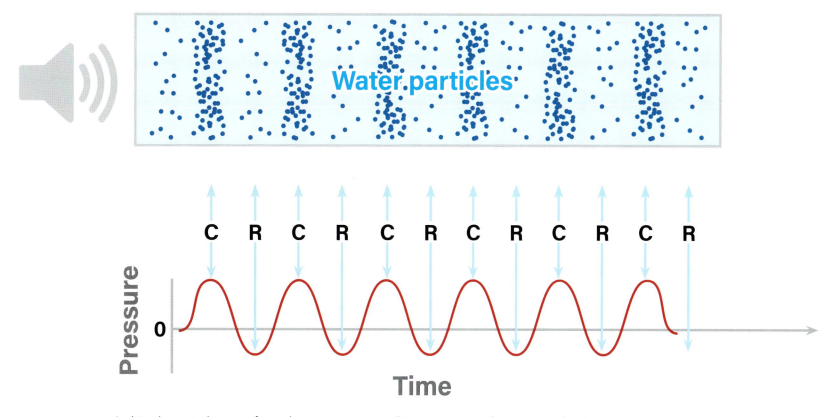

In this schematic diagram of sound as a pressure wave, C is compression, where water molecules are closer together, and R is rarefaction, where the particles are farthest apart. Sound travels significantly faster in water than in air because neighboring water particles more easily bump into one another.

To model sound in water, one of four methods is typically employed: ray tracing, wavenumber integration or spectral methods, normal modes, or parabolic equation-based simulations.

Ray tracing is an acoustic modeling technique used to calculate the path of sound through the ocean up to very large distances. The 3D map of transmission loss shows a ray trace simulation used to model the acoustic intensity of a source in one location in all directions, first by modeling the acoustic intensity in range and depth at a single bearing, and then repeating that simulation in all directions. This ray tracing method is called an N x 2D simulation, where N is the number of radials, and 2D refers to range and depth. The result is a 3D parabolic equation-based acoustic simulation that also accounts for the horizontal refraction of sound in seawater. The final output gives us the acoustic field at every voxel, or 3D pixel, in the ocean.

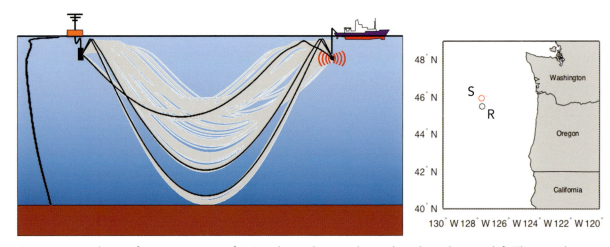

*A ray trace simulation of acoustic energy refracting due to the sound-speed gradient shown at left. The sound source (S) is the boat propeller that releases sound in all directions, and the hydrophone (Receiver -R) at right picks up the sound waves as they reach it at different times.*

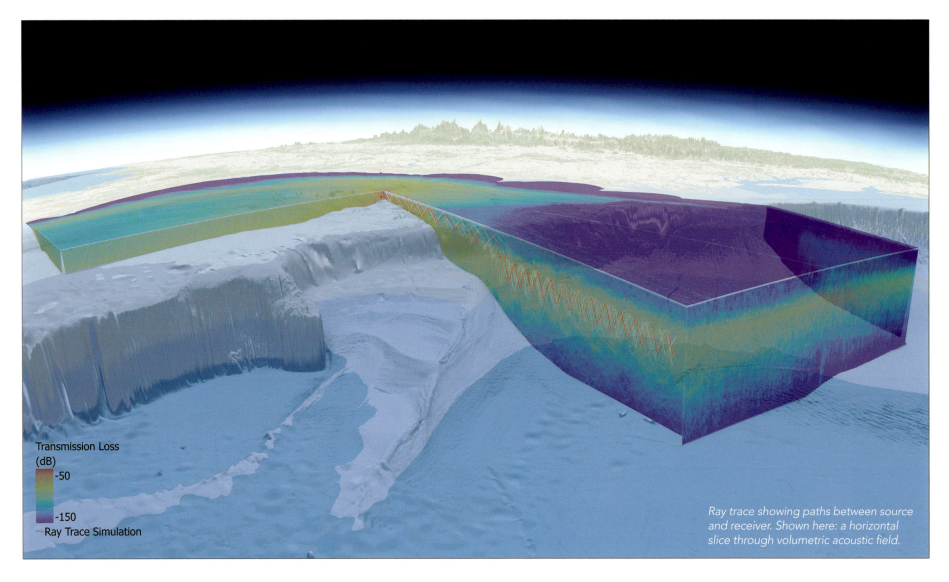

Transmission Loss
(dB)
-50
-150
— Ray Trace Simulation

*Ray trace showing paths between source and receiver. Shown here: a horizontal slice through volumetric acoustic field.*

Acoustic propagation modeling can be used to estimate or predict natural and anthropogenic sounds underwater. The natural soundscape consists of everyday noises, such as wind, waves, and marine mammal communication, and more unique sounds, such as cracking and ridging ice or earthquakes (pictured in map A is the March 1964 earthquake in the Gulf of Alaska, the strongest ever recorded in North America.).

Anthropogenic sounds also fall into two categories of common sounds and unusual events. The Comprehensive Nuclear-Test-Ban Treaty Organization (CTBTO) has a worldwide monitoring network in place to make sure that one of the most devastating unusual events (testing of nuclear weapons) doesn't happen, and if it does happen, the international community is aware of it immediately. Pictured in map B is the hydroacoustic portion of the CTBTO international monitoring system, with the red dots representing the acoustic moorings and the colored raster images illustrating the coverage of the monitoring network with the colors representing how loud a source would be in the network.

A

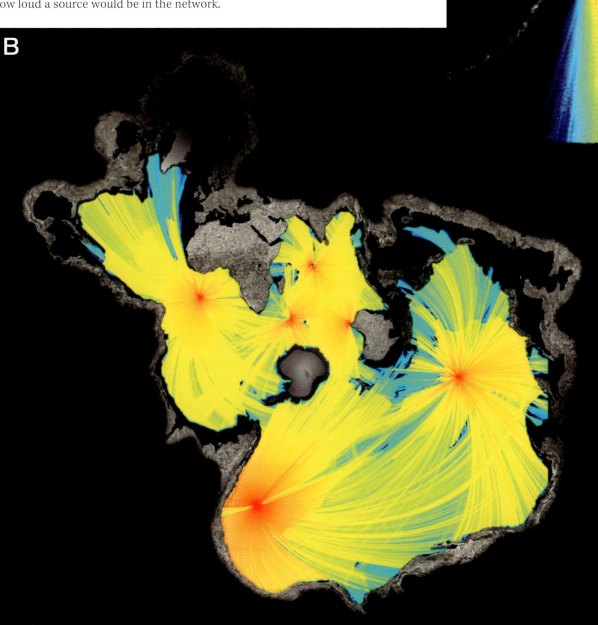

B

*Map A simulates the sound of the magnitude 9.2 earthquake in the Gulf of Alaska on March 27, 1964. The color red symbolizes received sound levels of more than 130 decibels. Earthquakes can sometimes be detected on hydrophones around the world, illustrating the importance of accounting for 3D propagation effects, which can be substantial over great distances. Map B represents the coverage of the CTBTO global monitoring array. Acoustic propagation is largely reciprocal, which means the same simulation software used to model how loud a source such as an earthquake would be all over the ocean also can show how loud a source in any position in the world would be on a network of hydrophones, such as shown here. At any location shown in light blue, yellow, or red, hydrophones would detect an explosion.*

# GIS WORKFLOW FOR SOUNDSCAPE MODELING

The six-step process of a typical ocean acoustic GIS workflow involves synthesizing a vast array of environmental input data and making that data available to an acoustic propagation model, such as a ray trace simulation or parabolic equation-based model. The workflow then takes the results of the acoustic simulation and uses analysis tools to answer questions ranging from how far away one bowhead whale can hear another bowhead to where sensors should be placed to detect a vessel fishing illegally in a closed area. This workflow resembles an hourglass, which is wide at both ends and narrow in the middle. The top of the hourglass represents the data management and processing required to feed the acoustic model what it needs. The bottom represents the wide range of analysis techniques that can be applied to the results of the acoustic simulation. The narrow middle represents the acoustic simulation itself. The wide parts—the top and bottom of the hourglass—are both GIS.

## Step 1: Gather data

Acoustic propagation is affected by an array of ocean properties. The data for each of these properties (and providers) includes: ocean temperature (map A) and ocean salinity (World Ocean Atlas); ocean bathymetry (the General Bathymetric Chart of the Oceans); bottom type (Bottom Sediment Type v 2.0 database); wind and wave height (National Oceanic and Atmospheric Administration); and ship location (Spire).

## Step 2: Dependent variables

Temperature, salinity, and pressure are used to calculate the speed of sound in seawater (map B). Bathymetry and information about bottom composition are used to determine how the sound will reflect off of and be absorbed by the seafloor.

## Step 3: Grid data

The manner in which the environmental data is discretized and gridded is critical for acoustic modeling. For example, the parabolic equation-based acoustic models used by this team require the acoustic field to be computed every one-half wavelength. A 50 Hz sound has a wavelength of 30 m.

## Step 4: Acoustic simulation

The sound speed, bathymetry, bottom characteristics, and source location are input into an acoustic propagation model to estimate acoustic transmission loss. This parabolic equation-based model conducts the simulation along radials at discrete bearings and then integrates the resulting vertical slices into a full 3D field using spatial interpolation. Horizontal refraction is also accounted for because subtle horizontal sound speed gradients deflect the sound horizontally. The result of the simulation is the channel impulse response (CIR), a mathematical term that represents how the acoustic signal changes between a source location and every voxel in the simulation area. The CIR can be used to reconstruct what a signal generated in one location might sound like in another location, or estimate acoustic transmission loss (map C). Transmission loss (TL) is a measure of how much energy is lost between source and receiver.

## Step 5: Repeat

At this point, researchers repeat the acoustic simulation for all sources of noise, including all ships, global wind, and waves. The acoustic intensity from each of these sources is summed with the transmission loss computed using the propagation model to determine how loud the ocean is in every location.

## Step 6: Analysis

Using this global ambient noise model can help answer questions related to acoustic communications, marine mammal stressors, and optimal acoustic sensor placement for various oceanographic sensing or defense applications.

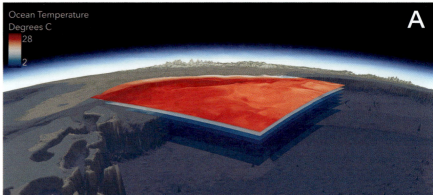

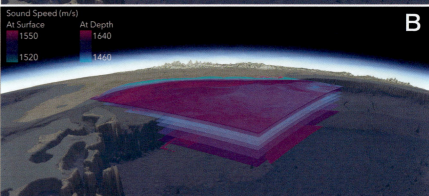

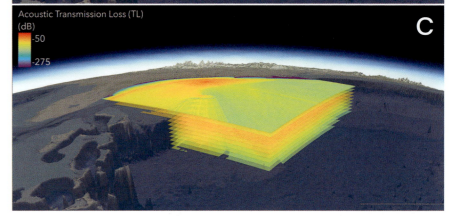

*The complexity of the ocean acoustics modeling challenge results from the data being volumetric. In the GIS workflow, the water "layers" are combined in a model to create layers that can describe the long-distance behavior of sound in the world's oceans.*

# GLOBAL SOUNDSCAPE MODELING

Applying the GIS workflow to global ship-tracking databases such as the Spire Global AIS dataset, along with global weather, bathymetry, and bottom composition, makes it possible to model the entire global soundscape in near-real time. This workflow application can help marine biologists understand the communication and sensing capabilities of marine mammals or help law enforcement and military personnel understand the performance of acoustic monitoring networks. Modeling the global soundscape presents four primary challenges:

1. **Computational expense**: An estimated 200,000 commercial ships operate in the ocean; capturing small-scale variability in noise from wind and waves requires global soundscape models at approximately 1 km resolution. And powerful computers must run efficient algorithms to model these sources in near-real time.

2. **Input database management**: Acoustic models require accurate information about seawater and bottom properties. As a rule, these environmental inputs should be known with more than one-half wavelength resolution. For 150 Hz acoustic simulations, this means 5 m voxels. These datasets also must be kept up to date. As with any computer simulation, effective management of input environmental data is key.

3. **Process optimization**: The mathematics involved in the parabolic equation-based acoustic simulations used for soundscape modeling have existed for decades, and sources such as Spire Global make obtaining global ship-tracking datasets fairly straightforward. The challenge becomes setting up a computational architecture to grid and discretize environmental data inputs, manage and filter source information, schedule computational load online and on physical computer servers, and then effectively store simulation results.

4. **Data analysis:** Once the acoustic simulations have been completed, further analysis is needed to understand the soundscape. Source levels must be applied to each ship, and based on international databases and historical empirical measurements. Then the data must be made available to organizations that need it through web-API or a data marketplace such as Esri's ArcGIS Marketplace or ArcGIS Online. The data must be in a format accessible to organizations that need it and communicated in a clear and accessible way.

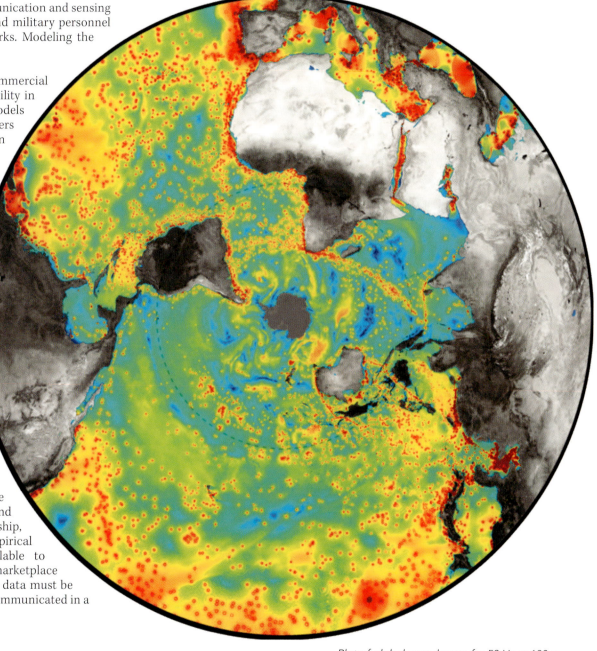

*Plot of global soundscape for 50 Hz at 100 m receiver depth simulated using global ship-tracking data and global wind data.*

# THE YEAR OF A QUIETER OCEAN

The COVID-19 pandemic, although tragic on so many levels, has provided scientists with a unique opportunity to study the global soundscape.

The increasing sound levels in our oceans in modern times are generally understood to be detrimental to certain sea life—in particular such creatures as whales, which communicate over long distance using sound. Baleen whales are the largest animals on earth and include gray, humpback, minke, right, blue, and fin whales. All but the gray and humpback whales are considered threatened. These whales emit extremely low-frequency vocalizations that travel surprisingly long distances underwater. The wavelengths of these calls can be longer than the bodies of the whales themselves.

Global decreases in ship traffic during the pandemic (maps A and B) led to a significant decrease in anthropogenic noise in the global ocean (maps C and D). The North Atlantic Ocean, for example, experienced an average decrease in noise levels from ships of 3.4 dB at 50 Hz, a frequency that is highly relevant for many large baleen whale species. A decrease of 3.4 dB might not sound like a lot, but dB are logarithmic, which means 3 dB represents a doubling of acoustic intensity.

Put a different way, a whale—which uses sound to communicate, locate food, and find other whales—could hear another whale over twice as far away in 2020 during the pandemic in the North Atlantic Ocean than during normal conditions in 2019.

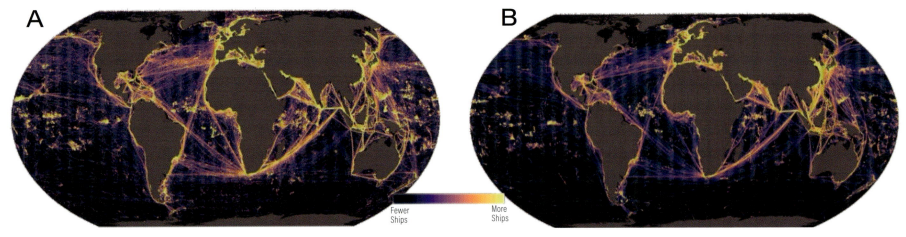

Global average shipping density during April 2019 (A) and April 2020 (B) estimated using Spire Global AIS tracking data. Shipping lanes are clearly identifiable. Subtle differences can be seen, such as the less bright shipping lane off Brazil.

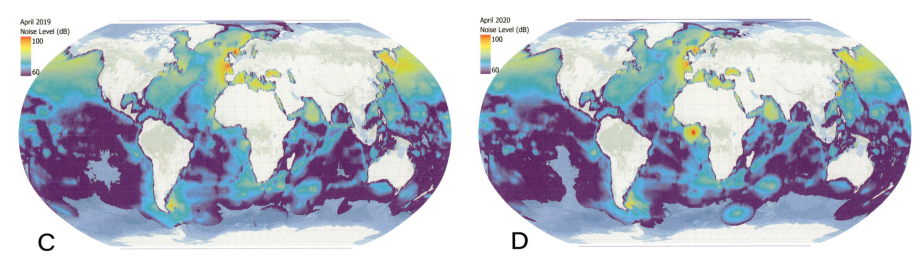

Average ambient ocean noise from ships at 50 Hz and 100 m receiver depth during April 2019 (C) and April 2020 (D). Differences at first glance appear subtle between 2019 and 2020, but by comparing the 2019 and 2020 maps, researchers can compare the ambient noise difference between the two years. In 2020, the global ocean averaged 2–4 dB quieter than 2019. Red, or positive, means that 2020 was quieter than 2019.

Although the changes in anthropogenic sound from 2019 to 2020 seem subtle on the global scale, they are much more obvious when one focuses on a specific location. Using Hawaii as a specific case study, we see in the maps of the Hawaiian Islands that a humpback whale is only detectable by other whales over a very small distance in 2019, due to the amount of ship noise in the area. But in 2020, with the lack of ships present, those whales are able to communicate over an area 10 times as large. The use of soundscape GIS in combination with propagation models and relevant data allows researchers to determine how many other whales will hear the call of the humpback and over what area that call is audible.

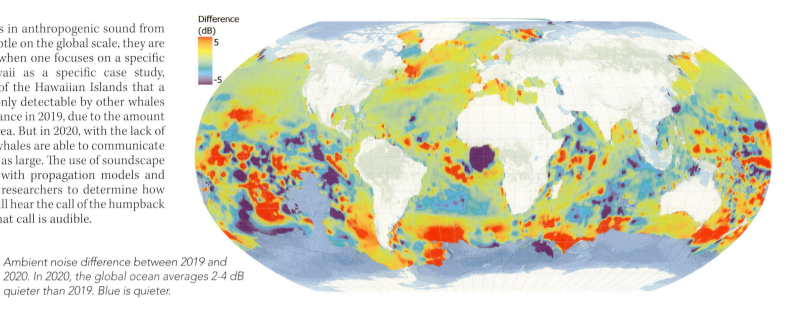

Difference (dB)

*Ambient noise difference between 2019 and 2020. In 2020, the global ocean averages 2–4 dB quieter than 2019. Blue is quieter.*

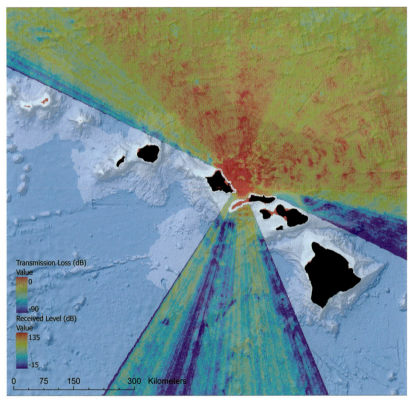

Transmission Loss (dB)
Value
0
-90

Received Level (dB)
Value
135
-15

0   75   150        300   Kilometers

*Plot of source level (how loud it will be) of a humpback whale call north of the island of Oahu in the Hawaiian Islands. To determine where that whale might be detected by other whales, researchers must determine the ambient noise levels (soundscape) and then subtract the soundscape from the source level to get the signal excess, or the detectable sound.*

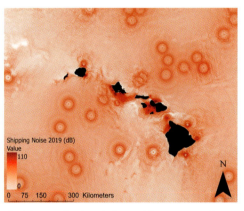

Shipping Noise 2019 (dB)
Value
110
0

0   75   150        300   Kilometers

*Ambient noise levels from ships on a random day in March 2019. A large number of ships are under way. As a result, the ocean is very loud. The noise will make it more difficult for whales to communicate with each other.*

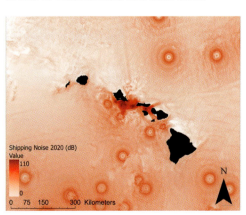

Shipping Noise 2020 (dB)
Value
110
0

0   75   150        300   Kilometers

*On the same random day and time in 2020, fewer ships are at sea. Consequently, the ocean is quieter than in 2019.*

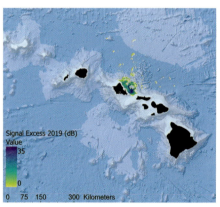

Signal Excess 2019 (dB)
Value
35
0

0   75   150        300   Kilometers

*An examination of the signal excess for the humpback whale call reveals that on this random day in 2019, the whale is only detectable by other whales over a very small area, approximately 4,025 km².*

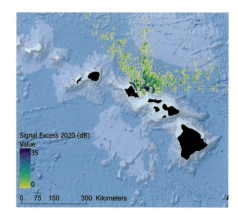

Signal Excess 2020 (dB)
Value
35
0

0   75   150        300   Kilometers

*Signal excess for the same humpback whale call in 2020 in the same location audible over an area of 42,240 km². This represents a tenfold increase in the area over which this whale can communicate in 2020 compared with 2019.*

# ARCTIC SOUNDSCAPE

The Arctic is a perfect example of the value of underwater acoustics in the study of oceanography and the challenges and complexities of its application. From an acoustics standpoint, the Arctic is a unique environment. Cold water near the surface leads to lower sound speeds at the surface. The result is a condition called surface ducting, which allows sound generated at the surface, such as noise from ships and breaking ice, to propagate great distances.

The consequence is that a single ship in the Arctic can ensonify an entire basin, whereas adding a single ship to the ocean in a different region might not even be noticed.

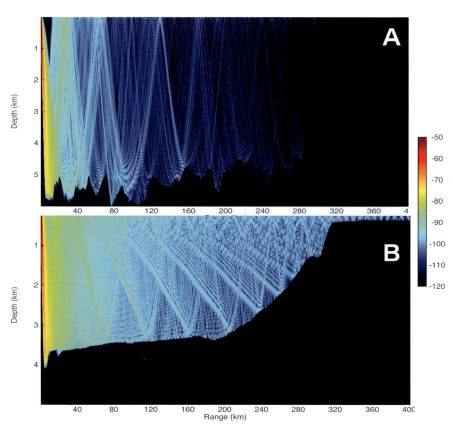

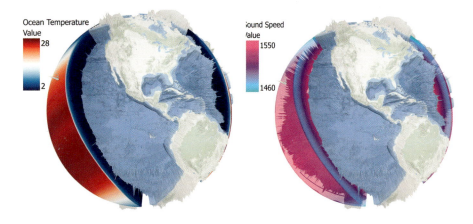

Ocean temperature and sound speed along meridional transects in the Pacific and Atlantic Oceans with corresponding acoustic ray paths getting shallower at the poles.

Typical vertical slice acoustic transmission loss simulation in the mid-latitude Pacific Ocean (A) compared with the Arctic transmission loss simulation (B). Transmission loss in dB plotted as a function of range and depth. Surface ducting keeps sound generated near the surface propagating at the surface over great distances.

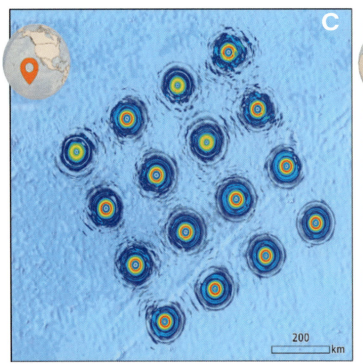

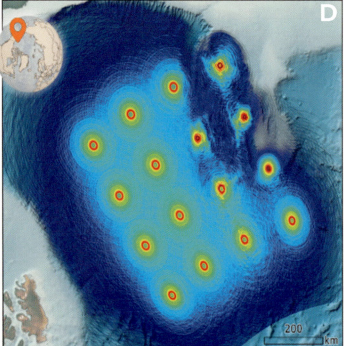

Map C shows acoustic simulation at 100 m depth of a fictitious grid of 16 vessels spaced 200 km apart in the South Pacific Ocean. The region is fairly quiet. Map D shows acoustic simulation at 100 m depth of the same grid of 16 vessels spaced 200 km apart in the Arctic Ocean. The entire region is ensonified.

## The effects of surface ice on sound propagation

In addition to the unique sound speed profile resulting from cold surface temperatures, which causes sound to be trapped in a duct at the surface, the presence of sea ice causes sound to scatter and sometimes be blocked entirely. Ice ridges that form when sheets of ice come together can extend 100 m beneath the surface, forming walls that block propagation paths. The presence of ice has a significant masking effect that contributes to the characterization of the pre-industrial Arctic soundscape as quiet or pristine. The combination of surface ducting, ice melting, and increased vessel traffic in the Arctic has led to the region today becoming much louder than it was 20 years ago.

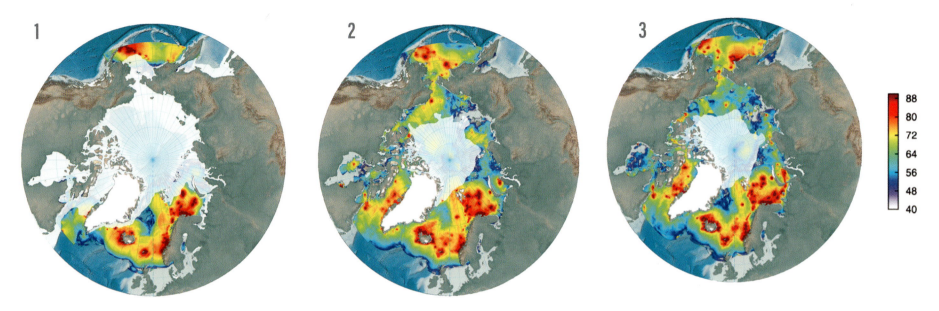

Pan-Arctic 25 Hz band weekly median sound exposure level (SEL) for 1) March 2015; 2) September 2015; and 3) September 2019 (SEL is in units of dB/μPa2). Sea ice is represented by white shading. Images from Underwater Noise Pollution from Shipping in the Arctic, Protection of the Arctic Marine Environment (PAME), 2021.

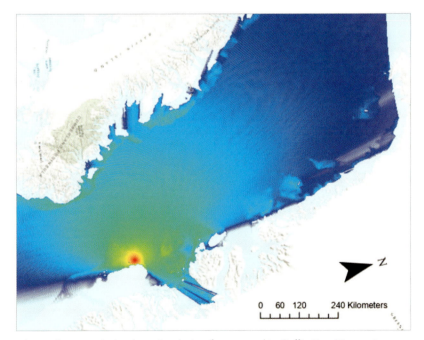

Acoustic transmission loss simulation for a vessel in Baffin Bay, Nunavut, Canada, during the summer, when ice is not present. A single ship ensonifies the entire region across the bay to the west coast of Greenland. Images from Underwater Noise Pollution from Shipping in the Arctic, PAME, 2021.

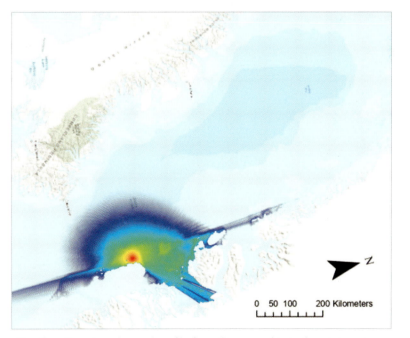

Significantly reduced sound profile from the same ship in the winter, when sea ice is present and scatters the sound as it reflects off the surface.

# WHY ACOUSTICS MATTER

**Marine mammals:** Acoustics matter a lot to whales and other marine mammals. Whales, such as this humpback, use sound to communicate, navigate, and sense their environment. When the ocean becomes noisier, many large baleen whale species vocalize louder and exhibit increased levels of stress hormones such as cortisol and aldosterone. During a brief hiatus in ship traffic following the terrorist attacks in New York City on September 11, 2001, researchers found that stress hormones in whales decreased. Researchers demonstrated that increased anthropogenic noise from ships causes physiological stress to marine mammals. Marine mammals use sound in ways we do not yet fully understand, and the input of additional sound into their environment can be damaging. For example, increased noise levels likely make it more difficult for whales to find mates and potentially interfere with their ability to navigate and find prey.

**Underwater communications:** Researchers have reasons to study the ocean soundscape beyond learning how noise impacts marine mammals. Because electromagnetic radiation such as radio waves do not propagate far underwater, acoustics is the primary mechanism for communicating underwater. Controlling underwater vehicles or extracting data from an underwater sensor is typically done with acoustics. But this type of communication is impossible if the ambient noise is too loud.

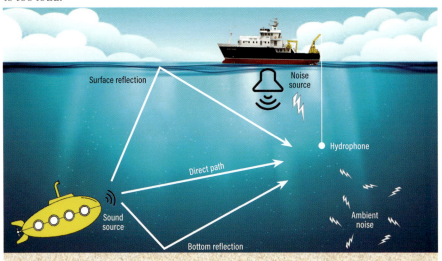

**Oceanographic and hydrographic sensors**: Oceanographers and hydrographers rely on acoustic sensors to measure and understand the marine environment. Sensors help measure the speed of ocean currents, and we use bottom profilers, multibeam, and side scan sonar to map the properties of the seafloor. We can even use acoustics to measure the temperature and salinity of the ocean using techniques such as acoustic tomography. This technique uses the travel time of sound along known paths to infer the properties of seawater that affect sound speed, including temperature and salinity. This process is accomplished by generating acoustic signals at specific frequencies and listening for their arrivals on distant sensor arrays or by passively gathering the various ambient noise signals received through multiple sensors. Recent work indicates that ocean acoustics could be used to measure the pH, or acidity, of the ocean. All of these sensing techniques rely on knowledge of the soundscape.

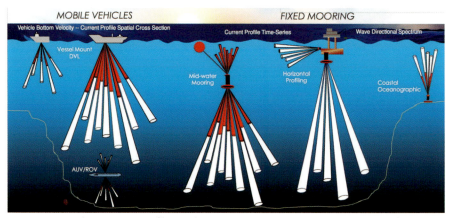

*Acoustic Doppler Current Profiler and its uses in oceanographic sensing. Credit: Teledyne RD and Rowe Technologies (acoustic device).*

**Military and law enforcement**: Finally, military and maritime law enforcement use acoustics to find surface and underwater vessels using passive and active sonar.

## Wind-generated noise

The applications presented earlier rely on knowledge of the soundscape. Here, we show the implications of changing the soundscape. In the 40 kn–70 dB map, we have a plot of the area in which a whale can be detected in Beaufort level 8 conditions, which are characterized by 40 kn wind, with seas between 12 ft and 18 ft. The wind noise at 1,000 m depth would be approximately 70 dB. The red dots in all the scenes show the detection volume, or the area with a positive signal excess, for a humpback whale, vocalizing at 50 Hz with a source level of 135 dB for the Atlantic Deepwater Ecosystem Observatory Network (ADEON) sensor network.

On a calmer day, when winds are less than 5 knots (kn) and seas are less than 2 feet, the detection volume changes dramatically. Bottom right map shows the detection volume for that same humpback whale, vocalizing at 50 Hz with a signal level of 135 dB. Everywhere marked in red is a voxel with a positive signal excess and therefore a place where a whale would be detected on the ADEON array. You can see that a whale would be detected over virtually the entire region, in this case the area offshore of Jacksonville, Florida. Even a small change in conditions, for example, a 5-kn decrease in surface winds, can impact the soundscape and detectability.

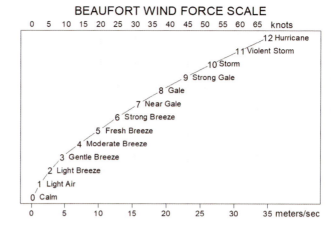

*The traditional Beaufort scale is an empirical measure that relates wind speed to observed conditions at sea or on land.*

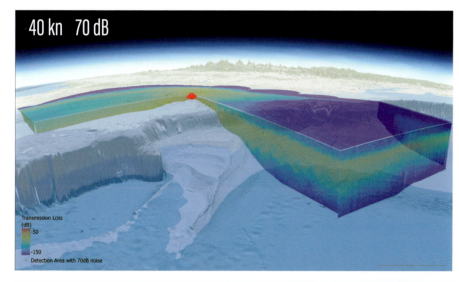

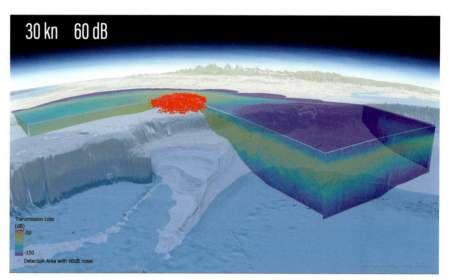

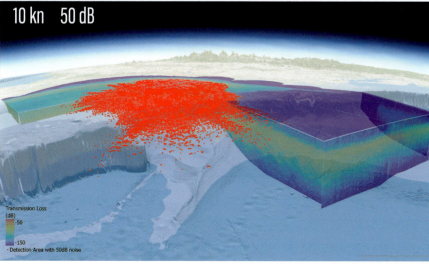

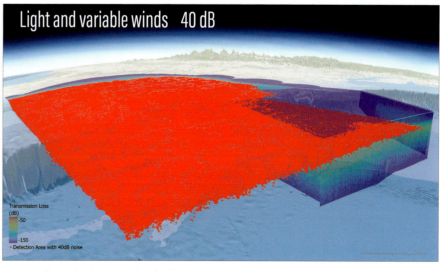

*Four panels showing signal excess, or detection volume, for a whale under various sea states. Voxels in which a humpback whale is detectable plotted in red. Bottom right: Noise from wind and waves at level 1.5 conditions, less than a light breeze on the Beaufort scale. On this calm day, a humpback whale is detectable virtually everywhere.*

# VALUE OF GIS

Researchers increasingly understand the importance of soundscape to marine mammals, humans, and planet Earth. As we have seen, the global soundscape can affect everything from a whale's ability to find a mate to a submarine's ability to stay hidden beneath the ocean's surface. GIS technology is one of the big reasons that research has progressed in the study and understanding of soundscape. Researchers use GIS to process, curate, and manage the 4D datasets required to do high-quality acoustic modeling. GIS allows researchers to quickly analyze the results of an acoustic simulation, such as those generated in the examples in this chapter, and turn it into something meaningful. Spatial analysis can help us answer questions about marine mammal stress and ecology and tell us where to deploy acoustic sensors to detect a vessel fishing illegally in a closed area. The scientific community needs tools that make this type of analysis available to everybody who needs it, from marine biologists and conservationists to law enforcement officials and military officers.

Applied Ocean Sciences (AOS), with its partners Spire Global and Esri, continues looking for innovative ways to make soundscape data and acoustic analysis tools available for wider use. We have created the capability to combine Spire's real-time and forecast ship-tracking and weather data with AOS's acoustic models, delivered through the ArcGIS Pro platform for the scientific community.

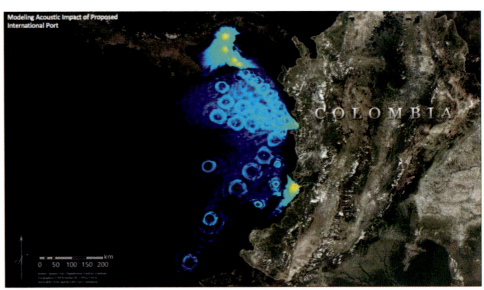

*GIS can model the potential acoustic impacts of a proposed port on marine life.*

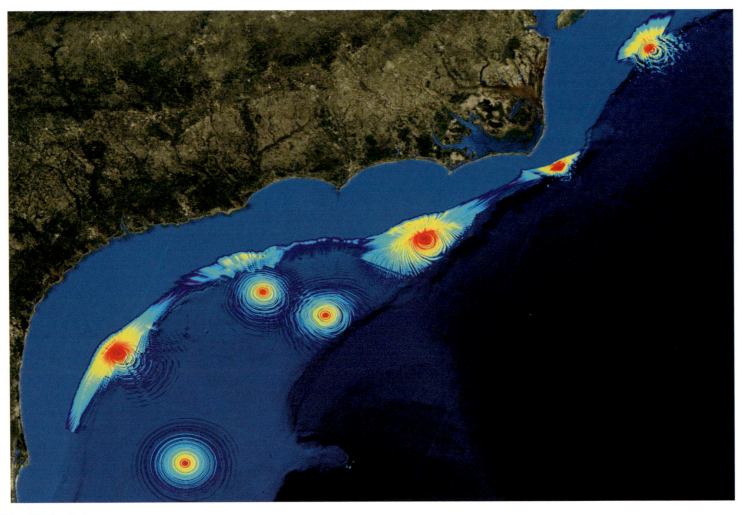

*Measuring the acoustic coverage of hydrophones off Maryland for the Atlantic Deepwater Ecosystem Observatory Network (ADEON).*

# REFERENCES

Behringer, D., T. Birdsall, M. Brown, B. Cornuelle et al. 1982. "A Demonstration of Ocean Acoustic Tomography." *Nature* (299): 121–125. https://doi.org/10.1038/299121a0.

Boyer, T.P., J.I. Antonov, O.K. Baranova, C. Coleman et al. 2013. "World Ocean Database 2013." *NOAA Atlas NESDIS 72: World Ocean Database 2013*, edited by S. Levitus. Silver Spring: National Oceanographc Data Center. https://doi.org/10.7289/V5NZ85MT.

Cornuelle, B.D. 1982. "Acoustic Tomography." *IEEE Transactions on Geoscience and Remote Sensing*. GE 20 (3): 326–332.

Heaney, Kevin, and Richard Campbell. 2019. "Modeling 3-Dimensional Sound Propagation Using the Parabolic Equation." *The Journal of the Acoustical Society of America* 146. 3035–3035. https://doi.org/10.1121/1.5137519.

International Maritime Organization. 2004. Regulation 19.2.4. Solas Chapter V– Annex 1–Automatic Identification Systems (AIS).

IOC, IHO, and BODC. 2003. "Centenary Edition of the GEBCO Digital Atlas." Liverpool: British Oceanographic Data Centre. Part of *General Bathymetric Chart of the Oceans*. Published on behalf of the Intergovernmental Oceanographic Commission and International Hydrographic Organization.

Jensen, F.B., W.A. Kuperman, M.B. Porter, and H. Schmidt. 2011. *Computational Ocean Acoustics* second edition. New York: Springer Science and Business.

Masters, Dallas, Timothy Duly, Stephan Esterhuizen, Vladimir Irisov et. al. 2020. "Status and Accomplishments of the Spire Earth Observing Nanosatellite Constellation." *SPIE Digital Library* (September 20): 11530, Sensors, Systems, and Next-Generation Satellites XXIV, 115300V. https://doi.org/10.1117/12.2574110.

McKenna, M.F., D. Ross, S.M. Wiggins, and J.A. Hildebrand. 2012. "Underwater Radiated Noise from Modern Commercial Ships." *The Journal of the Acoustical Society of America* 131, no. 1 (January): 92–103. https://doi.org/10.1121/1.3664100. PMID: 22280574.

Monterey Bay Aquarium Research Institute (MBARI). https://www.mbari.org/technology/solving-challenges/persistent-presence/mars-hydrophone.

Munk, W., and C. Wunsch. 1979. "Ocean Acoustic Tomography: A Scheme for Large Scale Monitoring." Deep Sea Research Part A. *Oceanographic Research Papers* 26, no. 2 (February): 123-161. https://doi.org/10.1016/0198-0149(79)90073-6.

Munk, W. and A. Baggeroer. 1994. "The Heard Island Papers: A Contribution to Global Acoustics." *The Journal of the Acoustical Society of America* 96: 2327–2329. https://doi.org/10.1121/1.411316.

"Acoustic Thermometry of Ocean Climate (ATOC)." 1995. In the *3rd Symposium on Integrated Observing Systems*.

Munk, W., P. Worcester, and C. Wunsch. 1995. *Ocean Acoustic Tomography*. New York: Cambridge University Press.

Ocean Networks Canada Data Archive. Neptune Array Acoustic Data. Oceans Networks Canada. University of Victoria, Canada. http://www.oceannetworks.ca. Downloaded in June of 2014.

PAME. 2019. *Underwater Noise in the Arctic: A State of Knowledge Report*. Roveniemi: Protection of the Arctic Marine Environment (May).

PAME. 2021. *Underwater Noise Pollution from Shipping in the Arctic*. Protection of the Arctic Marine Environment (PAME) Working Group. (May).

Porter, M.B. 2011. The "BELLHOP Manual and User's Guide." Preliminary draft.

Porter, M.B., H. Schmidt, F.B. Jensen, and W.A. Kuperman. 2011. "Computational Ocean Acoustics." *American Institute of Physics*, Woodbury, New York.

Rolland, R.M., S.E. Parks, K.E. Hunt, M. Castellote, P.J. Corkeron, D.P. Nowacek, S.K. Wasser, and S.D. Kraus. 2012. "Evidence That Ship Noise Increases Stress in Right Whales." *Proceedings of the Royal Society. Biological Sciences* 279 (1737): 2363–2368.

Ross, D. 1976. *Mechanics of Underwater Noise*, New York: Pergamon, 272–287.

Sabra K.G., B. Cornuelle, and W.A. Kuperman. 2016. "Sensing Deep-ocean Temperatures." *Physics Today* 69 (2): 32–38.

Sabra, K.G. 2016. "Passive Travel Time Tomography Using Ships as Acoustic Sources of Opportunity." *The Journal of the Acoustical Society of America* 140 (4): 3074–3074.

Stojanovic, M. 2003. "Acoustic (Underwater) Communications." *Encyclopedia of Telecommunications*, edited by John G. Proakis. New York: John Wiley & Sons.

Talley L.D., Pickard G.L., W.J. Emery, and J.H. Swift. 2011. *Descriptive Physical Oceanography: An Introduction*, 6th edition. Boston: Elsevier.

U.S. Coast Guard. 2012. *Nationwide AIS Database*. AIS Messages. U.S. Coast Guard Navigation Center.

Urick, R.J., and A.W. Pryce. 1954. *A Summary of Underwater Acoustic Data*. Part V. Background Noise. Arlington: Office of Naval Research. 1–61.

Wagstaff, R.A. 1981. "Low-Frequency Ambient Noise in the Deep Sound Channel—The Missing Component." *The Journal of the Acoustical Society of America* 69: 1009–1014.

Wenz, G.M. 1962. "Acoustic Ambient Noise in the Ocean: Spectra and Sources." *The Journal of the Acoustical Society of America* 34: 1936–1956.

Woolfe, K.F., S. Lani, K.G. Sabra, and W.A. Kuperman. 2015. "Monitoring Deep-Ocean Temperatures Using Acoustic Ambient Noise." *Geophysical Research Letters* 42: 2878–2884.

Wunsch, C. 2006. *Discrete Inverse and State Estimation Problems: with Geophysical Fluid Applications*. Cambridge: Cambridge University Press.

# PART 4
# TRAINING FUTURE GENERATIONS OF SCIENTISTS

Read the inspiring stories of students doing science in the greatest classroom of them all—outdoors in the field. From drone and satellite imagery to GPS and field observation, students of anthropology, conservation biology, forestry, glaciology, and many other disciplines are learning to use traditional and cutting-edge tools, technologies, and methods in their research. These students and their instructors and mentors show us "what's next" in training this next generation of scientists.

National Geographic Young Explorer Nathaniel Soon prepares for a nighttime dive near his home in Singapore.

# SPATIAL THINKING EFFECTS ON THE HUMAN BRAIN

The Geospatial Semester (GSS) partners Virginia high schools with the School of Integrated Sciences at James Madison University. In addition to launching hundreds of students into academic careers in GIS and geographic sciences, GSS uniquely measures the effects of geospatial thinking on brain function. Its work provides compelling evidence that a spatial approach to science, technology, engineering, and math effectively teaches students crucial skills to succeed in college and beyond.

By Bob Kolvoord, **James Madison University**

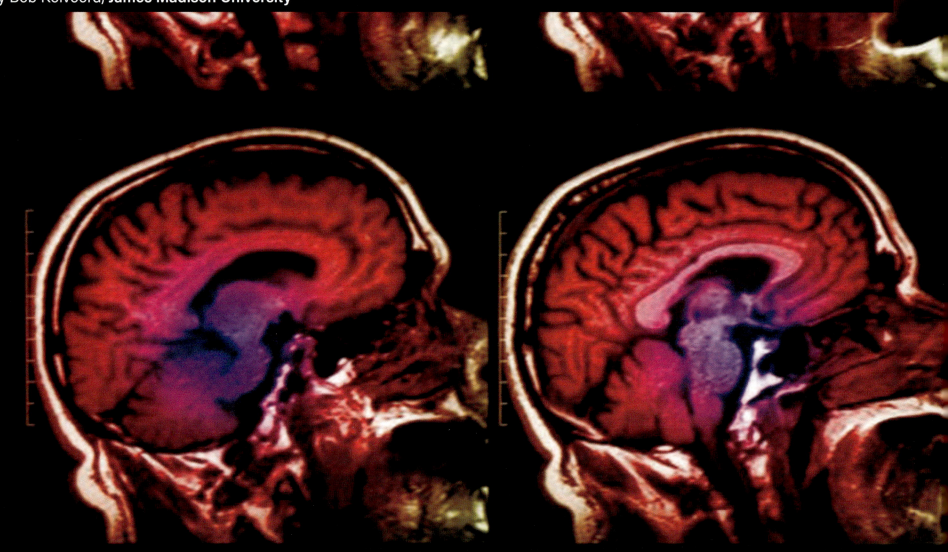

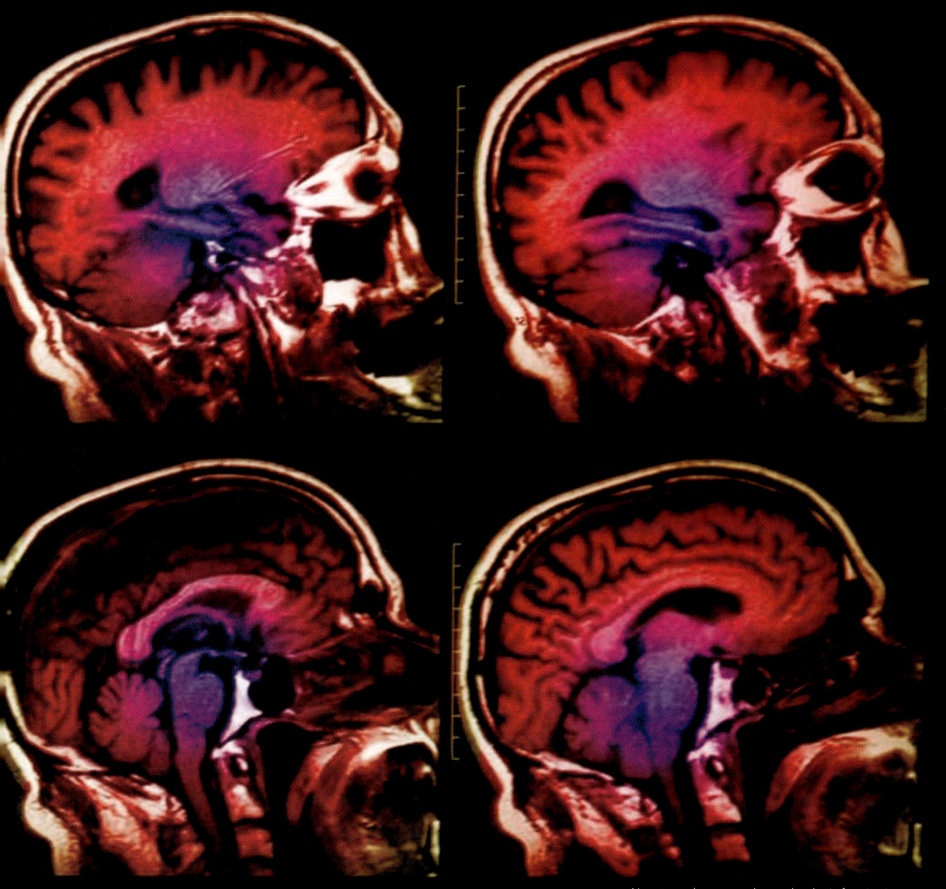

*New research suggests that student brain function is measurably altered by using GIS in spatial problem solving.*

# EVOLUTION OF STUDENTS' SPATIAL SKILLS

How does a student become a GIS professional, especially one who uses GIS as part of the scientific endeavor? Most children today don't grow up dreaming of making maps and performing spatial analysis for a living. In many U.S. schools, geography is almost an afterthought. But in Virginia, a unique partnership between high schools and James Madison University (JMU) is inspiring students to learn skills that prepare them for college and careers related to science, technology, engineering, and math (STEM). For more than 15 years, the program's geospatial approach has inspired young people to embrace STEM learning in college and beyond. Research about the impact of the Geospatial Semester (GSS), including brain scans of students in the program, has revealed promising results that warrant more study.

Two examples will help illustrate the success of this program. The first story is about Becky Schneider, who attended a suburban Washington, D.C. high school and originally wanted to be a choir teacher. The second is about Drew Mehfoud, who attended a high school south of Richmond, Virginia, and planned to become a pediatrician. Instead, both shifted their focus to geospatial technology and chose careers as geographic information system (GIS) professionals. Schneider today is a geospatial analyst with Dewberry Engineers, and Mehfoud is a GIS technician in the environmental division of Timmons Group.

Looking back, both credit their careers in large part to high school coursework that supported scientific inquiry using geospatial technologies. Schneider said that her high school has offered a course in geospatial technology since the mid-2000s. The idea of taking such a class intrigued her, but she had no idea what to expect. But once she began, Schneider was "amazed by the functionality of GIS software and the wide variety of career possibilities that would be plausible." Mehfoud's high school also offered a GIS-based course to seniors for more than a decade. His desire to get involved led him to pester guidance counselors so he could take the class a year early as a junior. He enjoyed it so much, he returned for a second year, serving as a teaching assistant to his classmates. Taking that class and an advanced placement environmental science course put him on the career path he's followed ever since.

These classes opened a new world of possibilities that led to geospatially focused college degrees for Schneider, at JMU, and Mehfoud at Virginia Tech. But what kind of high school class has this sort of impact on students, opening their eyes to possibilities once hidden in plain sight? And what can we learn about the impact of such a class on how students think and reason?

The class in question is called the Geospatial Semester (GSS), a joint effort between the School of Integrated Sciences at JMU and high school districts across Virginia.[1] The class introduces students to the array of opportunities provided by geospatial technologies. This chapter will 1) discuss the impact of this class on students' spatial and scientific problem-solving skills, 2) review the history and impact of the GSS on the thousands of students who have participated, and 3) present the results of a multiyear research study measuring how the class affected students' skills in spatial thinking and scientific reasoning.

## Larger context

What impact does the extended use of GIS have on students? Many of us have long advocated its use in K-12 schools, primarily because of intriguing anecdotes and our own experiences. Unfortunately, relatively little research has been done to date on this topic. Prior work focused either on student content knowledge[2,3] or on changes in standardized test scores.[4] As a result, we know a lot about why it can be challenging to bring these technologies to K-12 schools[5,6] but have relatively little research to show why it's worthwhile.

We wanted to see whether using geospatial technology in school supported students' thinking and scientific problem-solving skills. If we had this evidence, we could make a compelling case to include geospatial technology in the K-12 curriculum. It would also foster a broader discussion about the research agenda for geospatial technologies in education.[7] Working with researchers at Northwestern, Georgetown, Dartmouth, and American Universities, we've tried to do just that. Here's our story and what we know so far.

Collaborating with Michael Charles, a longtime professor at Pacific University in Oregon who died in 2020, we began by evaluating GSS student final projects to assess how well the projects demonstrated competence in GIS and applying spatial analysis. The next section presents sample student projects and illustrates the range of student topics, from the hyperlocal to the global. Many of the projects are science-based and driven by student interest. Few of them would fit into any standard science class currently offered in U.S. high schools. No project is perfect, but these projects suggest what high school students can accomplish when they are given the time and opportunity to follow their interests with a spatial perspective.

*The home page for the Geospatial Semester at James Madison University's School of Integrated Sciences, https://www.jmu.edu/sis.*

# Student project example 1

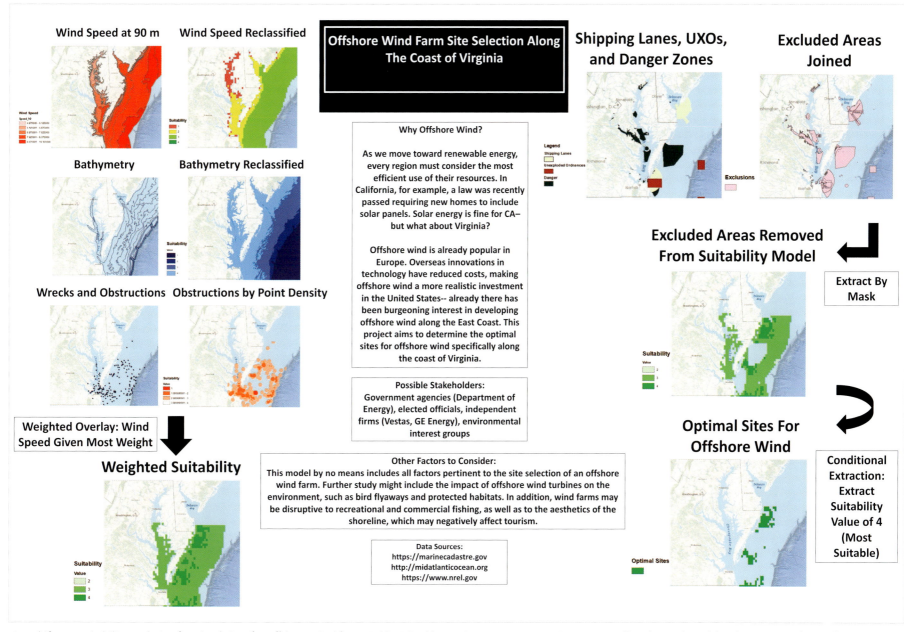

*A multifactor suitability analysis of optimal sites for offshore wind farms in Virginia. Alternative energy source siting, especially solar and wind, has been a popular student project topic for many years.*

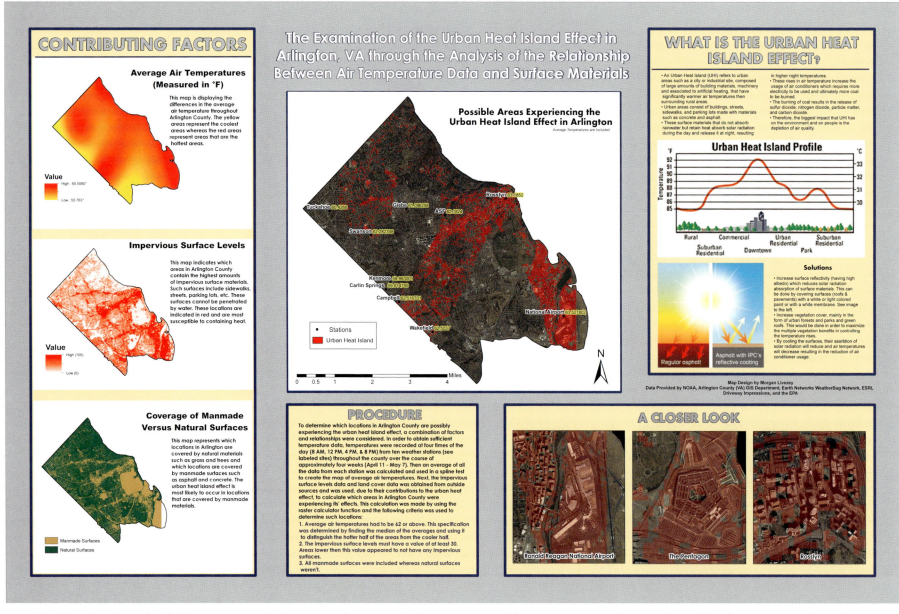

A study of the urban heat island effect in Virginia suburbs near Washington, D.C. This project exemplifies an environmental science theme that characterizes much student interest and a variety of student projects.

# An Analysis of the Change in the Amount of Ground-Level Ozone for the Metropolitan Region from 2013-2016

## Designed By: Natalie Warnke

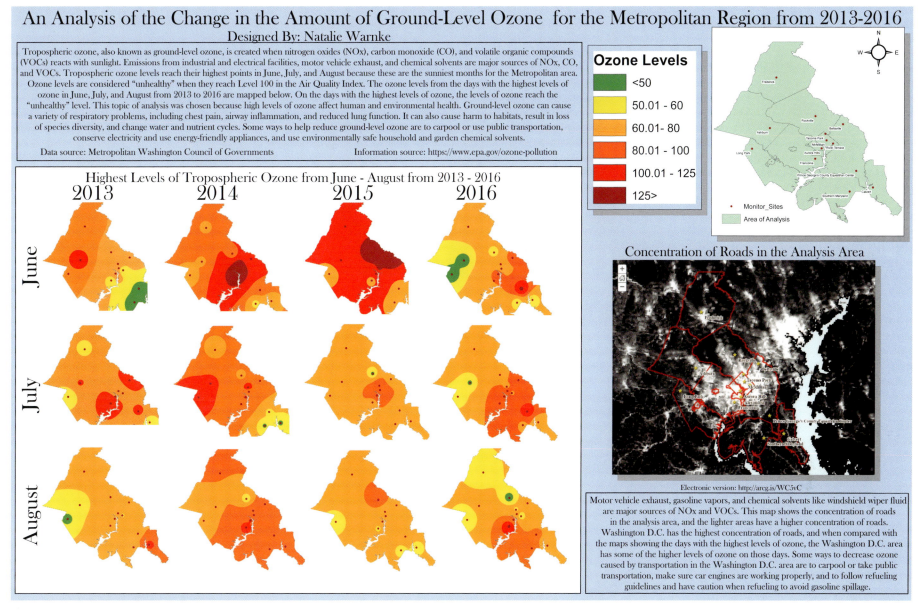

Tropospheric ozone, also known as ground-level ozone, is created when nitrogen oxides (NOx), carbon monoxide (CO), and volatile organic compounds (VOCs) reacts with sunlight. Emissions from industrial and electrical facilities, motor vehicle exhaust, and chemical solvents are major sources of NOx, CO, and VOCs. Tropospheric ozone levels reach their highest points in June, July, and August because these are the sunniest months for the Metropolitan area. Ozone levels are considered "unhealthy" when they reach Level 100 in the Air Quality Index. The ozone levels from the days with the highest levels of ozone in June, July, and August from 2013 to 2016 are mapped below. On the days with the highest levels of ozone, the levels of ozone reach the "unhealthy" level. This topic of analysis was chosen because high levels of ozone affect human and environmental health. Ground-level ozone can cause a variety of respiratory problems, including chest pain, airway inflammation, and reduced lung function. It can also cause harm to habitats, result in loss of species diversity, and change water and nutrient cycles. Some ways to help reduce ground-level ozone are to carpool or use public transportation, conserve electricity and use energy-friendly appliances, and use environmentally safe household and garden chemical solvents.

Data source: Metropolitan Washington Council of Governments     Information source: https://www.epa.gov/ozone-pollution

**Ozone Levels**

- <50
- 50.01 - 60
- 60.01- 80
- 80.01 - 100
- 100.01 - 125
- 125>

Monitor_Sites
Area of Analysis

## Highest Levels of Tropospheric Ozone from June - August from 2013 - 2016

2013  2014  2015  2016

June

July

August

## Concentration of Roads in the Analysis Area

Electronic version: http://arcg.is/WC5vC

Motor vehicle exhaust, gasoline vapors, and chemical solvents like windshield wiper fluid are major sources of NOx and VOCs. This map shows the concentration of roads in the analysis area, and the lighter areas have a higher concentration of roads. Washington D.C. has the highest concentration of roads, and when compared with the maps showing the days with the highest levels of ozone, the Washington D.C. area has some of the higher levels of ozone on those days. Some ways to decrease ozone caused by transportation in the Washington D.C. area are to carpool or take public transportation, make sure car engines are working properly, and to follow refueling guidelines and have caution when refueling to avoid gasoline spillage.

*This project focused on air quality in the metropolitan Washington, D.C., area and used ground-level ozone as a primary indicator. The variation during the summer months shows an interesting pattern.*

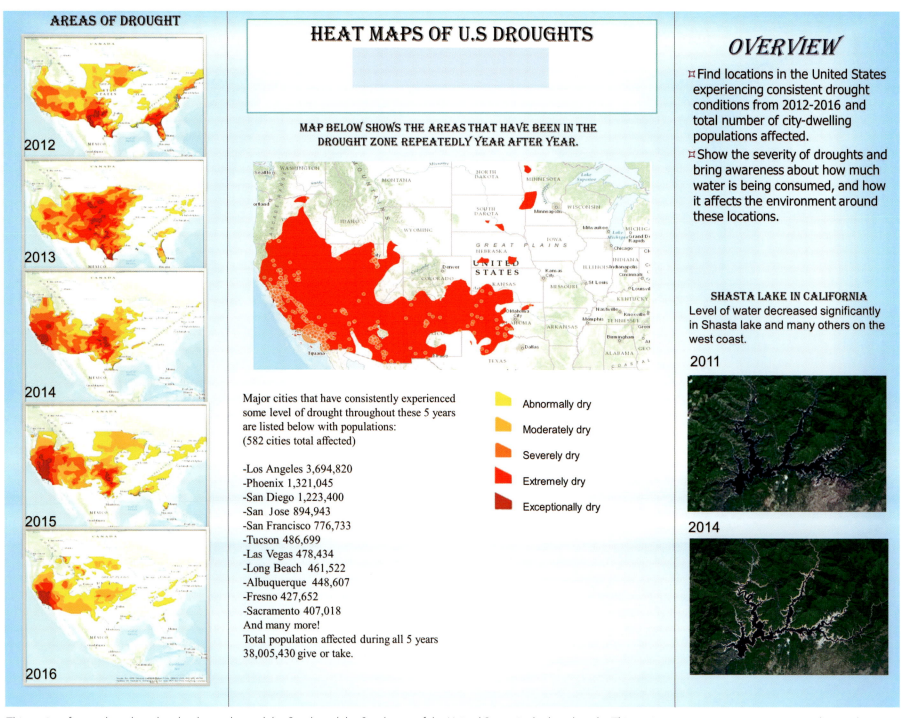

# AREAS OF DROUGHT

2012

2013

2014

2015

2016

# HEAT MAPS OF U.S DROUGHTS

MAP BELOW SHOWS THE AREAS THAT HAVE BEEN IN THE DROUGHT ZONE REPEATEDLY YEAR AFTER YEAR.

Major cities that have consistently experienced some level of drought throughout these 5 years are listed below with populations:
(582 cities total affected)

-Los Angeles 3,694,820
-Phoenix 1,321,045
-San Diego 1,223,400
-San  Jose 894,943
-San Francisco 776,733
-Tucson 486,699
-Las Vegas 478,434
-Long Beach  461,522
-Albuquerque  448,607
-Fresno 427,652
-Sacramento 407,018
And many more!
Total population affected during all 5 years
38,005,430 give or take.

Abnormally dry

Moderately dry

Severely dry

Extremely dry

Exceptionally dry

# OVERVIEW

¤ Find locations in the United States experiencing consistent drought conditions from 2012-2016 and total number of city-dwelling populations affected.

¤ Show the severity of droughts and bring awareness about how much water is being consumed, and how it affects the environment around these locations.

SHASTA LAKE IN CALIFORNIA
Level of water decreased significantly in Shasta lake and many others on the west coast.

2011

2014

*This project focused on droughts that have plagued the South and the Southwest of the United States in the last decade. This project connects to an environmental issue that continues to impact the United States. Students are not limited to local problems in their projects.*

# DRILLING INTO BRAIN SCIENCE

We discovered that most students at the end of the GSS demonstrated significant ability in using the GIS software, and some students showed strong spatial thinking and scientific reasoning skills.[8] While this was reassuring, it didn't give us any idea of what might be happening during the course and how students might be gaining these skills. Was it from the GSS, or was it just the way adolescents develop? We realized we needed to do more research.

With support from the Spatial Intelligence and Learning Center (SILC), a National Science Foundation-supported research center, we teamed with Dr. David Uttal and his colleagues at Northwestern University to dig deeper. We set up another study in which we identified two student groups, one taking the GSS and another taking a similar set of classes but not the GSS. We then followed both groups through a school year and interviewed the students at regular intervals (four times for the GSS group and twice for the non-GSS group). During interviews at the beginning and end of the year, we gave students in each group a scenario (e.g., "Your town needs a new landfill; how would you decide where to site it?") and asked them to think aloud as they tried to answer the question. We videotaped the interviews and transcribed the results. We then used a dictionary of spatial language and explored which group used more spatial language. We found no statistically significant difference between the groups at the beginning of the year, but by the end of the year, the GSS students used significantly more spatial words. We also analyzed their answers to look at scientific thinking and problem solving and discovered that the GSS students used more evidence, provided more complete claims, and showed better reasoning skills than the non-GSS students.[9]

The research suggested many potential benefits of geospatial learning for GSS students, but the results came from a pilot study. It's difficult to do this kind of research in schools because of all of the various factors that can impact a high school student throughout a year. Further, we could see substantial behavioral changes, but we didn't have any evidence about changes in cognition. We decided to dig deeper.

Dr. Adam Green and his laboratory at Georgetown University joined us, and we secured National Science Foundation funding to redo our earlier study in a more rigorous fashion and add a cognitive component to assess the impact of the GSS on brain function. This study was managed by Dr. Emily Peterson, now on the faculty at American University, and focused on two school districts in Virginia near Washington, D.C.

We revisited how we identified non-GSS students as a comparison group to GSS students. Using a technique called propensity score matching, we found non-GSS students who were a good match to GSS students across a number of variables (gender, age, SAT score, school grades, etc.). In this way, we tried to limit the number of confounding variables and provide a more robust comparison group. We repeated the behavioral interviews described above and added some new measures. The students also took several standard psychometric tests that evaluate spatial thinking, including the mental rotation and embedded figures tests. The students reported on the kinds of spatially related activities they participated in during their childhood and their interest in spatially related activities. Most importantly, we took a subset of students to Georgetown University and conducted functional magnetic resonance imaging (fMRI) scans of their brains at the start and the end of the school year. We required a parent to accompany them and surveyed the parents about their attitudes toward their child's spatial thinking abilities.

While the analysis of these data are ongoing, we've already discovered a great deal.

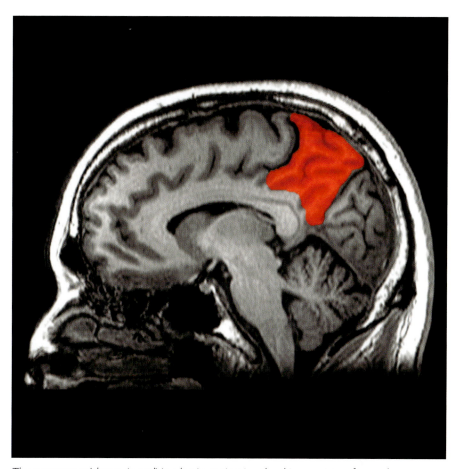

*The precuneus (shown in red) is a brain region involved in a variety of complex functions, including spatial thinking and human perception of the environment.*

As we observed the GSS students taking a more spatial approach to their problem solving, we also saw increased performance on the embedded figures task when compared to the non-GSS students. Our fMRI results show that GSS students had increased activation in their parietal cortex and caudate, a result we think is related to increased spatial habits of mind. GSS students also showed more efficient use of their frontal regions and increased connectivity between parts of the prefrontal cortex and motor and parietal areas. This shows a significant transfer effect because the GSS students didn't see or practice these psychometric tests in class, but they showed improvement because of an increase in their spatial habits of mind.

One of the biggest questions in our work is to understand how taking the GSS affects *transfer*, that is, how do GSS students deal with things they've never studied and what kinds of strategies do they use to solve novel problems? We explored this question through a deductive reasoning task. We don't teach reasoning strategies in GSS, yet GSS students reasoned more effectively than their non-GSS comparison group peers. More specifically, our neural data showed that GSS students had increased activation in the parietal cortex during deductive reasoning, and increased connectivity to the motor and parietal cortices. In other words, the GSS students likely adopted a more spatial strategy for solving reasoning problems than the non-GSS students.[10]

# THE GEOSPATIAL SEMESTER

Our demonstration of transfer from in-classroom spatial teaching to improvements in verbal reasoning (associated with the changes in spatial habits of mind and the recruitment and connectivity of spatial brain resources mentioned earlier) is a potentially important finding in STEM education. Spatial thinking strongly predicts achievement and persistence in STEM fields.[11,12] In some STEM fields, spatial ability contributes more to the prediction of achievement and attainment by scientists and engineers than SAT scores. Other research describes the evidence linking spatial ability and future STEM attainment as exceedingly strong, with more than 50 years of research and more than 400,000 participants.[13] Likewise, the ability to reason is perhaps the most important cognitive skill for achievement across academic and professional pursuits, including in STEM.[14,15] So, the connection of a real-world classroom approach (the GSS) with an associated neural mechanism (increased recruitment and connectivity of spatial brain resources) for improving spatial thinking and deductive reasoning has major implications.

One additional and particularly compelling finding in our study is that students in the GSS class closed the gender gap in spatial performance from the start of the school year to the end, while students in the comparison group did not. So, beyond the improvements in spatial thinking and scientific reasoning described earlier, the GSS has potential to contribute to greater inclusion in the STEM workforce, a critical national priority.

Our work helps bridge the significant gap between real-world classrooms and the laboratories that study cognitive neuroscience. The use of the fMRI has allowed us to determine that cognitive changes occurred with learning and that we can develop inferences about how these changes occur. To our knowledge, ours is the first project to measure how learning in a real-world high school class drives changes in the brain over time.

Capturing neural measures of change has provided compelling, physical evidence of the effectiveness of the GSS, which is likely to support greater implementation of a spatial approach to STEM learning and more geospatial use in the classroom. We'll next describe the GSS course that provided these dramatic impacts to participating students.

To appreciate the learning value of the GSS, some background knowledge of its history and evolution is instructive. The GSS is a unique concurrent enrollment program in which high schools in Virginia partner with JMU to offer yearlong classes to high school seniors (and the highly motivated juniors such as Mehfoud) that feature geospatial technologies and provide students the opportunity to engage in extended spatial projects that match their interests, all while earning college credit from JMU.

The program, inaugurated in 2005, has served more than 5,000 students and has introduced many students,[1] such as Schneider and Mehfoud, to the world of geospatial technologies. We developed the program to be a viable alternative to the test-heavy and project-light world of many high schools and to introduce geospatial technologies in a variety of contexts. Starting with four schools and a few dozen students in 2005, the GSS has grown to more than two dozen schools and more than 500 students in the 2020-2021 school year.

From the outset, the GSS has been a collaboration between college faculty and high school teachers to create a project-based curriculum to help students transition from high school to college, the military, or the workplace, and to a world where academic disciplines weren't the primary organizing principle. We also wanted to give students exposure to cutting-edge technologies that might capture their

*Heritage High School students from Loudoun County, Virginia, representing the GSS at the Esri Federal GIS Conference in 2020.*

interest and open new career horizons. Lastly, we wanted to create opportunities for all students, not just students who had learned how to flourish in a test-crazed environment.

Our recipe was simple—provide teachers with software and support, work together to learn the latest technology, and then visit them and their students at regular intervals. Working with Charlie Fitzpatrick and the Esri Education team, we provided school site licenses for ArcGIS and helped teachers get the software installed (later, some schools migrated to ArcGIS Online). Our regular visits gave us the opportunity to share new tools and techniques with the classes and the chance to build relationships with the students. We also offered teachers a lot of flexibility in the curriculum as we helped assess the students' performance with oral midterms and presentations of their final projects.

*Students from Washington-Liberty High School in Arlington, Virginia, presenting their projects to 16,000 people at the annual Esri User Conference several years ago in San Diego.*

The final projects deserve particular attention because they represent the culmination of the class for the students, and in several instances, the first time students had chosen and pursued projects of their own interest without the tight constraints of an academic discipline. We work with students to help them craft projects that are important to them and that can be done within the time available with existing data. Many students chose an area of scientific inquiry that's amenable to the application of geospatial technologies. They can hone their scientific reasoning and problem solving while using cutting-edge technology.

The work in Virginia has led to other adoptions of the GSS. Pacific University in Oregon and the Chicago Public Schools have created versions of the GSS in hopes of offering their students similar opportunities. We've been in discussion with schools in Beijing and Redlands, California, about starting programs.

The stories from the GSS are similar to stories from several other programs that have featured GIS during the last two decades. Programs such as EAST (https://www.eastinitiative.org/) and 4-H have offered students in-school and after-school hands-on programs that provide exposure to GIS software and data. However, the GSS now has behavioral and cognitive data to support the anecdotes about its effectiveness.

## SUMMARY

We've taken an in-depth look at the impact of extended high school coursework using GIS to address a variety of interesting problems. We observed dramatic changes in student behavior and cognition after a yearlong experience. These findings lead to more research questions, and we're now embarking on a new project to try to determine how much exposure to geospatial technologies will provide similar behavioral and cognitive results.

Ultimately, it is the students' experiences that motivate and propel them into scientific and geospatial careers. When asked what she might say to a high school student in light of her experience, Becky Schneider offered the following: "GIS is so much more than IT or computer science, and the applications of this type of knowledge and background are endless. You can take this knowledge and apply it anywhere from environmental science to intelligence analysis, work in either the public or the private sector, and have so much fun along the way!"

When Drew Mehfoud was asked what advice he would give to schools considering the GSS, he replied, "I would tell any schools considering the class to do everything they can to add it. The class is interesting, challenging, teaches a highly demanded skill, and is worth college credits. The class changed my mind on my future aspirations and could do the same for a student at any other school. I highly recommend, if able, adding a GIS class to the list of classes available for students to take."

Many of the thousands of students who've participated in the GSS share these sentiments. Through their experience, they've built their critical thinking and spatial reasoning skills and been exposed to the broader world of geospatial technologies and their applications. We hope our work will support the adoption of the GSS and similar programs more broadly around the world and open this experience to a range of students who'll be prepared and eager to make a difference in the many scientific disciplines described in this book.

This chapter is dedicated to the late Dr. Mike Charles of Pacific University, a long-time collaborator and supporter of GIS in K-12 education.

# NOTES

1. B. Kolvoord, K. Keranen, and S. Rittenhouse, "The Geospatial Semester: Concurrent Enrollment in Geospatial Technologies," *Journal of Geography* 18, no. 1 (2019): 3–10, https://doi.org/10.1080/00221341.2018.1483961.

2. A.T. Bodzin and V. Kulo, "Efficacy of Educative Curriculum Materials to Support Geospatial Science Pedagogical Content Knowledge," *Journal of Technology and Teacher Education* 20, no. 4 (2012): 361–386.

3. A.M. Bodzin, Q. Fu, V. Kulo, and T. Peffer, "Examining the Effect of Enactment of a Geospatial Curriculum on Students' Geospatial Thinking and Reasoning," *Journal of Science Education and Technology* 23, no. 4 (2014): 562–574, https://doi.org/10.1007/s10956-014-9488-6.

4. D. Goldstein and M. Alibrandi, "Integrating GIS in the Middle School Curriculum: Impacts on Diverse Students' Standardized Test Scores," *Journal of Geography* 112, no. 2 (2013): 68–74, https://doi.org/10.1080/00221341.2012.692703.

5. J. Kerski, "Implementation and Effectiveness of Geographical Information Systems Technology and Methods in Secondary Education," *Journal of Geography* 102, no. 3 (2003): 128–137, https://doi.org/10.1080/00221340308978534.

6. T.R. Baker, A.M. Palmer, and J. Kerski, "National Survey to Examine Teacher Professional Development and Implementation of Desktop," *Journal of Geography* 108, no. 4 (2009): 174–185.

7. T.R. Baker, S. Battersby, S.W. Bednarz, A.M. Bodzin, B. Kolvoord, S. Moore, and D. Uttal, "National Survey to Examine Teacher Professional Development and Implementation of Desktop," *Journal of Geography* 114, no 3 (2015): 118–130.

8. M. Charles and R.A. Kolvoord, "Geospatial Semester: Developing Students' 21st Century Thinking Skills with GIS: A Three Year Study," *Pyrex Journal of Educational Research and Reviews* 2, no. 6 (December 2016): 67–78.

9. E.W. Jant, D. Uttal, R. Kolvoord, K. James, and C. Msall, "Defining and Measuring the Influences of GIS-Based Instruction on Students' STEM-Relevant Reasoning," *Journal of Geography* 119, no. 1 (2019): 22–31, https://doi.org/10.1080/00221341.2019.1676819.

10. Robert C. Cortes, Emily G. Peterson, David J. M. Kraemer, Robert A. Kolvoord et al., "Spatial Education Improves Verbal Reasoning: Classroom Neural Change Predicts Transfer Beyond Tests or Grades." Submitted to *Science* (2021).

11. D.L. Shea, D. Lubinski, and C.P. Benbow, "Importance of Assessing Spatial Ability in Intellectually Talented Young Adolescents: A 20-year Longitudinal Study," *Journal of Educational Psychology* 93, no. 3 (2001): 604–614, http://dx.doi.org/10.1037/0022-0663.93.3.604.

12. J.D. Wai, D. Lubinski, and C.P. Benbow, "Spatial Ability for STEM Domains: Aligning Over 50 Years of Cumulative Psychological Knowledge Solidifies its Importance," *Journal of Educational Psychology* 101, no. 4 (2009): 817.

13. D.H. Uttal and C.A. Cohen, "Spatial Thinking and STEM Education. When, Why, and How?" *Psychology of Learning and Motivation—Advances in Research and Theory* 57 (2012): 147–181, https://doi.org/10.1016/B978-0-12-394293-7.00004-2.

14. A.R. Duschl, A.H. Schweingruber, and W.A. Shouse," *Taking Science to School: Learning and Teaching Science in Grades K-8* (Washington, D.C.: The National Academies Press, 2007).

15. M.C. Linn, D. Clark, and J.D. Slotta, "WISE Design for Knowledge Integration," *Science Education* 87 (2003): 517–538.

Support for this research was provided by the Spatial Intelligence and Learning Center and the National Science Foundation (DRL 1420600).

# FUELING CURIOSITY TO FOSTER A HEALTHY PLANET

Geography opens the door for learners to better understand the interconnected world. It gives young people the insight to draw connections, measure how individual actions can change the world, assess costs and benefits, and seek solutions to the many complex questions about our planet. Operating at the intersection of exploration, documentation, and mapping, young people are positioning themselves to confront the future's looming problems.

By Vicki Phillips, **National Geographic Society**

*School of jack fish. Photo by Nathaniel Soon.*

# AT THE HEART

*" Thanks to generations of curious, daring, intrepid explorers of the past, we may know enough, soon enough, to chart safe passage for ourselves far into the future."*
—Sylvia A. Earle

Without the ocean, human life on Earth would be impossible. The ocean covers more than 70 percent of our planet and is one of our most critical life-support systems. It generates the oxygen we breathe and provides the food we eat. It regulates our climate and weather, and anchors livelihoods for billions of people. While the ocean's resources are vast, they are not inexhaustible. Increasing global threats, such as overfishing, pollution, and global warming, are pushing the ocean to a tipping point beyond which we may never recover. The loss of biodiversity, together with climate change, are the most urgent global crises affecting our planet today. We cannot protect our planet—or our own health and well-being—without protecting the ocean.

It's easy to feel discouraged when thinking about the enormity of Earth's most pressing problems, but we are not without hope. As oceanographer and National Geographic Explorer-at-Large Sylvia Earle states, "Knowing is the key to caring, and with caring there is hope that people will be motivated to take positive actions. They might not care even if they know, but they can't care if they are unaware."

In this context, young people are one of the most powerful resources for hope.

Nathaniel Soon, a student at Yale-National University of Singapore (NUS) College, works tirelessly to ensure the planet's blue heart is protected for future generations. Motivated by an innate curiosity about the ocean, Soon combines his passion for documentary filmmaking, photography, and ocean conservation to cultivate understanding and inspire others to protect the sea. It is why he founded Our Seas, Our Legacy, a documentary collective to engage Singaporeans about the area's marine environments and biodiversity, and educate them on ways to create a healthier ocean.

Research shows young people are among Earth's most passionate defenders. Data shows they are most willing to act on issues and have high potential to influence others. Why? They are curious, have abundant energy, and bring new perspectives. They also have the most at stake and want to ensure our planet thrives.

Addressing the pressing problems of our time requires us to support young people as forces for meaningful change. We can do that by fueling their curiosity, supporting their energy and optimism, amplifying their ideas, and lifting up their leadership.

Educators, experts, mentors, and families can all serve as essential role models to help cultivate an explorer's mindset and launch them into a life of purpose and hope. This community, combined with immersive learning experiences and innovative tools, can empower and equip the next generation to reverse harmful environmental trends, forge solutions, and scale bold ideas.

At the heart of this monumental pursuit is geography.

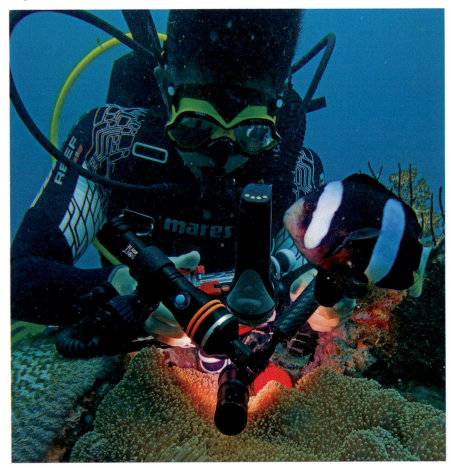

*Nathaniel Soon at work filming along Abana Reef in Brunei.*

*A broadclub cuttlefish (Sepia latimanus) makes its way along the reef just off the coast of Pulau Jong island, Singapore.*

# A MODERN VIEW OF GEOGRAPHY

Although geography helps us understand our interconnected world, this field of science is sometimes overlooked or taught in an outdated way. Geography is a complex discipline. It has never been about just finding places on a map or memorizing the names of rivers or capitals; it's about skill and insight and vision.

Geography is the art and science of understanding how everything affects everything else. It's exploring why things are where they are and why things are the way they are. It's comparing two places and discovering why they're different, or looking at places at two points in time and explaining the change. Geography is studying how things came to be the way they are, with a special focus on how Earth shapes our way of life and how our way of life affects everything on Earth.

Having the tools and insights to learn how things came to be the way they are allows us to figure out what human activities are helpful or harmful and what will happen if trends don't change.

The ability to think like a geographer is a crucial skill in this era because this is the first time human beings can see, within our lifetimes, the impact of the way we live. At its most pointed, geography is about seeing and changing the future if we don't like what we see.

Consider the ocean. Protecting the right places in the ocean offers humanity many benefits: more fish to catch, improved livelihoods, new economic opportunities, and an ocean that mitigates climate change.

So, a high-stakes test of geographical knowledge today is not, "Can you find a river on a map?" but, "Can you see the patterns and connections?" "Can you see the connection between a polluted river in one country and threatened food supply in another?" Geography unlocks insights to make informed decisions and strategic solutions. With these insights, can you be part of the solution and inspire others to act?

Because we are part of the natural world, our impacts on the world also impact us. To have a command of geography is to understand the interconnected forces of science, economics, history, culture, and beyond. This sentiment was reinforced for Nathaniel Soon during his studies: "I found a lot more meaning when I combined my curiosity for understanding human dynamics with the natural world. Understanding both the environment and humanity required an appreciation of their interdependence."

These are among the many reasons why geography is particularly well suited to solve the challenges of the 21st century.

Modernizing geography also means making it as diverse as the world in which we live and elevating historically unacknowledged or excluded stories and voices. Among the young, passionate advocates for this vital change is Francisca Rockey, a geography student at York St John University in England. As she studies for a career in biogeography and conservation, Rockey is simultaneously focused on bringing equity and diversity to geography, making it more inclusive of people of color. "People have an idea of what a scientist looks like, and it's not me," she said.

In the spring of 2020, Rockey posted a single tweet that changed her life.

*Francisca Rockey conducting field work.*

The response was huge, and helped spark an idea that evolved into a powerful movement. Rockey and her peers launched the group Black Geographers to diversify the curriculum and raise the work of Black geographers. They've conducted research, started a scholarship with help from Esri, and established a mentorship program. Together, they are working to ensure Black students are empowered and encouraged to find work and have representation in the field.

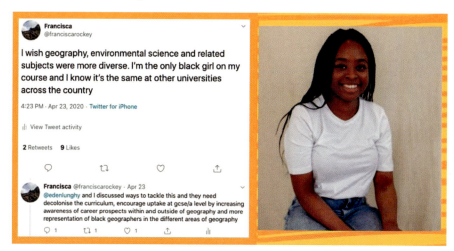

*Francisca launched a movement for Black people in geography with a simple statement expressed in a tweet.*

# AN EXPLORER'S MINDSET

As young geographers, Soon and Rockey exemplify what it means to have an explorer's mindset. Cultivating this mindset begins with persistent curiosity, a passion for finding out, and a restless quest for answers. As young people explore the world around them, curiosity becomes the energy that drives their learning. Curiosity is energy. And fueling that curiosity is one of the most effective ways that educators can help their learners develop the attitudes, skills, and knowledge to become architects of change.

## Fueling curiosity

Educators in the broadest sense are instrumental in helping young people develop the essential attitudes, skills, and knowledge to:

- lead, seek solutions, and tell stories.
- become informed, curious, and capable individuals committed to making the world a better place.
- have a sense of responsibility and respect for other people, cultures, and the natural world.
- make a difference, pursue bold ideas, and persist in the face of challenges.
- observe, document, and engage with the world around them.
- create and foster a global community committed to a sustainable future.
- support diversity, equity, and inclusion in their communities and respective fields.

An array of tools helps educators to engage learners in asking questions, investigating, collecting data, thinking critically, and developing solutions. The use of these tools is a striking shift from the older methods of giving students chapters to read and facts to remember. Students take on some of the most perplexing problems of the world and use tools such as National Geographic's own MapMaker, ArcGIS Online, ArcGIS Survey123, and ArcGIS StoryMaps to design and communicate solutions.

How can communities combat plastic pollution? How can human activities help animal migration? How can we actively save species from extinction? From collecting and visualizing data to studying evidence and forming hypotheses, no learners are too young to start thinking like geographers.

Rockey says she wishes she had access to GIS earlier in her schooling before she was a university student. Just recently, she used GIS to map the spatial distribution of students categorized by educational qualifications around the city of York, England. She says providing children with earlier access to geographic tools can help them address challenges in their community and inspire them to embark on careers in which they can put their problem-solving skills to use.

Lindsay Hyneck, a kindergarten teacher at Oak Harbor Elementary School on Whidbey Island in Washington State, created a lesson to foster an explorer's mindset in her 5-year-old students. "This meant pushing the boundaries of their curiosity, their observational skills, and their knowledge with a hands-on approach," Hyneck said. She brought her learners outside to map the community around the school through photography. As they documented the area, Hyneck asked questions with the goal of inspiring critical thinking: "Why are certain buildings located in particular places? What civic resources are important to be located near a school? And how does city design lead to healthy communities?"

Back in the classroom, students studied their photos and created a floor-sized map of their community. Using construction paper, paint, and other supplies, they built schools, businesses, and parks and placed them on the map. They discussed how the map looked from above and whether all communities looked like theirs. With that in mind, Hyneck then led her students in a lesson about other island communities and asked them to compare Whidbey Island to the Isle of Man in the Irish Sea. "They started locally and observed globally," the teacher said. By mapping their community and examining essential questions, they drew connections to another island on the other side of the world.

*Oak Harbor Elementary students map their Whidbey Island environs in 3D.*

While Hyneck's kindergartners are perhaps not quite ready to use mapping tools such as ArcGIS Online, the applied learning they are doing is a first step in that direction. It develops their spatial thinking, helps them understand their place in the world, and empowers them to see themselves as explorers, even without leaving their small island.

*Observation notes showing that Whidbey Island, Washington, and Douglas, Isle of Man, aren't that different after all.*

# PREPARING FOR CAREERS IN THE SCIENCES

Through inquiry-based work, students sense they can do something about the challenges facing the world. When students start with a pressing question and then receive the data, resources, and guidance they need to explore options, and present solutions, then they can take charge of their own learning. Alicia (Ali) Pressel uses this approach to great success in her environmental science classes at Creekside High School in St. Johns, Florida, where she was honored as 2021 Teacher of the Year.

Students in Pressel's class embrace the learning process as a form of exploration. "Too often, that's a missing link in our education system. We're not doing enough to let students ask their own questions and inquire why things are happening," Pressel said. Cultivating a young explorer's mindset requires that teachers model it. "You and your students both have to try new things," she said. "You have to be willing to challenge yourself, and you have to be willing to fail. Students have to know if you're going to be an explorer, things are not going to work every time."

Pressel challenges her students to think through questions such as, "How does land development and habitat fragmentation influence our community watershed? How does it influence biodiversity in the state?" She guides them in using ArcGIS, survey tools, and storytelling resources to research, form, and test hypotheses; create action plans; and communicate the issues and solutions to the community.

The teacher is inspired by the work of her current and former students to improve lives beyond their own. She recently connected with a former student who is working in the Dominican Republic using GIS to identify where fresh water is scarce, but direly needed. The former student works with local engineers to provide fresh well water access to underserved communities in mountainous areas of the country. "I don't remember her as outdoorsy, but she's trekking in the highlands with all this equipment to install these pumps so impoverished communities can have clean water," Pressel said.

Another former student, Jule Campbell, is a graduate student studying conservation biology and environmental sustainability at the University of Central Florida. During a visit to Creekside High, she encouraged Pressel's students to use their skills to help people outside the classroom. Her own dream is to become an environmental educator. "You can't make change without education," Campbell said. "Teachers are there with you since kindergarten. They're consistently making an impact on your life. That's true with me and Ms. Pressel. She's literally the reason I'm on the path I'm on, and to be able to give that back to students is something I'm really excited about."

*These photos show environmental science classes under way in St. Johns, Florida.*

# NATIONAL GEOGRAPHIC EXPLORERS

Nothing inspires young people's curiosity and actions more than seeing their work come alive through National Geographic Explorers. One of these explorers, Agustina Besada, discovers and documents the world's wonders while also striving to protect it, often through her teaching. Several years ago, Besada traveled by small sailboat from New York to Gibraltar, and then to Africa's west coast, and finally, to Brazil. During the voyage, she collected and mapped water samples to test for microplastics. As expected, she found plastics in the most pristine areas of the oceans, thousands of miles from human habitats.

Besada explains that there is no *away* when someone throws plastic away. It returns to us as litter and waste—on our shores, in our food, and in the marine life we lose when species get caught in plastics or consume microplastics, which enter the food chain and eventually, our bodies. Recycling can help, but it's not a solution. She teaches that we have to redesign our relationship with single-use plastic and use it only sparingly.

Besada spends much of her time working with students and schools to create projects that respond to ocean plastic waste by finding alternatives. She encourages the students to examine plastic use in their communities, draw connections to how local habits influence global waterways and biodiversity, and draft action plans. One team she worked with noticed that cookie packaging was a big source of plastic at school. So the team surveyed students to determine their favorite cookie, wrote to the cookie maker, and asked, "You make our favorite cookie, but we don't want to use plastic, can you sell them to us in bulk?"

The company didn't respond, so they partnered with a local bakery and purchased cookies without plastic packaging. Another group found that much of the plastic use came from students and others leaving campus to buy food and returning with lunch. The group responded by supplying reusable bags in the lobby. Now at lunchtime, they grab a reusable bag and return it.

The value of the project goes far beyond reducing the use of plastic. It prompts students to think like explorers, and once they begin thinking that way, it becomes part of their mindset. The ability to seek and find solutions and bring them to fruition is valuable in business, government, and life. These projects change the way students think and live in the world—and that can change the future.

*Agustina Besada at sea.*

*Besada uses time with students to create local ocean plastic research.*

*Students are encouraged to work on solutions.*

# USING GIS TO CULTIVATE EMPATHY FOR THE EARTH

Can we care about something that we never see or experience? Studying manta rays allows students to learn about the species, such as their place in the food web and importance to the marine ecosystem, but does it inspire them to protect manta ray habitats?

How different would the learning experience be if students traveled to the Great Barrier Reef with National Geographic Explorer Erika Woolsey and to see hundreds of manta rays feasting on plankton together? And then map their migration as they seek food supplies and warm water temperatures? This is the power of geospatial technologies.

More than 20 years ago, Sylvia Earle raised a question that remains just as relevant today: "Where does GIS come into all of this? I'll put it another way. Where doesn't GIS come into the understanding of the ocean? After all, marine ecosystems, just as those on the land, are geospatial, and therefore, so are the solutions that we must craft as we go forward."

GIS takes learners and solution seekers to any part of the globe and helps them visualize information in ways that reveal patterns and trends that can unlock new insights about Earth. Modern tools help learners develop geographic thinking and problem-solving and storytelling abilities so they can confront challenges. Introducing these resources to children early in their school years helps them become digital and geospatial learners.

## Architects of change

The greatest accomplishment of human civilization, by far, will not come from the genius of one scientist, inventor, activist, or moral leader. It will come from the combined efforts of people helping to build a culture that uses its energy, genius, innovation, inclusivity, and idealism to create a thriving planet.

When we consider that 42% of the world's population is under the age of 25, young people are key in seeking solutions now and sustaining a healthy planet in the future. They have the collective talent, energy, and vision, but equipping them with the necessary skills requires cultivating learning environments that spark their curiosity, engage them with dynamic and contextualized learning experiences, and empower them with geospatial concepts and tools to make a difference.

Nathaniel Soon, for example, is part of a large, diverse community of young solution seekers and planetary stewards—a generation known as GenGeo. This global community organizes peer meetups, serves as mentors, models leading with empathy, and channels collective knowledge into ideas, action, and real-world solutions.

As Soon progresses through college, he champions ocean conservation globally through his visual storytelling work, collaborations with partners, and other social enterprises. He is an Ocean and Climate Youth Ambassador with Peace Boat, an

*Fabien Cousteau, Sylvia Earle, and Nathaniel Soon aboard the* Peace Boat.

international NGO based in Japan focused on human rights and ocean sustainability. It was during one of these Peace Boat programs that Nathaniel met two of his heroes: Earle and aquanaut and documentary filmmaker Fabien Cousteau (grandson of Jacques Cousteau).

In exploring the world and sharing their journey, explorers and scientists such as Earle model curiosity and inspire generations of young people to follow in their footsteps. It is fitting then that Soon was recently named a National Geographic Young Explorer—a testament to his explorer's mindset in action.

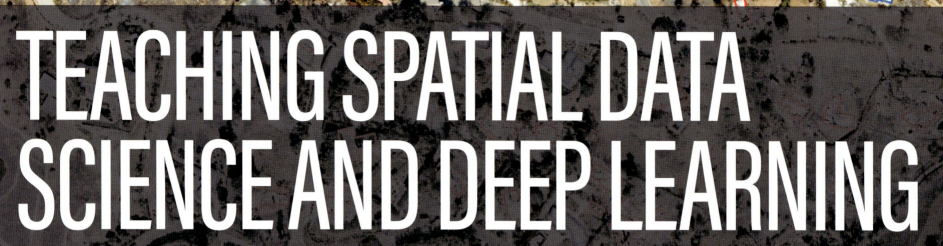

# TEACHING SPATIAL DATA SCIENCE AND DEEP LEARNING

Fueled by massive GIS data repositories, deep learning and neural networks enable data scientists to apply machine learning techniques to a growing range of real-world problems. Education leaders at the University of California, San Diego, are training a new generation of geo-literate scientists. This chapter explores the short history of deep neural networks, describes the teaching approach at UCSD, and examines several practical applications at the intersection of GIS and AI.

By Ilya Zaslavsky, **UCSD;** and Dmitry Kudinov, **Esri**

*Results of an automated damage assessment of homes after the devastating 2018 Woolsey Fire in Southern California. Red outlines show damaged structures; blue undamaged.*

# TEACHING PRACTICAL DEEP LEARNING SKILLS

On September 30, 2012, AlexNet, a neural network model designed and trained by Alex Krizhevsky, achieved a top-5 error of 15.3% in the ImageNet 2012 Challenge—an annual competition that has been instrumental in advancing computer vision and deep learning research. This achievement came in more than 10.8 percentage points lower than that of the runner-up and launched a new era of convolutional neural networks in image processing. Relying on GPU parallel computations, convolutional neural networks since then have improved state-of-the-art results across the entire spectrum of computer vision and achieved superhuman levels of accuracy in image classification, object detection, segmentation, etc.

The community of deep learning researchers and the number of practical applications and profitable businesses relying on deep neural networks are growing rapidly, reaching today far beyond the computer vision domain and making groundbreaking leaps in natural language processing, content generation, robotics, time-series analysis, computational physics and biology, partial differential equations, and beauty industry, to name a few. Not surprisingly, geospatial data—the original "big data"—is now coming into focus in the AI world.

The GIS world is not falling behind. Just a few years after the AlexNet introduction, deep neural networks have transformed and keep delivering new spatial analytic capabilities at an accelerating pace. For example, the ArcGIS API for Python's Learn module effectively tripled in size during 2020, bringing the count to about 30 ready-to-use deep neural networks in its November release.

Deep learning as a field has its own detailed and specific workflows covering data preparation, training and validation of neural networks, and deployment of trained models. The ArcGIS Platform seamlessly integrates these tasks into traditional GIS workflows through a set of tailored geoprocessing tools (APIs) that offer enhanced and distributed functionality. The result has been a more streamlined integration of deep learning capabilities into research and production systems, and growing interest from the academic community.

Deep learning can be thought of as an evolution of machine learning. Where traditional machine learning requires explicit and careful preparation of input features (training examples) by the programmers and data scientists, deep learning relies on large graph structures and corresponding algorithms, making them less sensitive to noise and correlations in the input data.

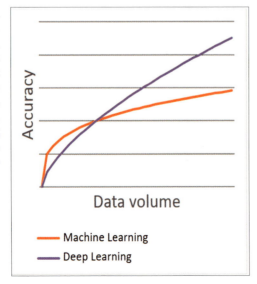

*The full advantage of deep learning versus machine learning comes with large volumes of training data.*

ArcGIS has multiple machine learning tools covering often the same tasks, for example, clustering, pixel classification, point cloud segmentation, etc., so when to use machine learning and when to use their deep learning counterparts is an important question.

There are multiple aspects to consider, but the main one can be illustrated in this simple chart: the full advantage of deep neural networks comes with big data.

Therefore, an important aspect to remember in the case of deep learning, is that for efficient training of deep neural network models, one or more modern GPUs with a significant amount of video memory is required.

And, of course, there are cases in which no machine learning models, or deterministic algorithms exist to automate the task, and where organizations historically rely on manual labor. Here, too, a deep learning solution can be a great option to try—the notion of searching the solution space by training a neural network, not by writing more code.

A good example here is the task of labeling specific classes of objects in lidar point clouds. There are existing statistical algorithms that allow the system to automatically label noise, ground, buildings, and a few other classes of points. But what if we need to label distinct classes of railroad switches? Or specific types of utility poles, types of powerline attachments, or different classes of transformer, etc.? Writing and maintaining unique algorithms may be prohibitively expensive, whereas training and deploying a neural network to do this job is a matter of a few days.

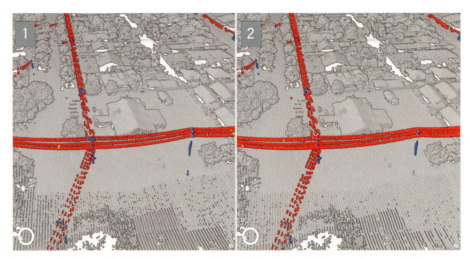

*Wires and utility poles labeled 1) manually and 2) by PointCNN neural network. Upon examination, one can see that the neural network replicated the performance of a human analyst doing the same classification.*

# DEEP LEARNING LAB AT UC SAN DIEGO

Given the importance of the deep learning workflows and availability of cost-efficient GPU offerings by multiple cloud providers, we have successfully offered a set of deep learning exercises to the data science students at UCSD since 2019. In 2020, we used Amazon EC2 G4 instances equipped with NVIDIA T4 GPUs, which worked well for running these exercises. Specifically, the g4dn.XL instances come with affordable hourly rates, allowing us to stay within a small budget for classes of 40–50 students.

Each student (or pair of students) was given a designated cloud virtual machine with a preconfigured software stack (Microsoft Windows, ArcGIS Pro, ArcGIS API for Python, Doccano), step-by-step PDF instructions for each exercise, initial ArcGIS Pro project files, data, and semicomplete Python notebooks. The student ArcGIS Online accounts (or Portal for ArcGIS accounts) used to sign in to ArcGIS Pro were provisioned with the ArcGIS Image Analyst extension.

Students were asked to complete four exercises during the limited time window of three days. Each exercise required the students to complete the full deep learning cycle: create and export training data, train a specific deep neural network, validate the resulting model, run the model on unseen data and, finally, write a short essay documenting the student's observations of the training process, hyperparameter search, and reasoning about further potential improvements of the models.

The exercises cover four scenarios with distinct workflows:

## 1. Feature classification in world imagery

Forest fires can spread quickly and destroy thousands of acres of land. In California, where many urban areas infringe on forestlands and native habitat, wildfires threaten people's lives and homes. The 2018 Woolsey Fire affected almost 100,000 acres of forest and threatened nearly 50,000 people in Ventura and Los Angeles counties. During a disaster, time is of the essence for response and recovery efforts. Unfortunately, the lack of real-time data and need to interpret data manually can cause critical response delays. In this exercise, with the help of deep learning, students quickly and accurately identify urban structures that have been destroyed by fires and structures that are still safe.

*Damaged structures have a distinct digital signature that the neural network learns to recognize through training data.*

## 2. Land cover classification

Land cover classification is an important prerequisite for performing regional planning exercises, prevention of land degradation, forest and ecological conservation, and change detection. It is labor-intensive and often cost-prohibitive to produce high-resolution land cover maps.

Deep learning provides the opportunity to take advantage of previous manually produced high-resolution land cover data to train an AI model to automatically perform this task. In this exercise, students use a portion of 1 m resolution land cover data from the Chesapeake Conservancy project and train a deep learning pixel classification model, which then gets applied across a large geographical area to produce land cover classification raster automatically.

*Land cover classification from satellite, aircraft, and imagery from drones—a historically labor-intensive task—can be successfully automated using deep learning methods.*

## 3. Detect objects in imagery

In Tonga, an archipelago in the South Pacific, inventorying each palm tree and building on Kolvai's landed estate would take a lot of time and require a large workforce. To automate the process of geographic feature extraction for this location on the island of Tongatapu, students train a deep learning model to identify tree species of interest and houses in the imagery collected by a drone.

*The process of building training data begins with circling all the known palm trees in a small sample area.*

*This procedure is repeated for several areas of a set of images. The same training process is run for buildings.*

*An example of detections produced by a student-trained neural network after approximately one hour of training.*

## 4. Recognize and extract entities from unstructured text

It is a labor-intensive task to read through thousands of documents and extract information useful for further GIS analysis. The EntityRecognizer module in ArcGIS API for Python uses deep learning to learn from just a few hundred labeled text reports.

In this exercise, students create a training set from unstructured police reports, train an EntityRecognizer model, and then apply it to extract geographic features automatically from thousands of similar text documents in seconds.

*Entity-recognizing software interface used to extract geocodable addresses, times, dates, names, and other facts from text documents.*

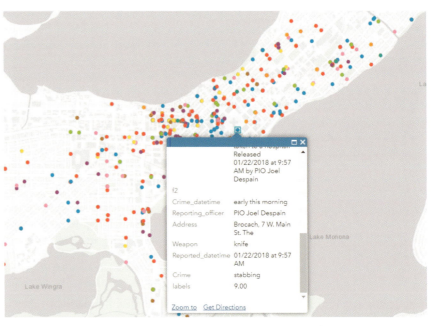

*The extracted features are then published as an ArcGIS Online feature service and added to a web map, ready for further spatial analysis.*

The exercises are organized in a progressive disclosure form, taking students first through a simple workflow of training a neural network using ArcGIS Pro user interface tools, and then diving deeper into Python code organized into visually rich, convenient Jupyter Notebooks annotated with detailed technical instructions and recommendations.

This way, students gain an insight into the deep learning training process: grasp the concepts of overfitting and underfitting, importance of validation-train splits, training data augmentation, optimal learning rate search, etc. The allotted virtual machine time also allows for experimenting with neural network architectures and inviting students to compete with each other for the best metric accuracy.

# TEACHING SPATIAL DATA SCIENCE TO DATA SCIENCE STUDENTS

As demand for data science specialists grows and graduates with these skills command higher average salaries, universities have rushed to create data science classes and programs. Enrollments in such programs often eclipse enrollments in geography, environmental science, urban planning, and other more traditional homes for geographic information science classes. Spatial location is integral to the increasingly large datasets that scientists use to train machine learning models. As such, spatial analysis is a key skill needed in such disciplines as environmental science, public health, urban planning, chemistry, business analytics, and neuroscience.

In a more traditional GIS curriculum, undergraduate and graduate students normally take GIS programming, data science, and machine learning courses near the end of their programs. These courses are based on previous coursework, which typically includes foundations of cartography, spatial analysis, and experience with a desktop or web-based GIS. In new data science programs, GIS and spatial analysis is a completely new topic, as lower-division training largely focuses on the theory and practice of data science, data structures and algorithms, statistics, and Python coding. For this large new cohort of undergraduate students, a path to spatial data science is also necessary. Most importantly, this different path must be practical and project-based, letting students work with real data and develop solutions to real challenges posed by local governments and communities or identified in research projects.

Students encounter these key principles in Spatial Data Science and Applications, an upper-level elective course that the new UCSD Data Science Program has offered in collaboration with Esri since spring 2019. While several remarkable spatial data science programs have been created at the graduate level, this class was among the first of its kind at the undergraduate level, focused on a combination of machine learning and Python coding with spatial analysis and GIS.

When UCSD students take this, they are proficient with Python and several key data science libraries, such as Pandas, Matplotlib, Seaborn, scikit-learn, and Keras/TensorFlow. They have learned key machine learning and statistical analysis techniques, and many of them have taken image processing and similar classes. With a few exceptions (typically, graduate students from other departments), the students had no previous classes with elements of mapping and spatial data management and analysis.

The program structured the course as a sequence of modules, gradually introducing spatial analysis concepts, techniques, and platforms, building on what students know at the start of each module:

1. *Foundations of spatial data, online mapping, and analysis*, using open-source Python packages (GeoPandas, Shapely, etc.). Getting started with this environment was simple for students trained in Python and Pandas and proficient with Jupyter notebooks they developed on the UCSD Data Science and Machine Learning Platform. This familiarity helped us avoid technical distractions as we explored foundational concepts: layered organization of spatial information, key topological and distance-based operations, spatial join, and techniques for online mapping, projections, spatial data structures, spatial data quality, and metadata.

2. *Advanced feature management and analysis*, with ArcGIS API for Python and ArcGIS Online. In this module, students learn components of the ArcGIS ecosystem for spatial data management, and experiment with more advanced concepts that are especially useful for data science projects: geoenrichment to generate variables for classification and regression models, operations for joining and aggregating spatial information from various data layers, geocoding, geoprocessing, and network analysis.

3. *Raster analysis and modeling*, with ArcGIS API for Python on ArcGIS Enterprise. This module introduces raster data structures and operations over grids of various types, common indexes computed over Landsat and Sentinel imagery, map algebra operations, and rules of map combination for suitability modeling.

4. *Spatial statistics*, primarily with ArcGIS API for Python and PySAL: including understanding of spatial weights, geographically weighted regression, computing hot and cold spots in spatial distributions, and point pattern analysis.

5. *Machine learning and deep learning with spatial data*. While machine learning examples appear in all modules of the course, this module summarizes techniques and tools for feature engineering based on spatial relationships to improve model accuracy, and experiment with deep learning models.

Each lecture in the 10-week course includes Jupyter notebooks showing spatial analysis concepts and coding recipes. The idea is to demonstrate how these new concepts help address practical challenges that researchers encounter as they work with spatial information. For example, projections are introduced once students try to plot a map with data from various sources in GeoPandas, and the results are not what they expect. Results of binary spatial operations often appear incorrect because of geometric imperfections or different spatial resolutions of input layers. These results lead to a discussion of spatial uncertainties and formal data quality descriptions in spatial metadata, while the need to ensure topological integrity triggers a discussion of spatial data structures.

## Class projects

Learning new concepts and understanding caveats of using spatial data and spatial analysis techniques by resolving practical roadblocks is the most important part of the course. Students learn these skills through a series of mini-projects culminating with a final project at the end of the quarter. Unlike typical assignments in previous classes, these projects are open ended. Students are given a general prompt, for which they are expected to find appropriate datasets, apply spatial operations and machine learning models, and critically discuss their findings using Jupyter notebooks.

Teams of two students work on each of the mini-projects (except the first one, which is intended to establish individual baseline proficiency with spatial operations) so that they can discuss design choices and limitations of their approach. For final projects, some student teams choose ArcGIS Pro to develop and edit data layers or ArcGIS StoryMaps to present their results. Students received no instructions on the use of these Esri products but often found them easy to use after experiencing GIS through the lens of Python packages. The next section presents a selection of example student projects.

## Examining food deserts

The U.S. Department of Housing and Urban Development (HUD) designated selected high-poverty communities as Promise Zones. In these areas, government agencies work with local partners to increase economic activity, improve educational opportunities, enhance public health, reduce crime, and otherwise improve community well-being. One such zone is established in southeast San Diego.

The student teams for this mini-project examined whether the Promise Zone in San Diego was also a food desert—an area with limited access to affordable healthy food options. They used the Promise Zone boundaries delineated by HUD, local census tract boundaries and social-economic statistics from the San Diego Association of Governments (SANDAG), and business tax certificate listings from the City of San Diego. The latter included names and North American Industry Classification System (NAICS) codes of various food-related businesses such as grocery stores, restaurants, and fast-food establishments. To create a dataset for regression analysis, students needed to join and aggregate this data by census tracts. While this task appeared initially straightforward, students had to address several caveats. The Promise Zone boundaries, while generally following census tract boundaries, were generalized, which rendered results of geometric operations (such as which census tracts are in the Promise Zone) incorrect. Students had to discover and discuss the geometric misalignments and plan to programmatically create the most representative selection of census tracts within and beyond the Promise Zone. An additional source of bias was relying on NAICS codes to identify which businesses provided healthy and unhealthy food options.

As is often the case with real data science projects, regression modeling results (using scikit-learn, a popular machine learning Python library that students are familiar with) predicting ratios of healthy and unhealthy food options by census tracts initially appeared counterintuitive and often inconclusive. They showed a slightly higher ratio of healthy food options in the Promise Zone, because of restaurants clustered mostly elsewhere in the city. This unexpected result made students reconsider the problem formulation and their assumptions, including redefining "healthy" and "unhealthy" food options, re-examining the spatial joins, and considering additional independent variables.

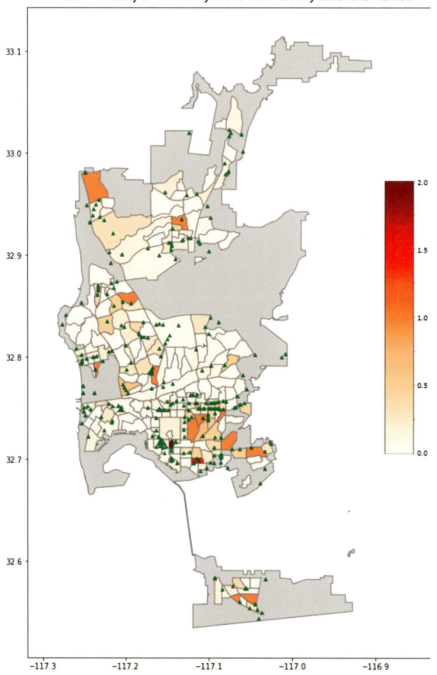

A snapshot of a map generated in GeoPandas showing a ratio of healthy and unhealthy food options by census tracts, and healthy food business locations.

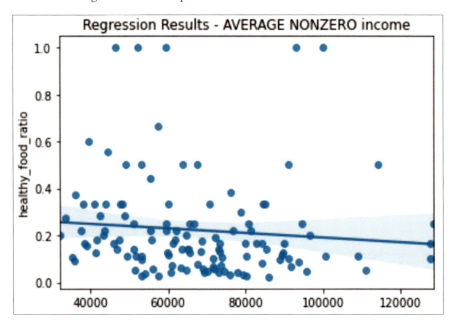

Regression modeling of the ratio of healthy food options versus average income initially appeared counterintuitive.

## Is it safer to ride on bike paths?

The next assignment focused on a well-known challenge: improving bicycle safety. The number of bicycle fatalities has reached a 30-year high in the United States, because of more bicyclists on the road, cars driving at faster speeds, distractions from cell phones, and the lack of bike path infrastructure. The task was to examine whether bicyclists were more likely to be injured if they were on a bike path. The assignment asked the teams to describe severity of injuries depending on several factors, including characteristics of the zip code where the accident occurred, and accident-related factors, such as whether alcohol was involved. Students retrieved bike path and other spatial data layers from SANDAG, accident information from the Statewide Integrated Traffic Records System at UC Berkeley, and business data from the City of San Diego. Students also generated additional variables through the ArcGIS geoenrichment service.

The likelihood of getting into a collision on a bike path depended on how bike paths were defined in the first place, and how the accident locations were recorded (as either geocoded addresses, or GPS coordinates, or both). In each case, students had to choose more accurate coordinates and define a distance buffer around bike paths so that they could approximately place accidents on a bike path or outside of it. As before, these design choices resulted in different modeling results, often forcing students to reconsider their approach.

Using exploratory analysis, students found that biking accidents are more likely to occur in zip codes where more alcoholic beverages are sold and that accidents on bike paths generally resulted in fewer severe injuries. Students chose the data science algorithms for their analysis. In most cases, the choices about how to define spatial features and compute spatial relationships for inclusion

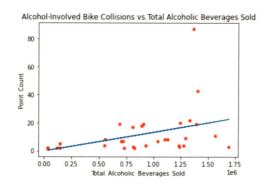

in machine learning models had a larger effect on the accuracy of results than a selection of a machine learning technique. That was an important takeaway from this exercise and from the course as a whole.

## Suitability modeling

In this next mini-project, students tried to find and organize raster data in ArcGIS Living Atlas of the World, converted from vector data of their choice or published as an ArcGIS raster image collection from scenes downloaded from the U.S. Geological Survey. The raster layers were clipped to a study area and remapped for inclusion in a suitability or a risk model, which students implemented using raster analytics functions in ArcGIS API for Python on the ArcGIS Enterprise installation at UCSD Library. Besides demonstrating the machinery of raster analytics and map algebra functions, this exercise showed that model results differed depending on the map combination techniques used by students, such as exclusionary screening or a weighted linear combination of factors expressed as raster layers.

As with the previous mini-projects, the students were expected to discuss how their choices—such as the raster layers they selected, the remapping rules for each layer, the map algebra and map combination techniques, and the assigned

weights—affected the results. Within these general expectations, topics of student mini-projects ranged from determining suitable locations for soccer fields in San Diego and highlighting wildfire or desertification risks, to finding best sites for rock climbing, mountain biking or camping, locating wildfire alarm outposts and drive-through restaurants. Grading such open-ended assignments is time-consuming but the reward is better student learning as they creatively apply spatial analysis techniques to intriguing problems.

## Why spatial is special in machine learning

The key takeaway from this course: knowing how to find, interpret, and efficiently use spatial information through spatial APIs in a common coding environment 1) lets users create more comprehensive datasets by integrating data from different sources, 2) makes data science projects more visually attractive, and 3) helps improve performance of machine learning models. One example is more accurate prediction of childhood asthma hospitalization rates—information that would help hospitals better allocate personnel and other resources, and help families with affected children. This class explored a random forest regression and computed asthma hospitalization rates with and without spatial variables, resulting in improved model accuracy using spatial data.

Asthma hospitalization data is sometimes incomplete. Besides asthma hospitalization rates by some (but not all) Connecticut census tracts, additional variables reflect educational attainment, unemployment, other health indicators, and income levels. Using these variables, obtained through geoenrichment, students built a predictive model to impute the missing asthma hospitalization values. Using random forest regression in scikit-learn, we experimented with various model parameters and derived the best model, which gives us a prediction accuracy of 0.73 on the testing set. Then, we can try to predict asthma hospitalization rates, but with additional spatial data layers, such as distance to toxic release points and to primary and secondary roads. Using the Forest-based Classification and Regression tool in ArcGIS.Learn (an ArcGIS Pro machine learning tool not to be confused with the training site Learn.arcgis.com) increased the prediction accuracy and allowed the model to be applied to a similarly geoenriched layer at a finer spatial granularity.

The next snapshot shows a fragment of a Jupyter notebook with a map of predicted child hospitalization rates by block groups, as computed by the model.

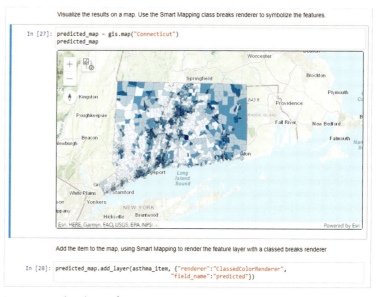

*Jupyter notebook interface.*

# FINAL PROJECTS

The self-directed final projects give students a chance to apply any combination of spatial analysis and machine learning techniques to a problem they select and present their work to peers and a group of experts from local government, industry, nonprofits, and UCSD faculty.

Topics may range from detecting wildfires, predicting habitat change, and analyzing spatial patterns of crime and vehicle traffic, to modeling housing prices and affordable housing strategy, assessing hospital accessibility, and predicting success of sports teams depending on their travel schedules. When students took the course remotely during the pandemic, many projects focused on the coronavirus disease 2019 (COVID-19), including modeling indoor transmission risks, locating testing facilities, predicting infection rates by zip codes, and examining the effects of COVID-19 on home values in various parts of California.

## Wildfire prediction

For a project that focused on predicting California wildfires, the student team used exploratory spatial data analysis to identify spatial patterns of historical wildfires. Students then built a machine learning model of wildfire frequency based on several raster layers (rainfall, topography, biomass, etc.), geoenrichment, and identification of "under-serviced" areas that would benefit from additional fire stations and cameras to monitor remote locations. Improving wildfire analysis and evaluating prevention strategies is critical for a state that spends $2.5 billion annually fighting wildfires and where the 2020 wildfire season alone caused more than $12 billion in direct damage.

## Finding equitable polling locations

Another interesting student project explored polling stations in San Diego during the 2020 election year and compared them with optimal locations generated through a multistep spatial analysis workflow implemented in a Jupyter notebook using location-allocation capabilities through ArcGIS API for Python. The process involved these steps:

1. Filter out precinct polygons categorized as "Vote-by-Mail."
2. Map current in-person polling stations and remove duplicate locations.
3. Identify potential areas for polling locations based on land-use codes and on the number of in-person voters.
4. Use these areas to assign candidate locations for the polling stations.

With the centroid of each precinct considered *demand points*, weighted by the share of in-person voters, the students could then run location-allocation analysis for five regions within San Diego to determine the most equitable polling locations. In many cases, the model's suggested polling locations differed from the official 2020 polling stations.

This project showed that GIS and data science together could help develop equitable polling place locations, especially when more voters cast ballots by mail and when the relative shift to mail revealed different voting patterns for different socioeconomic groups.

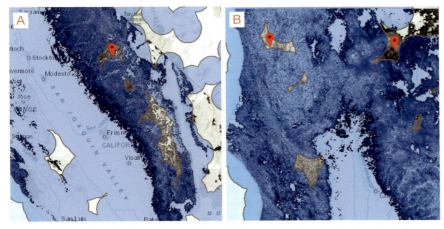

*These images show highest-risk areas for wildfires and under-serviced national forests, with potential locations for additional fire stations in A) Stanislaus National Forest and B) Klamath National Forest and Modoc National Forest.*

This is the output of all 5 of the location-allocation sub regions combined into an interactive webmap.

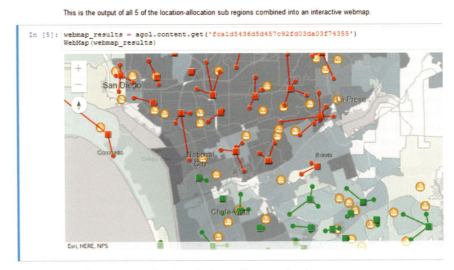

*A fragment of Jupyter notebook with the polling results shown.*

# CONCLUSION

## Spatial analysis and higher learning

The most important outcome of this introductory spatial data science course is that students could interpret challenges they encountered in data science practice as spatial data science problems and solve them using techniques learned in class. Many topics worth studying remained outside the 10-week course format. Data science programs are still being refined. Because these programs are packed with undergraduate coursework, they often don't have room to include a dedicated spatial data science track at that level. Nevertheless, many students remained interested in spatial analysis and GIS, and continued their work through independent research or through a senior capstone course that produced remarkable projects. One project created a COVID-19 dashboard showing infection rates by zip codes, which are automatically updated daily from county sources (the list on the left and the central map frame); a forecast of infection rates (right map frame); and a range of socioeconomic, demographic, and health indicators for the selected zip codes (with data from the American Community Survey). The dashboard project won first place in the Urban and Regional Information Systems Association (URISA) 2020 University Student and Young Professional Digital Competition.

## COVID-19 transmission in elementary schools

Another interesting example is a spatially explicit agent-based model of COVID-19 transmission in elementary schools. In this project, students used real school floor plans, digitized with ArcGIS Pro and organized as shapefiles, to simulate viral transmission and accumulation in classrooms and other school areas under various scenarios (learning at individual desks, group activities, recess, lunch, etc.). The simulations helped identify school areas and activities with the highest risk of spreading the virus and demonstrated how a combination of non-pharmaceutical interventions, such as mask wearing, reducing class sizes, moving lunch from the cafeteria to classrooms, adding desktop dividers, staggering attendance, and adding ventilation, could reduce the risk of infection.

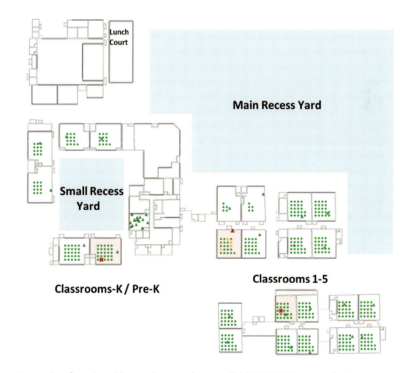

*Example of a school floor plan used to model COVID-19 transmission.*

*A student created this COVID-19 dashboard to show infection rates by zip code, a forecast of infection rates, and socioeconomic, demographic, and health indicators.*

# ACKNOWLEDGMENTS

The authors wish to acknowledge the many contributions of the UCSD students and tutors in DSC 170, Spatial Data Science and Applications, in particular, Kaushik Ganapathy, Johnny Lei, Eric Yu, Bailey Man, Xiangchen Zhao, Renaldy Herlim, Siddhi Patel, Nathan Roberts, Cameron Shaw, Akshay Bhide, Peter Larcheveque, Jiali Qian, and Songling Lu, whose projects are used as examples in the chapter. The authors also thank Kaushik Ganapathy, Bailey Man, and Miles Labrador for their comments on the draft of this text.

# PART 5
# TECHNOLOGY SHOWCASE

This book has already shown how science goes hand in hand with technology (engineering). One of the most exciting trends of the modern age is how science uses the exponential power and assistance of artificial intelligence (AI) to help address the unprecedented challenges facing humanity and the planet, including climate change, water scarcity, global health crises, food security, and loss of biodiversity. GIS technology is no different; it extends our minds by abstracting our world into knowledge objects that we can create, replicate, and maintain. These knowledge objects include data, imagery, and models that explain processes and workflows, as well as maps that communicate and persist in apps. Enjoy this section of vignettes on GIS technologies that help create new systematic frameworks for scientific understanding.

An ArcGIS web application illustrates vessel traffic monitored by a navigation device called the Automatic Identification System (AIS). By assembling reams of breadcrumb-like information, AIS helps identify and resolve space use conflicts and much more.

• CARGO

# DRONE DATA AUTOMATION WITH SITE SCAN FOR ARCGIS

Lauren Winter, Esri

Environmental researchers face significant barriers from the technological complexity and time-consuming nature of data collection, processing, and management. For time-based research such as marsh erosion, it's crucial to ensure that processes are repeatable for comparative analysis.

To address this challenge, many environmental scientists now use drones equipped with optical, multispectral, laser, thermal, and radiation sensors to collect data. These sensors and drones rely heavily on the same technologies used in mobile phones, so capability is rapidly increasing while cost is decreasing. The use of drones is clearly a boon for the quality of scientific instrumentation and the amount of data that can be collected. However, this capability can quickly turn into a mountain of unmanageable data and processing. For imagery, each drone flight commonly produces 500 to 1,000 4k photos that require complex image processing configurations and leave little time for environmental science. Also, these complexities are prone to human error, such as inconsistent drone flight patterns for image capturing, which makes it difficult to accurately compare and analyze results over time.

Site Scan for ArcGIS is an industrial software as a service (SaaS) product that enables calculated and automated drone flights to capture large sets of precisely posed images that are uploaded to the cloud for parameterized processing into 2D orthomosaic, 3D point cloud, and digital elevation model (DEM) data products. Site Scan organizes the data and a set of workflow tools for analysis, measurement, planning, and comparison. Researchers require this type of cost-effective system automation, data management, precision, and repeatability for tasks such as daily project planning, progress reports, and built-as-designed quality assessment.

Consider the task of measuring marsh erosion and the formation of surge channels over time, which is one of the San Francisco Estuary Institute (SFEI) programs. The institute discovered that using a drone carrying a high-quality camera is the perfect tool. It can be flown over a large area without significantly disturbing the study area and performed with fewer people and in less time than doing the work on foot.

To monitor the evolution and condition of a 90-acre marsh near Corte Madera, California, the team used Site Scan to program a drone flight that took about 45 minutes. The 640 high-resolution images, together with the processed orthomosaic image and DEM, provided detailed and measurable information for the study.

*Thanks to Pete Kauhanen, San Francisco Estuary Institute, and Ross Robinson, Esri.*

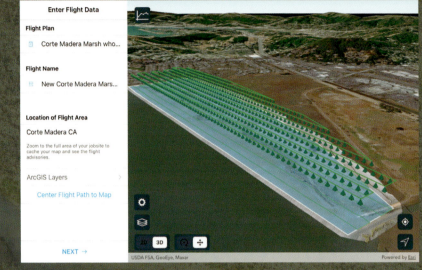

*Site Scan automatically completes the steps of drone flight planning, sensor setup, flight, image transfer to the cloud, image processing, data storage, and data analysis, saving valuable time and producing scientifically significant results.*

*Images and results are automatically managed in the cloud and can be shared with other scientists and used publicly in ArcGIS Online.*

Aerial view of marsh near
Corte Madera, California.

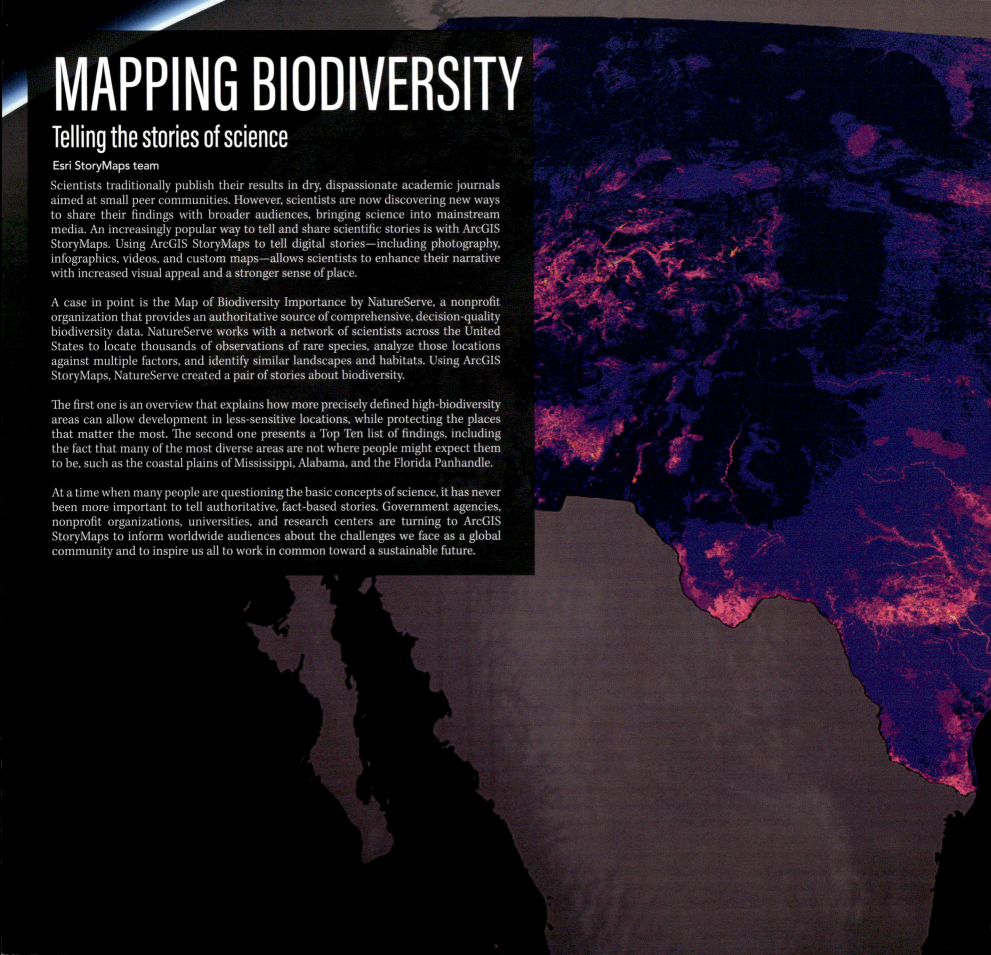

# MAPPING BIODIVERSITY
## Telling the stories of science

Esri StoryMaps team

Scientists traditionally publish their results in dry, dispassionate academic journals aimed at small peer communities. However, scientists are now discovering new ways to share their findings with broader audiences, bringing science into mainstream media. An increasingly popular way to tell and share scientific stories is with ArcGIS StoryMaps. Using ArcGIS StoryMaps to tell digital stories—including photography, infographics, videos, and custom maps—allows scientists to enhance their narrative with increased visual appeal and a stronger sense of place.

A case in point is the Map of Biodiversity Importance by NatureServe, a nonprofit organization that provides an authoritative source of comprehensive, decision-quality biodiversity data. NatureServe works with a network of scientists across the United States to locate thousands of observations of rare species, analyze those locations against multiple factors, and identify similar landscapes and habitats. Using ArcGIS StoryMaps, NatureServe created a pair of stories about biodiversity.

The first one is an overview that explains how more precisely defined high-biodiversity areas can allow development in less-sensitive locations, while protecting the places that matter the most. The second one presents a Top Ten list of findings, including the fact that many of the most diverse areas are not where people might expect them to be, such as the coastal plains of Mississippi, Alabama, and the Florida Panhandle.

At a time when many people are questioning the basic concepts of science, it has never been more important to tell authoritative, fact-based stories. Government agencies, nonprofit organizations, universities, and research centers are turning to ArcGIS StoryMaps to inform worldwide audiences about the challenges we face as a global community and to inspire us all to work in common toward a sustainable future.

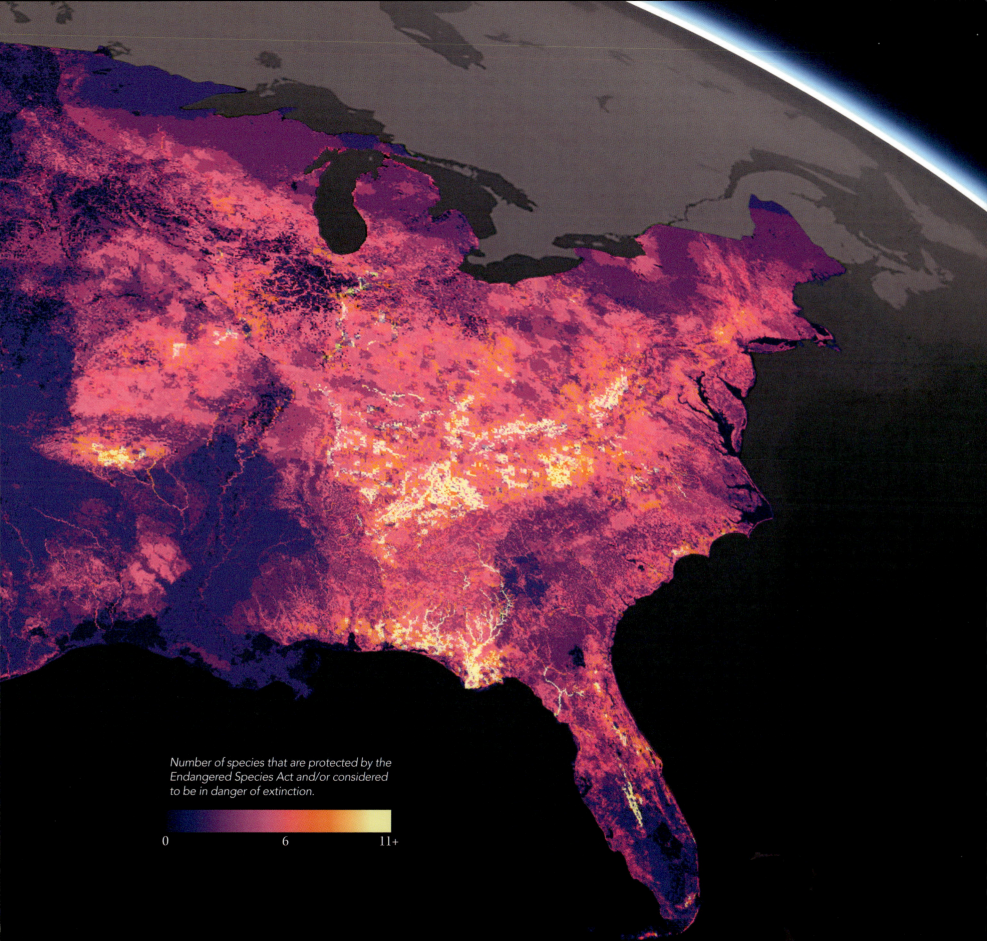

Number of species that are protected by the
Endangered Species Act and/or considered
to be in danger of extinction.

0          6          11+

# MODELING GLOBAL STREAMFLOW

## A worldwide service for forecasting water volume

Steve Kopp, Esri

During the last 20 years, issues related to flood, drought, and water affected more than 3 billion people, caused nearly $700 billion in economic damage, and killed an estimated 166,000 people. In most places in the world, people don't know how much water will be in their rivers next week or have reliable historic information to understand past floods or droughts. To help address these problems, Esri, in collaboration with Brigham Young University (BYU), the European Center for Medium-Range Weather Forecasts (ECMWF), the GEO Global Water Sustainability (GEOGloWS) program, and the World Bank, embarked on a plan to provide global streamflow forecasts and historic streamflow information with easy-to-use web maps. These maps and a related web service now provide access to forecast and historic flow for more than 1 million rivers around the world.

The backbone of the project is the ECMWF global runoff forecast and historic modeled runoff. The global runoff forecast is an 18 km resolution, 51-member ensemble model. The historic modeled runoff is a 31 km resolution, 40-year estimate of historic streamflow. BYU and Esri together transformed the surface runoff measurements into the quantity of water flowing in a stream. This transformation is done through a series of GIS steps, computing watersheds from digital elevation models (DEMs), aggregating runoff into watersheds, transforming land runoff into streamflow, and using the Routing Application for Parallel computation of Discharge (RAPID) model to route the flow downstream. The daily processing workflow, which was piloted as a research project at BYU, was successfully transitioned to ECMWF. Now, each day, a new forecast is published and made available through a REST API hosted by ECMWF and a web service hosted by Esri.

Knowing about local water is important to public safety, health, food, energy, and more. By providing free and open web services, an API, and easy-to-use configurable web applications, streamflow forecasting will become as common as a weather forecast. As an example, when Latin America was struck by two Category 4 hurricanes, Eta and Iota, only two weeks apart, the Central American disaster prevention and preparedness organization (CEPREDENAC) used the streamflow service to improve situational awareness and emergency management efforts by understanding how much water would be in the rivers each day as Hurricane Iota approached and moved through the region.

*Thanks to our project collaborators: Brigham Young University, European Center for Medium-Range Weather Forecasts, GEO Global Water Sustainability, National Aeronautics and Space Administration, National Oceanic and Atmospheric Administration, SERVIR, World Bank, U.S. Agency for International Development, European Commission, Copernicus, AquaVEO, and International Centre for Integrated Mountain Development.*

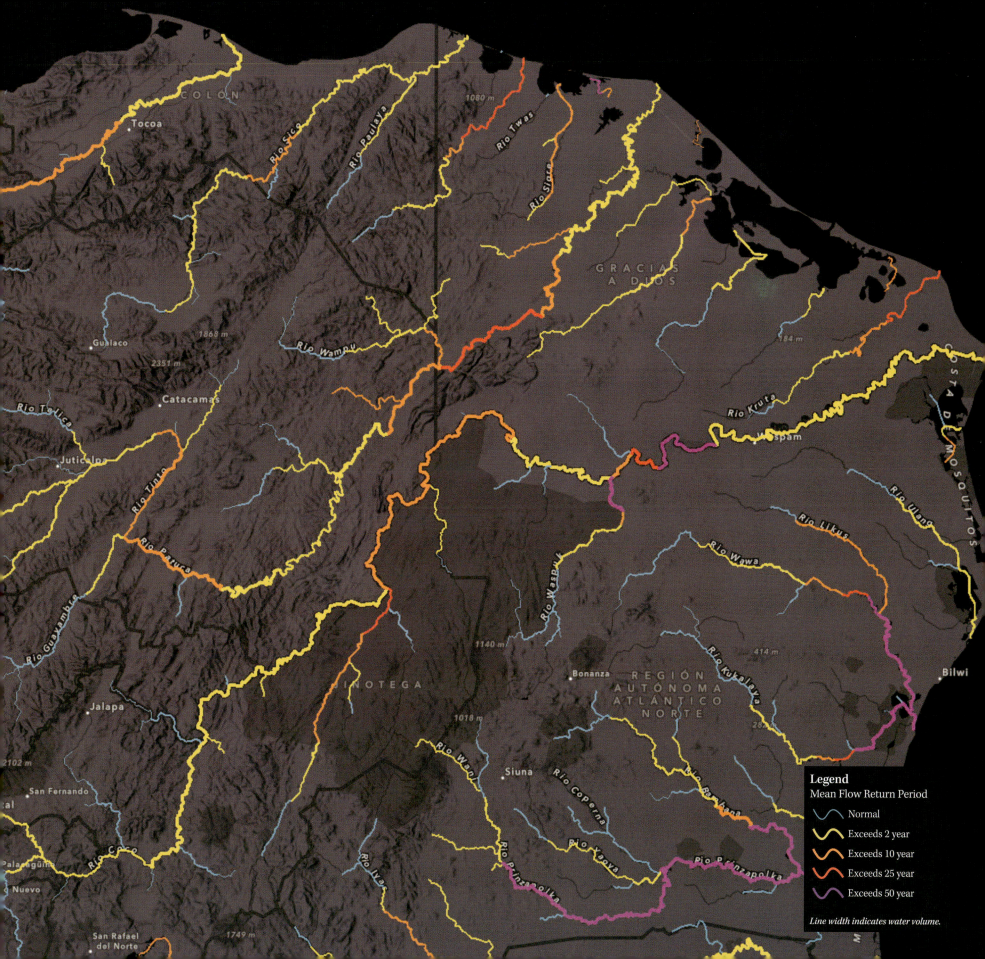

COLON

Tocoa

Rio Sico

Rio Paulaya

Rio Twas

Rio Sigra

1080 m

GRACIAS A DIOS

184 m

Gualaco

1868 m

2351 m

Rio WamPU

Catacamas

Rio Talica

Rio Kruta

Waspam

Juticalpa

Rio Tinto

Rio WasBuk

Rio Llikus

Rio Ulang

Rio Wawa

Rio Patuca

Rio Guayambre

COSTA DE MOSQUITOS

1140 m

Rio Kukalaya

414 m

Bonanza

REGIÓN AUTÓNOMA ATLÁNTICO NORTE

Bilwi

Jalapa

INOTEGA

1018 m

Rio Wani

Siuna

Rio Coperna

Rio Ra Aena

264

2102 m

San Fernando

Rio Kaova

al

Palanguin

Rio Coco

Rio Iyas

Rio Pinzapolka

Pio Prinzapolka

Nuevo

M

San Rafael del Norte

1749 m

**Legend**
Mean Flow Return Period

〰 Normal
〰 Exceeds 2 year
〰 Exceeds 10 year
〰 Exceeds 25 year
〰 Exceeds 50 year

*Line width indicates water volume.*

# CLIMATE DATA FOR THE GIS COMMUNITY

## Mapping a global network of weather stations

Charlie Frye, Esri

A global network of high-tech weather stations measures conditions every hour or even every few seconds—and stores and archives those measurements in databases. The data gathered by this network is the basis for how we know the normal daily temperatures and the record highs and lows. We also use this data to determine the best time to plant wheat or when to watch for frost or freezing conditions that could harm blooms on fruit trees. We depend on this data for many other reasons, including human health, cost of living, risk analysis of extreme weather, and more.

Historically, weather and climate data have been difficult to use in GIS. Often, weather data was only presented as tables, without station location information. With GIS, location is integrated with the weather and climate data, allowing each weather station to serve as an individual database.

As a result, annual and monthly summarizations of global climate records are starting to appear in GIS. This map and related graphs are examples of presenting National Oceanic and Atmospheric Administration (NOAA) historical climate data with GIS.

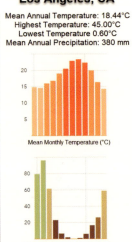

**Los Angeles, CA**

Mean Annual Temperature: 18.44°C
Highest Temperature: 45.00°C
Lowest Temperature 0.60°C
Mean Annual Precipitation: 380 mm

Mean Monthly Temperature (°C)

Mean Monthly Precipitation (mm)

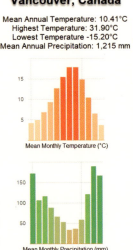

**Vancouver, Canada**

Mean Annual Temperature: 10.41°C
Highest Temperature: 31.90°C
Lowest Temperature -15.20°C
Mean Annual Precipitation: 1,215 mm

Mean Monthly Temperature (°C)

Mean Monthly Precipitation (mm)

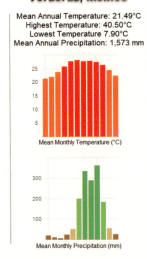

**Veracruz, Mexico**

Mean Annual Temperature: 21.49°C
Highest Temperature: 40.50°C
Lowest Temperature 7.90°C
Mean Annual Precipitation: 1,573 mm

Mean Monthly Temperature (°C)

Mean Monthly Precipitation (mm)

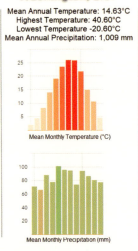

**Washington, D.C.**

Mean Annual Temperature: 14.63°C
Highest Temperature: 40.60°C
Lowest Temperature -20.60°C
Mean Annual Precipitation: 1,009 mm

Mean Monthly Temperature (°C)

Mean Monthly Precipitation (mm)

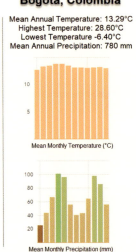

**Bogotá, Colombia**

Mean Annual Temperature: 13.29°C
Highest Temperature: 28.60°C
Lowest Temperature -6.40°C
Mean Annual Precipitation: 780 mm

Mean Monthly Temperature (°C)

Mean Monthly Precipitation (mm)

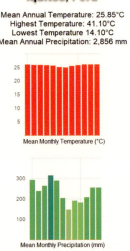

**Iquitos, Peru**

Mean Annual Temperature: 25.85°C
Highest Temperature: 41.10°C
Lowest Temperature 14.10°C
Mean Annual Precipitation: 2,856 mm

Mean Monthly Temperature (°C)

Mean Monthly Precipitation (mm)

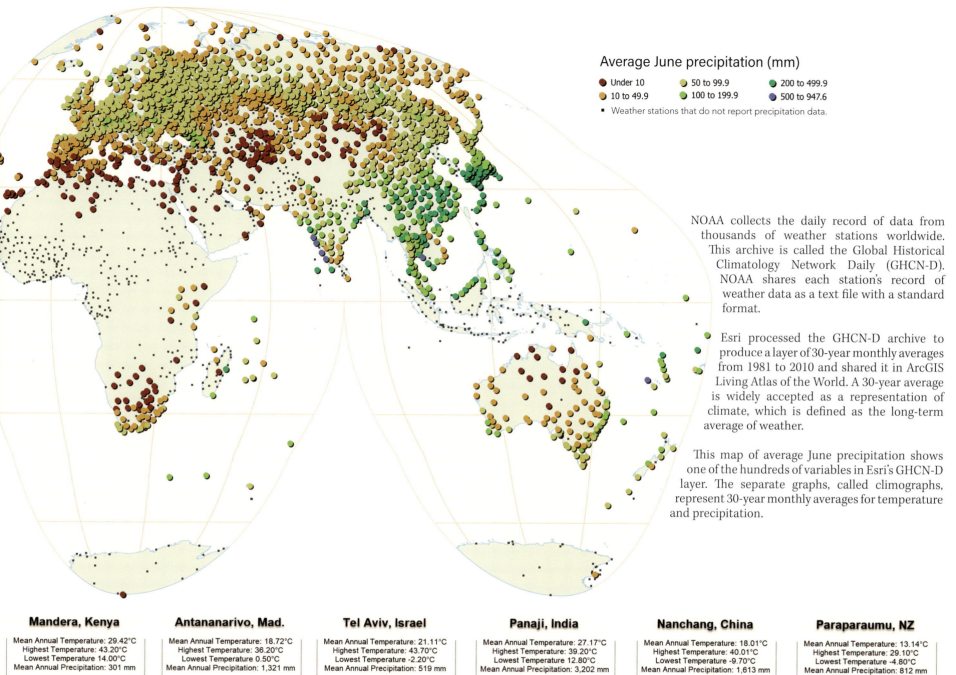

## Average June precipitation (mm)

- Under 10
- 10 to 49.9
- 50 to 99.9
- 100 to 199.9
- 200 to 499.9
- 500 to 947.6

▪ Weather stations that do not report precipitation data.

NOAA collects the daily record of data from thousands of weather stations worldwide. This archive is called the Global Historical Climatology Network Daily (GHCN-D). NOAA shares each station's record of weather data as a text file with a standard format.

Esri processed the GHCN-D archive to produce a layer of 30-year monthly averages from 1981 to 2010 and shared it in ArcGIS Living Atlas of the World. A 30-year average is widely accepted as a representation of climate, which is defined as the long-term average of weather.

This map of average June precipitation shows one of the hundreds of variables in Esri's GHCN-D layer. The separate graphs, called climographs, represent 30-year monthly averages for temperature and precipitation.

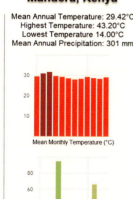

### Mandera, Kenya

Mean Annual Temperature: 29.42°C
Highest Temperature: 43.20°C
Lowest Temperature 14.00°C
Mean Annual Precipitation: 301 mm

Mean Monthly Temperature (°C)

Mean Monthly Precipitation (mm)

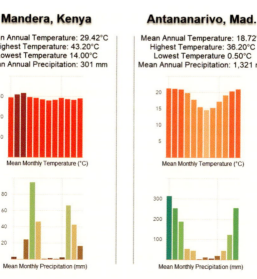

### Antananarivo, Mad.

Mean Annual Temperature: 18.72°C
Highest Temperature: 36.20°C
Lowest Temperature 0.50°C
Mean Annual Precipitation: 1,321 mm

Mean Monthly Temperature (°C)

Mean Monthly Precipitation (mm)

### Tel Aviv, Israel

Mean Annual Temperature: 21.11°C
Highest Temperature: 43.70°C
Lowest Temperature -2.20°C
Mean Annual Precipitation: 519 mm

Mean Monthly Temperature (°C)

Mean Monthly Precipitation (mm)

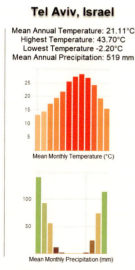

### Panaji, India

Mean Annual Temperature: 27.17°C
Highest Temperature: 39.20°C
Lowest Temperature 12.80°C
Mean Annual Precipitation: 3,202 mm

Mean Monthly Temperature (°C)

Mean Monthly Precipitation (mm)

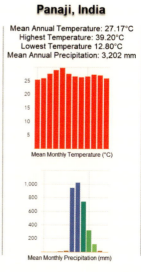

### Nanchang, China

Mean Annual Temperature: 18.01°C
Highest Temperature: 40.01°C
Lowest Temperature -9.70°C
Mean Annual Precipitation: 1,613 mm

Mean Monthly Temperature (°C)

Mean Monthly Precipitation (mm)

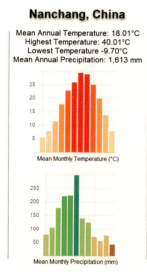

### Paraparaumu, NZ

Mean Annual Temperature: 13.14°C
Highest Temperature: 29.10°C
Lowest Temperature -4.80°C
Mean Annual Precipitation: 812 mm

Mean Monthly Temperature (°C)

Mean Monthly Precipitation (mm)

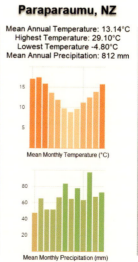

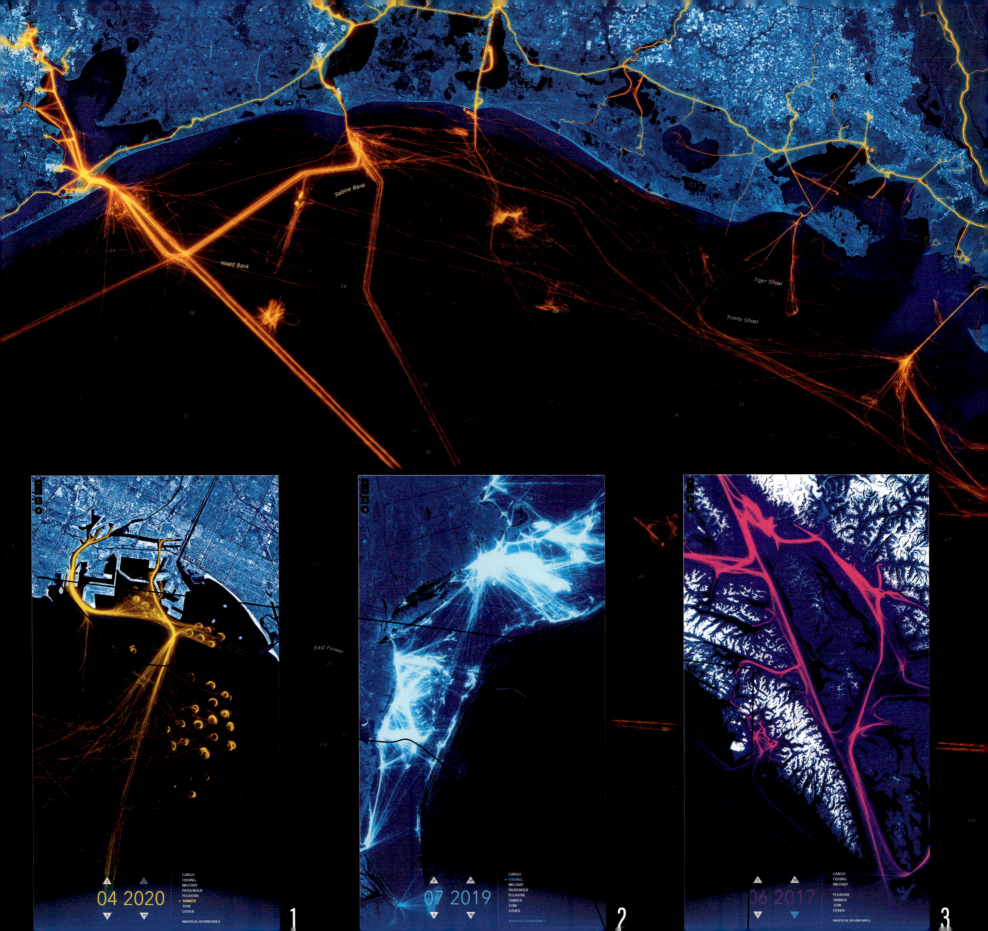

Sabine Bank

Heald Bank

Tiger Shoal

Trinity Shoal

East Flower

04 2020

CARGO
FISHING
MILITARY
PASSENGER
PLEASURE
TANKER
TOW
OTHER

1

07 2019

CARGO
FISHING
MILITARY
PASSENGER
PLEASURE
TANKER
TOW
OTHER

NAUTICAL BOUNDARIES

2

06 2017

CARGO
FISHING
MILITARY
PLEASURE
TANKER
TOW
OTHER

NAUTICAL BOUNDARIES

3

# VISUALIZING VESSEL TRAFFIC

## Analyzing spatial patterns and trends over time

Keith VanGraafeiland and John Nelson, Esri

Visualizing the where and when of vessel traffic provides an invaluable resource for understanding maritime space. The Automatic Identification System (AIS) included in ArcGIS Living Atlas of the World is just such a resource. AIS supports marine radar in allowing vessels to broadcast their locations to each other and avoid collisions. AIS helps assemble vast amounts of information to help identify and resolve potential space-use conflicts.

Using the Create Big Data Connection and Reconstruct Tracks geoprocessing tools, ArcGIS Pro users can connect to and analyze massive data sources, such as AIS data. For example, comma-separated value (CSV) files of raw vessel traffic positions and attributes from marine cadastre AIS archives can be aggregated into manageable and highly visual vessel track information products. Vessel tracks are crucial for additional analysis and understanding of key users in maritime space.

Increasing access to vessel traffic information gives specialists, researchers, and general users manageably sized and geographically and temporally specific vessel traffic information.

The U.S. Vessel Traffic application allows users to download AIS products for analysis and discovery. The Vessel Traffic application is organized into different vessel groups: cargo, fishing, military, passenger, pleasure, tanker, tow, and a catch-all "other" category. Users can browse these categories by date and time and download predefined areas at various geographic scales as a file database for further spatial analysis.

Visualizing AIS data can reveal the spatial patterns and trends of vessel traffic over time. For example, ringed patterns near bustling ports trace the locations of waiting cargo and tanker ships anchoring offshore. Fishing vessels in search of seasonal catches can reveal the migratory patterns of fish species and the otherwise unseen underwater terrain. The network of maritime transportation among coastal communities shows the connections of human relationships and economy, in some cases echoing the prevailing forces of national governance and epidemiology.

*Thanks to MarineCadastre.gov, the U.S. Coast Guard, National Oceanic and Atmospheric Administration, and the Bureau of Ocean Energy Management.*

*This map shows patterns and trends of tow vessel traffic 1) near the Port of Long Beach, California, 2) fishing vessel traffic along the northeastern coastline of the United States, and 3) passenger vessel traffic along the Gulf of Alaska.*

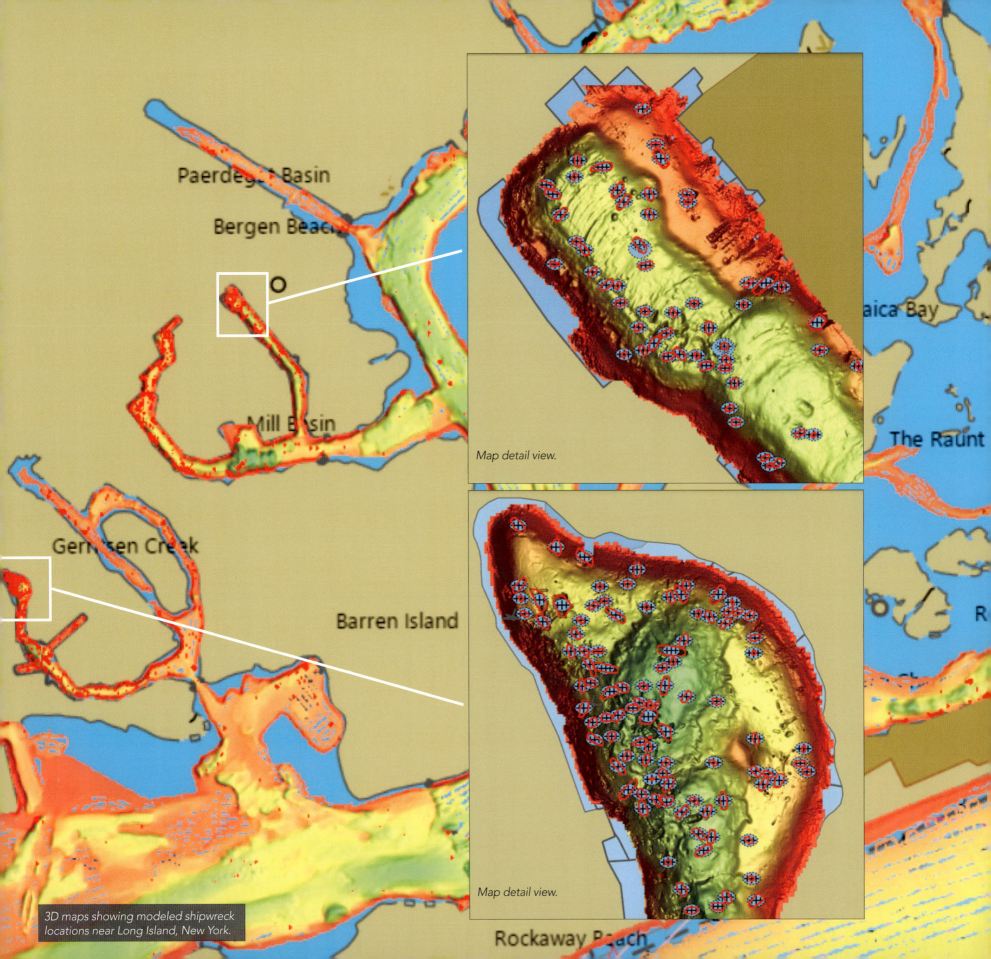

Paerdeget Basin

Bergen Beach

Mill Basin

Gerritsen Creek

Barren Island

Jamaica Bay

The Raunt

Map detail view.

Map detail view.

3D maps showing modeled shipwreck locations near Long Island, New York.

Rockaway Beach

# REVEALING SUNKEN SHIPS WITH GEOAI

## Deep learning detection extracts shipwreck locations

Madhu Hosuru, Vinay Viswambharan, Guneet Matreja, and Craig Greene, Esri

Geospatial artificial intelligence (GeoAI) is the combination of using location intelligence and GIS with the deep learning capabilities of AI. While GeoAI is often used for object and image classification, it can also be used for object detection, such as finding shipwrecks at the bottom of the ocean. The benefits of locating thousands of shipwrecks include:

- Updating navigational hazards for maritime mapping and charting
- Preserving maritime heritage, such as the National Oceanic and Atmospheric Administration database of shipwrecks
- Supporting programs such as the Remediation of Underwater Legacy Environmental Threats project, which identifies shipwreck locations and response efforts
- Introducing shipwrecks as potential artificial reefs to benefit ocean ecosystems

ArcGIS Pro deep learning–based object detection models were used to extract shipwreck information from bathymetric survey data and automate the task of detecting shipwrecks. Deep learning is a type of machine learning that relies on multiple layers of nonlinear processing for feature identification and pattern recognition described in a model.

The survey data was in Bathymetric Attributed Grid (BAG) format. BAG data is two-band imagery in which one band represents elevation and the other band represents uncertainty in the elevation values.

The model was trained on a ResNet-50 backbone containing more than a million images, 50 layers deep, and set for 80 epochs, the number of times the dataset is passed forward and backward through the neural network. Model training achieved an average precision score of 0.94, meaning the model was well trained.

The resulting trained model and model definition file were used to perform inferencing. In this case, the inference process searched for and identified shipwrecks in the BAG dataset.

Finally, the ArcGIS shaded relief function was applied, as shown in this map and inserts, to provide a 3D representation of the terrain and differentiate shipwrecks from the background.

# THE ART OF FREQUENCY AND PREDOMINANCE

## Mapping weather advisories, watches, and warnings

Emily Meriam and Kevin Butler, Esri

How frequently does the threat of severe weather occur in your neighborhood? The answer to this question is not only helpful, but it can help identify areas where "warning and watch fatigue" may be occurring. Experiencing too many notifications about weather warnings and watches raises the risk of people becoming desensitized. NOAA concluded that this desensitization might have contributed to the deaths of

more than 150 people from the 2011 Joplin, Missouri, EF-5 tornado. The majority of Joplin residents did not immediately seek shelter when the tornado warning was issued. In the United States, the National Weather Service (NWS) provides weather forecasts, which can include watches and warnings of hazardous weather to the public for protection, safety, and general information. Watches are issued when conditions

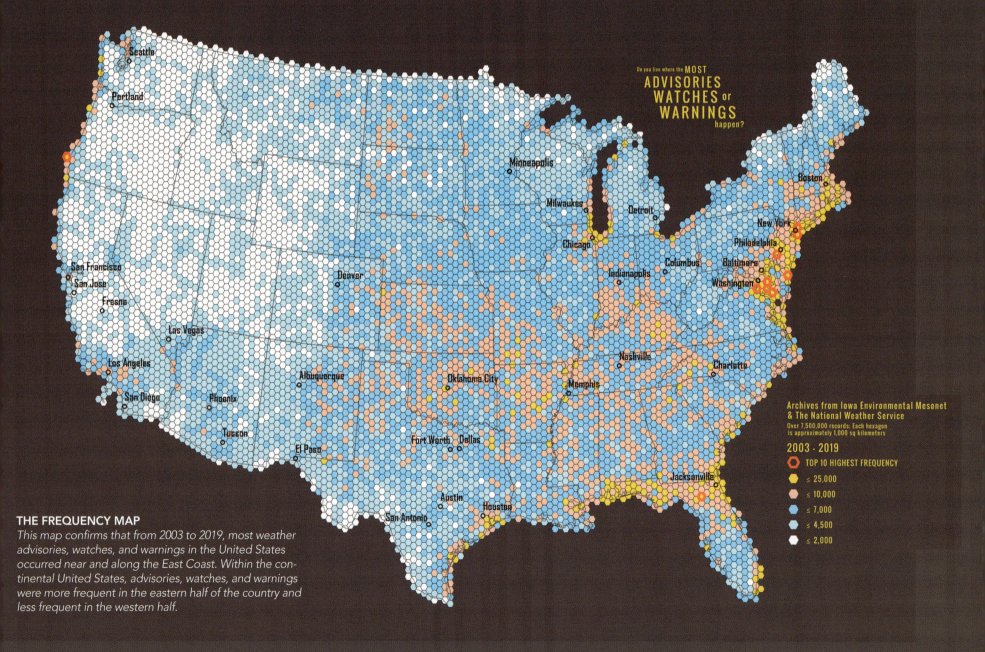

Do you live where the MOST **ADVISORIES WATCHES** or **WARNINGS** happen?

Archives from Iowa Environmental Mesonet & The National Weather Service

Over 7,500,000 records. Each hexagon is approximately 1,000 sq kilometers

**2003 - 2019**

- ⬤ TOP 10 HIGHEST FREQUENCY
- ⬤ ≤ 25,000
- ⬤ ≤ 10,000
- ⬤ ≤ 7,000
- ⬤ ≤ 4,500
- ⬤ ≤ 2,000

### THE FREQUENCY MAP

*This map confirms that from 2003 to 2019, most weather advisories, watches, and warnings in the United States occurred near and along the East Coast. Within the continental United States, advisories, watches, and warnings were more frequent in the eastern half of the country and less frequent in the western half.*

for a specific type of severe weather are favorable or expected but not occurring or imminent. Warnings are issued when conditions conducive to a particular kind of severe weather are occurring or impending. These warnings and watches are the primary means of communicating the risk of severe weather to the public.

In analyzing data from January 1, 2003, through July 31, 2019, our goal was to find areas that have high numbers of watches and warnings. The typical GIS workflow would be to intersect the 7.5 million storm watch and warning polygons. The solution: create a grid of hexagons (approximately 10,000 hexagons, each one about 1,000 km2) covering the United States, then intersect the hexagon grid with the 7.5 million warning and watch polygons using the ArcGIS Pairwise Intersect tool. Pairwise Intersect distributes the intersection work across multiple logical computer cores and employs an efficient

overlay algorithm. The final step is to use the Sort and Summary Statistics tools to calculate the number of warning and watch polygons under each hexagon and the predominant warning or watch event type.

Because the NWS issues 19 different advisories, watches, and warnings, using an intuitive color palette is necessary (e.g., winter events are blue, wind events are orange and pink, and floods are mostly in muddy water hues). These colors must look cohesive and yet differentiate from one another without individually dominating the map (e.g., dense fog, represented by the color purple, must not appear to outcompete high surf, represented by the color turquoise). This need is particularly true with severe thunderstorms. A significant portion of the map has this predominance, but by selecting a navy blue instead, the color is present on the map without dominating it.

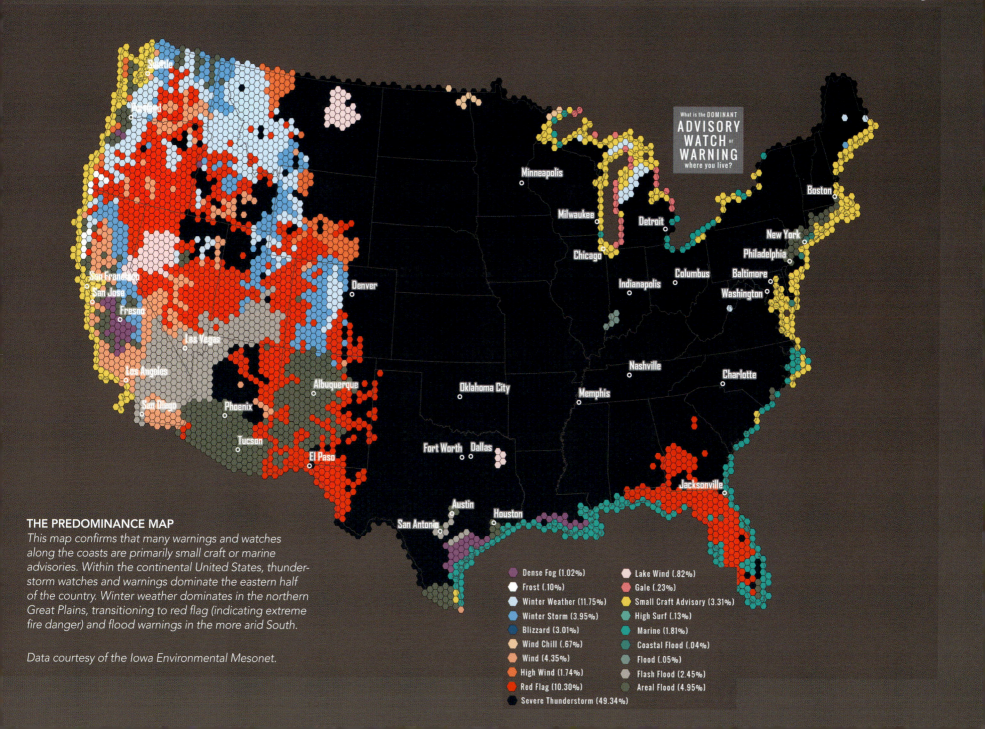

What is the DOMINANT
ADVISORY
WATCH or
WARNING
where you live?

THE PREDOMINANCE MAP
*This map confirms that many warnings and watches along the coasts are primarily small craft or marine advisories. Within the continental United States, thunderstorm watches and warnings dominate the eastern half of the country. Winter weather dominates in the northern Great Plains, transitioning to red flag (indicating extreme fire danger) and flood warnings in the more arid South.*

*Data courtesy of the Iowa Environmental Mesonet.*

- Dense Fog (1.02%)
- Frost (.10%)
- Winter Weather (11.75%)
- Winter Storm (3.95%)
- Blizzard (3.01%)
- Wind Chill (.67%)
- Wind (4.35%)
- High Wind (1.74%)
- Red Flag (10.30%)
- Severe Thunderstorm (49.34%)
- Lake Wind (.82%)
- Gale (.23%)
- Small Craft Advisory (3.31%)
- High Surf (.13%)
- Marine (1.81%)
- Coastal Flood (.04%)
- Flood (.05%)
- Flash Flood (2.45%)
- Areal Flood (4.95%)

# UNDERSTANDING THE PATTERNS OF COVID-19

## Clustering and detecting outliers in confirmed cases

Jie Liu, Cheng-Chia Huang, and Xiaodan Zhou, Esri

On January 22, 2020, King County, Washington, reported the first confirmed case of coronavirus disease 2019 (COVID-19) in the United States. Since then, several waves of outbreaks have surged across the country and around the world. Given the extreme risk to public health, local health departments collect detailed information on the number of confirmed cases, hospitalizations, and recovery rates. Using these data together with ArcGIS Pro, we can explore how the disease spreads in space and time and potentially forecast its future pattern.

### Data preparation

For each county in the contiguous United States, we acquired the cumulative number of confirmed COVID-19 cases from January 22, 2020, to December 13, 2020, from USAFacts, a not-for-profit provider of policy, economic, and demographic data. We converted the raw data to cumulative cases per 100,000 residents and daily new cases per 100,000 residents. To account for the lack of reporting on weekends in some locations, we calculated the seven-day average of daily new cases and removed the first six days of the original data.

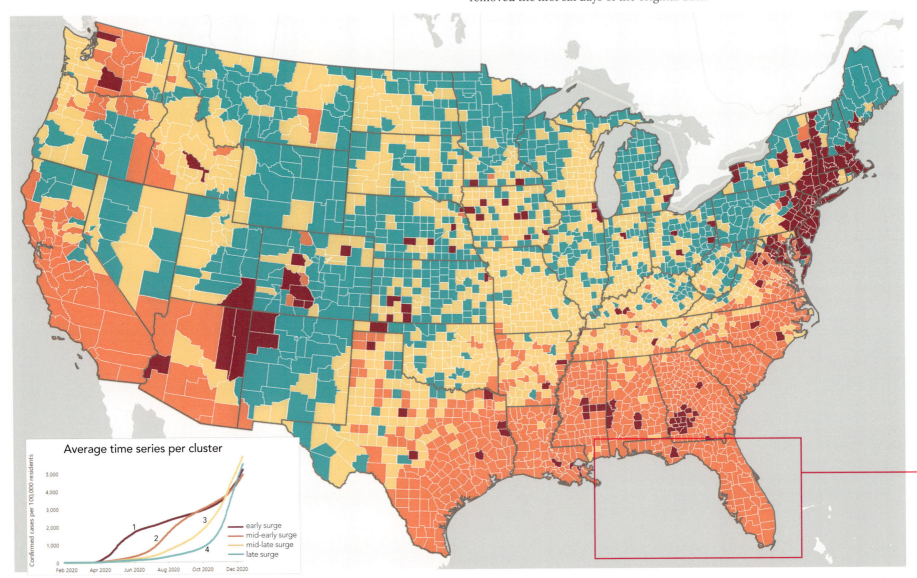

*Figure 1: This map shows four distinct temporal clusters of confirmed COVID-19 cases per 100,000 residents for the contiguous United States from January 22, 2020, to December 13, 2020: 1) early surge, 2) mid-early surge, 3) mid-late surge, and 4) late surge. The inset graph shows confirmed cases per 100,000 residents trending over the same time period.*

## Finding patterns of COVID-19 spread

Identifying temporal clusters of COVID-19 cases across the United States can help us understand which areas have had similar disease patterns or intervention strategies. The ArcGIS Time Series Clustering tool can easily find these clusters, a difficult task for the human eye. Using this tool, we grouped the U.S. counties based on their time-series profiles. Counties with similar patterns of increase or decrease in their cumulative cases per 100,000 appear in the same group (figure 1).

## Forecasting COVID-19 surges

The ArcGIS Time Series Forecasting toolset provides three ways to model the future spread of the disease. The Curve Fit Forecast tool applies simple linear, parabolic, exponential, or S-shaped curves to the time series. The Exponential Smoothing Forecast tool can incorporate seasonal patterns into the forecasts by decomposing the time series with season and trend. The Forest-Based Forecast tool uses machine learning to train and forecast the time series with moving time windows. We applied all three forecasting methods to the data and used the Evaluate Forecasts by Location tool to find the optimal forecast at each location.

## Detecting outliers in COVID-19 reporting

To explore extraordinarily high or low reports of new COVID-19 cases per 100,000 residents, we used the detect outlier option of the Time Series Forecasting tools and several visualization techniques (see figures 2, 3, and 4). The charts revealed several outliers in Miami-Dade County, Florida. Extraordinary high numbers of new confirmed cases were reported on August 12, September 1, and November 27, 2020. Low numbers of cases were reported on August 3 and November 26, 2020. Focusing on Florida, we generated a bar chart to show the counts of each type of outlier on each date that revealed many counties experienced similarly low reports of new cases on November 26, 2020, but high reports on November 27, 2020, likely related to the Thanksgiving holiday.

Spatial data scientists have been at the forefront of modeling and understanding the virus's spread and impacts. The Time Series Clustering and Time Series Forecasting tools available in ArcGIS Pro provide multiple traditional statistical and machine learning approaches that can help understand the COVID-19 pandemic.

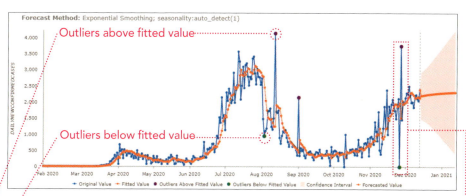

*Figure 3:* This chart highlights the outliers with extremely high values (in purple) and low values (in green) in Miami-Dade County, Florida.

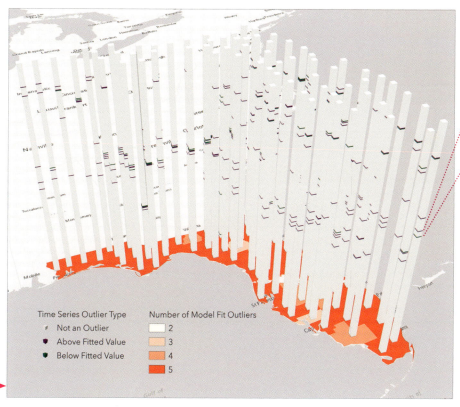

*Figure 2:* A 3D view of time-series outliers of new daily COVID-19 reports for Florida counties in 2020.

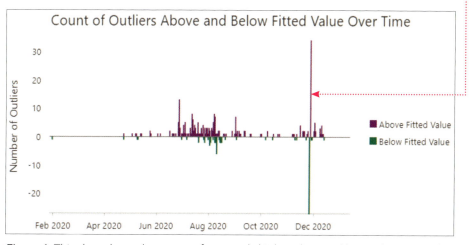

*Figure 4:* This chart shows the counts of extremely high outliers and low outliers on each date for all Florida counties.

# MONITORING GLOBAL SNOW COVER

## Using multidimensional raster analysis

Hong Xu and Nawajish Noman, Esri

Snow cover is a critical component of the climate system and the hydrological cycle because it regulates the exchange of heat between Earth's surface and the atmosphere. Studies show that the changing climate and increasing temperature have significantly decreased snow cover, leading to a warmer planet. Since changes in snow cover affect the ecosystem and access to water resources, the decrease could severely impact our lives.

Monitoring the extent and duration of snow cover provides valuable information to planners, decision-makers, and stakeholders. The Moderate Resolution Imaging Spectroradiometer (MODIS) snow cover data from the National Snow and Ice Data Center (NSIDC) provides long-term time-series analysis and understanding of the climate system. This study used Cloud Raster Format (CRF) data in ArcGIS Pro to analyze snow cover data over 20 years to monitor global, seasonal, and regional trends.

CRF is a raster format optimized for parallel and distributed computation in ArcGIS. The advantages of using CRF for multidimensional raster analysis in ArcGIS over other scientific data formats include simultaneous reading and writing, parallel and distributed computing, efficient extraction and processing of time series data through the creation of transpose, efficient compression using Limited Error Raster Compression (LERC) while maintaining accuracy, and the ability to store analysis templates for on-the-fly analysis.

Using MODIS monthly global snow cover data from 2000 to 2019, the study combines multidimensional CRF and multidimensional raster analysis tools in ArcGIS Pro to map the snow cover percentage and seasonal snow cover percentage data of all years to analyze the spatial variation and trend of snow cover within these years.

The snow cover data is available in Hierarchical Data Format (HDF). A mosaic dataset is created using the Snow_Cover_Monthly_CMG variable from 234 HDF files. A raster function is added to the mosaic dataset to extract only the monthly snow cover percentage and to mask out other categories such as cloud, night, and water as NoData. The mosaic dataset is converted to a multidimensional CRF and then used for further analysis. The overall and the seasonal snow cover percentages of the entire study period are derived using the Aggregate Multidimensional Raster tool. The yearly snow cover trend along with the corresponding statistics are calculated using the Generate Trend Raster and Zonal Statistics as Table tools, and raster functions.

The snow cover percentage map shows that most of the snow coverage is in the Northern Hemisphere. It occurs more frequently above 45 degrees latitude, whereas below 45 degrees, only the tall mountains such as the Himalayas in the Tibet Plateau, Mount Ararat in the Anatolian Plateau, and the Rocky Mountains in western North America experienced frequent snow coverage.

The seasonal characteristic of snow cover can be clearly observed by the seasonal maps of the Himalayas, where winter is January–March, spring is April–June, summer is July–September, and fall is October–December.

*A 3D representation of snow cover in the Himalayas during winter. Darker blue colors indicate a higher percentage of snow cover.*

*A 3D representation of snow cover in the Himalayas during spring. Darker blue colors indicate a higher percentage of snow cover.*

*Data provided by Hall, D. K., and G. A. Riggs. 2015. MODIS/Terra Snow Cover Monthly L3 Global 0.05Deg CMG, Version 6. Boulder, Colorado, USA, and NASA National Snow and Ice Data Center Distributed Active Archive Center.*

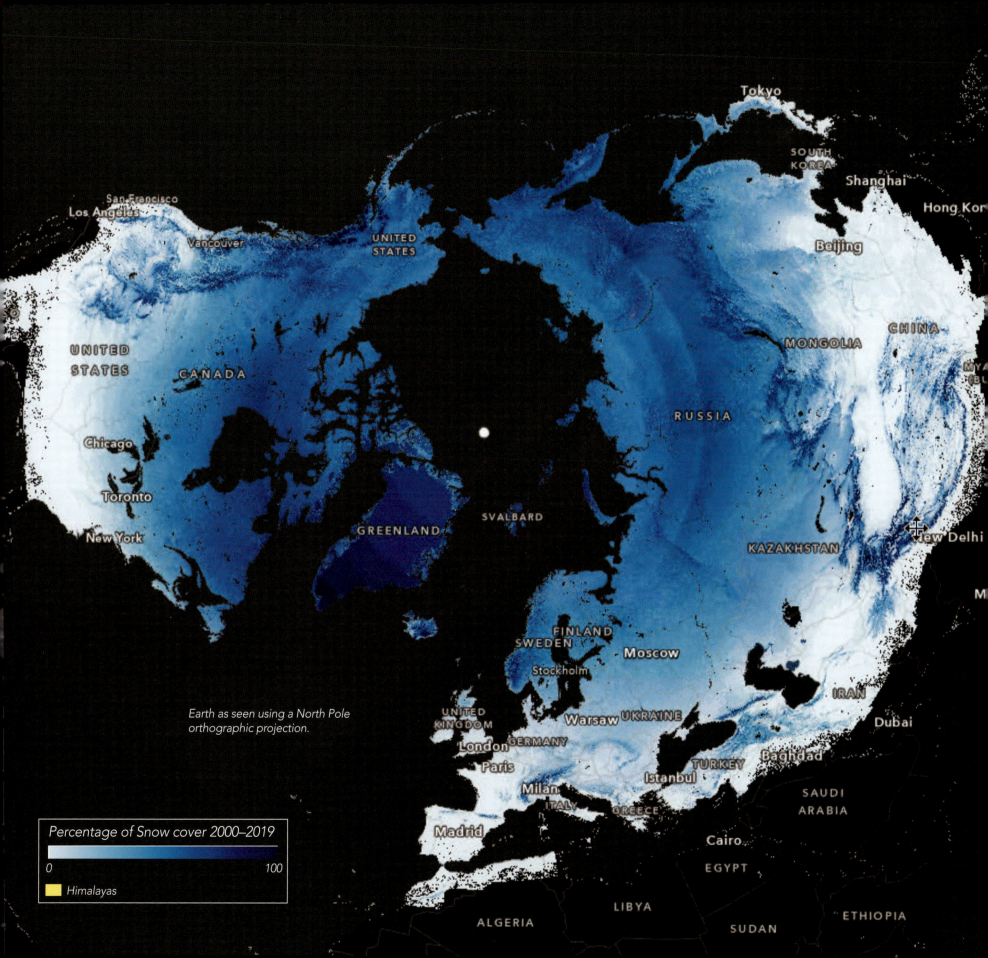

Percentage of Snow cover 2000–2019

0                    100

Himalayas

*Earth as seen using a North Pole orthographic projection.*

# PEOPLE FOR THE PEOPLE

## A new way to engage with politics

Whitney Kotlewski and Raynah Kamau, Esri

People for the People (P4TP) is a GIS community focused on educating and empowering individuals to engage with civic issues. The P4TP team is made up of volunteers from diverse backgrounds, ethnicities, thought processes, and even political party affiliations. This collective of about 150 individuals answered a call to action led by Black Girls M.A.P.P., which connects and empowers women of color in the field of GIS, and came together to provide information that would impact voters in the 2020 election and the broader GIS community.

Leading up to the 2020 U.S. presidential election, P4TP used GIS to reimagine how we engage with politics and politicians by breaking down the complexities of politics into digestible maps, apps, and stories that live on after the 2020 election. P4TP created geospatial solutions to recenter and focus political issues on people—our friends, family, and neighbors—while elevating untapped communities.

Among the deliverables P4TP made were 34 web maps, 19 stories created with ArcGIS® StoryMaps, 8 apps created with ArcGIS® Dashboards and ArcGIS® Web AppBuilder, 13 pages created using ArcGIS® Hub℠, and 7 apps created with the ArcGIS® Experience Builder application.

When identifying how communities can use GIS as an analytical tool, the team explored ways to analyze sentiment data and created two exemplary products to visualize sentiments related to social and political activity during the election.

## Sentiment mapping

The sentiment mapping application by P4TP brings together ArcGIS® API for JavaScript™, ArcGIS® Survey123, and Experience Builder to help crowdsource public feelings on sensitive social issues. The apps showcase a quantitative analysis of public feelings via a word cloud and provide additional context to explain those feelings in the map's pop-ups. Through charting, the app highlights the distribution of race to provide additional insight into the positioning of those responses. The responses are grouped to emphasize shared sentiments about social issues across the nation.

> "Educating and empowering people
> to engage with civic issues"
>
> —People4thepeople.org

## Sentiment analysis

In another example of visualizing sentiment data, P4TP used Twitter's API, ArcGIS Dashboards, and Python to help people stay current on the tweets of candidates in their area. This capability is important in the political landscape because many representatives use social media—Twitter especially—as a way to connect with and relate to their (potential) constituencies. In turn, these social media posts can give community members a more personal view of a candidate's priorities and affiliations that are not immediately apparent on a campaign website.

The Twitter sentiment application allows users to answer questions such as "Which district in my state represents the majority of Twitter activity?" or "Which party seems to dedicate more time to broadcasting their opinion on Twitter?" From there, community members can explore how a specific district, party, or candidate has appeared on Twitter during the past week.

## Leading by example

P4TP approached its work with the intention to empower others. The team remained steadfast for two months, building information products with hopes of empowering communities through GIS. Their work modeled what unity looks like when people work together, appreciate their differences, and get involved for positive change.

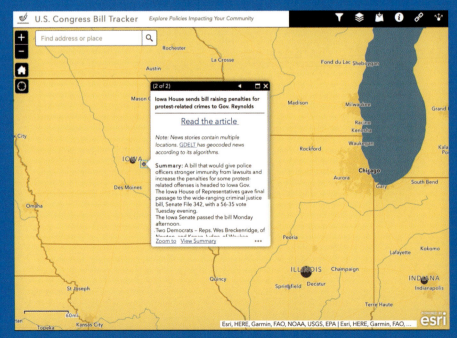

The U.S. Congress Bill Tracker app helps users monitor multiple news stories by location about legislation and political activities, such as protests.

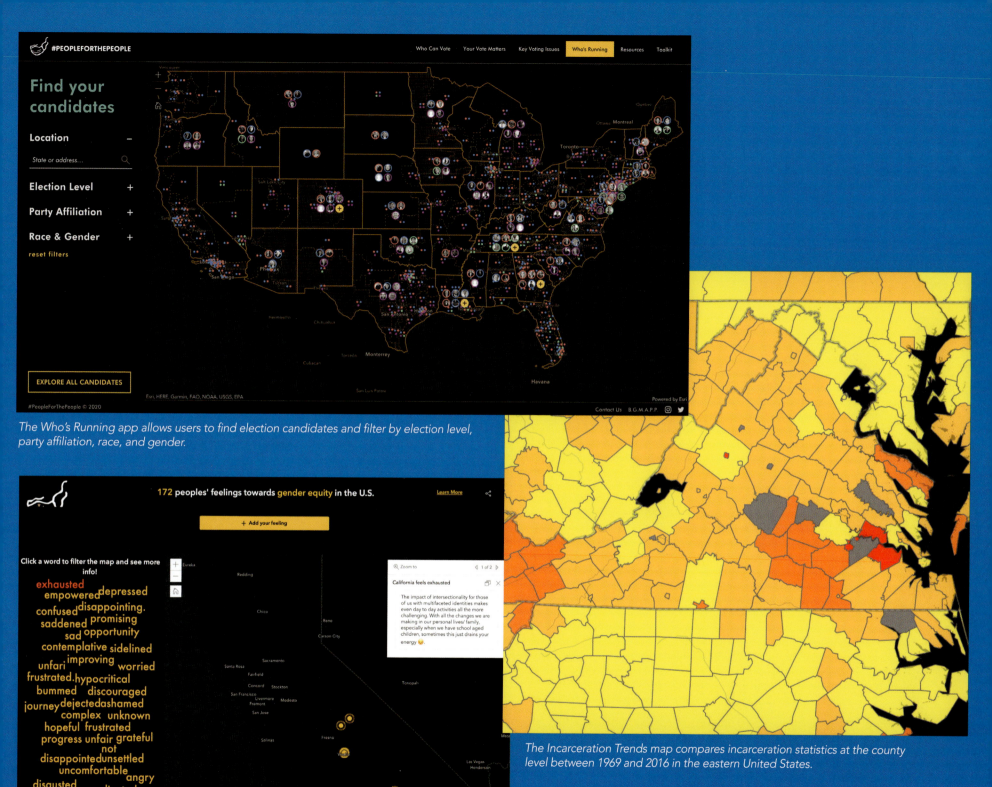

The Who's Running app allows users to find election candidates and filter by election level, party affiliation, race, and gender.

The Incarceration Trends map compares incarceration statistics at the county level between 1969 and 2016 in the eastern United States.

The Twitter Sentiment app shows how people are feeling about gender equity in California.

# ANALYZING GLOBAL WATER QUALITY OVER TIME

## A blueprint for Sustainable Development Goals

Emily Smail, GEO Blue Planet; Dany Ghafari, UNEP; and Keith VanGraafeiland, Esri

Eutrophication is a process driven by nutrient enrichment of water, especially compounds of nitrogen or phosphorus, according to the European Commission. The process leads to increased growth, primary production, and biomass of algae, resulting in adverse changes in the balance of organisms and water quality. Monitoring changes in chlorophyll provides information about biomass changes that may be related to nutrient enrichment. The United Nations Environment Program (UNEP), GEO Blue Planet, and Esri partnered to develop a new chlorophyll index to support UN Sustainable Development Goal (SDG) Target 14.1, which aims to prevent and significantly reduce marine pollution of all kinds, particularly from land-based activities, including marine debris and nutrient pollution. The statistical approach uses satellite data and GIS.

## Why is SDG Target 14.1 important?

Many countries depend on their coastal ecosystems to drive essential sectors of their economies, such as tourism, fisheries, and natural resources. These economies provide sustainable food to their populations. Fertilizers and other chemicals used on land disperse into the near-coastal ocean and cause blooms in marine algae that disrupt ecosystems and human health. For instance, some of these algal blooms release toxins that kill fish and other marine life and significantly impact humans with breathing difficulties. Other types of algal blooms may result in dead zones, where oxygen becomes so depleted in the water that marine life suffocates and dies. SDG Target 14.1 is part of a larger goal that aims to conserve and sustainably use the ocean, seas, and marine resources. In all, the United Nations has adopted 17 SDGs that aspire to address climate change, improve health and education, reduce inequality, and achieve world peace and prosperity.

## Knowledge transfer

Using ArcGIS Pro, a workflow was developed to globally identify and quantify the number and severity of eutrophication events in nearshore waters. A group of ocean color and statistics experts formulated the initial workflow. They provided the data and proposed a method for quantifying eutrophication based on satellite-derived chlorophyll measurements. Esri provided insight on how to scale and execute its method using GIS. This experience is a textbook example of collaborative science powered by GIS, more directly involving Esri in the SDG process.

The United Nations adopted and publicized this GIS-based method in the most recent update of the *Global Manual on Ocean Statistics*.

## Repeatable science

ArcGIS Pro is effective for creating a repeatable and understandable workflow essential for good science. Esri turned to a visual programming language within ArcGIS Pro called ModelBuilder to build geoprocessing workflows that automate and document spatial analysis and data management processes. The program helped Esri produce and report results for sub-indicator one and sub-indicator two. Sub-indicator one focuses on annual reporting, beginning in 2005, and compares each year to a baseline to identify potentially anomalous eutrophication events or deviations from the baseline. Sub-indicator two focuses on monthly reporting and classifies anomalous values as moderate, high, or extreme based on the 90th, 95th, and 99th percentiles, respectively. Results are broken down for exclusive economic zones in each coastal country to understand how chlorophyll-a deviations and anomalies (as a proxy for water quality) change over time.

## Community engagement

The project uses ArcGIS Hub, a community engagement platform that organizes people, data, and tools through information-driven initiatives to disseminate the results and engage with the community. Collaborative team members manage and update the ArcGIS Hub site to keep the public informed as the project progresses and results are published.

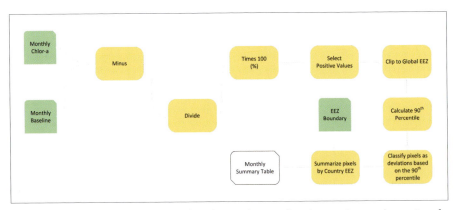

*This ArcGIS Pro workflow globally identifies and quantifies the number and severity of eutrophication events in nearshore waters.*

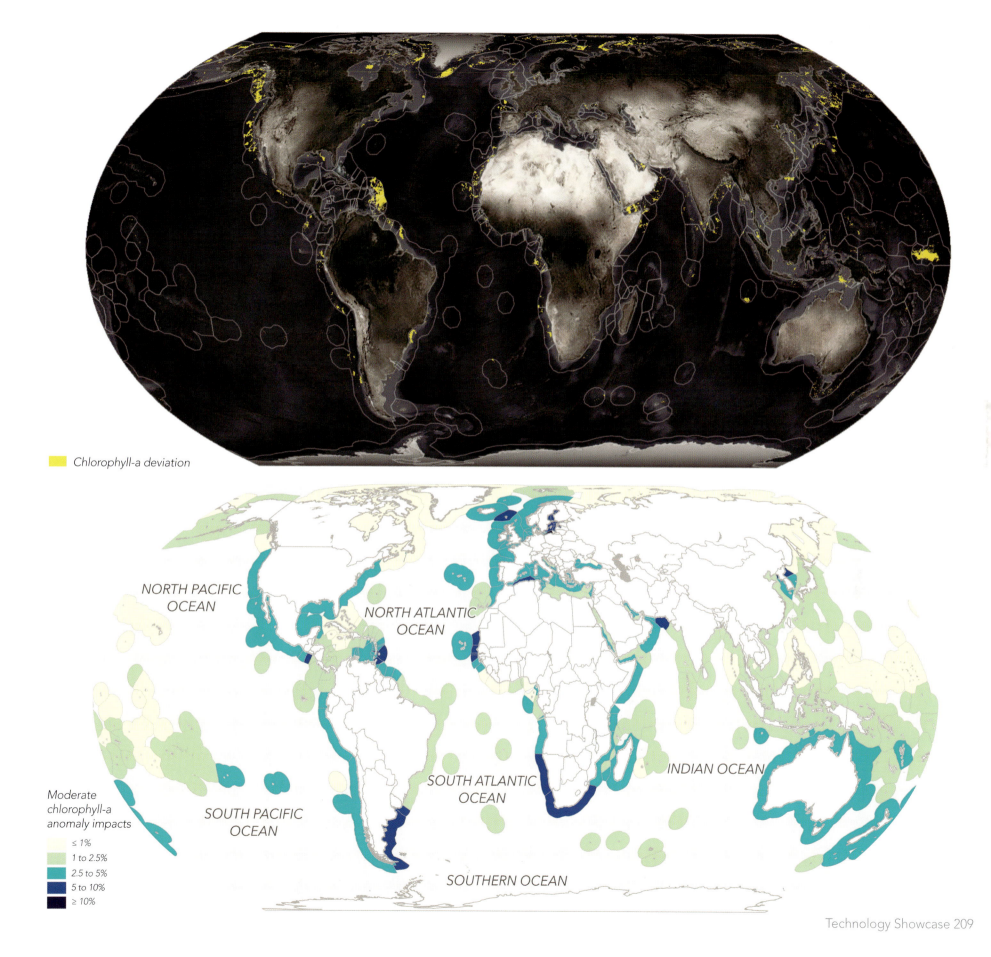

Chlorophyll-a deviation

NORTH PACIFIC
OCEAN

NORTH ATLANTIC
OCEAN

SOUTH PACIFIC
OCEAN

SOUTH ATLANTIC
OCEAN

INDIAN OCEAN

SOUTHERN OCEAN

Moderate
chlorophyll-a
anomaly impacts

≤ 1%
1 to 2.5%
2.5 to 5%
5 to 10%
≥ 10%

# GROWING DEGREE DAY MODELS
## Forecasting the future of viticulture

Julia Lenhardt and Elvis Takow, Esri

Observed climate change is impacting food security because of rising temperatures, changing precipitation patterns, and more frequent extreme weather events. The vulnerability of crop health to climatological variables requires adequate tools for forecasting those variables so that our planet can support food systems and the livelihoods that depend on them. The increasing use of growing degree day (GDD) models helps agricultural experts understand more about biological processes in plants and their response to rising temperatures.

## GDD models

GDD models are an integral tool in understanding plant phenology. All GDD models implicitly assume that plant development directly relates to time and temperature. As a result, numerous bioclimatic indices can measure crop suitability and are mostly developed based on climatic variables. GDD is historically and currently the most used measure of climatic suitability for viticulture, the study of grape cultivation.

GDD is calculated as the average of the daily minimum and maximum temperatures compared to a base temperature, which is assumed to be the minimum temperature at which plant growth occurs. In the case of grapevines, the base temperature is typically set to 50°F/10°C:

$$GDD = \frac{Tmax + Tmin}{2} - 10$$

## Grapevine suitability regions

The idea of a heat summation above a base temperature defining grapevine growth and grape maturation was first observed by Swiss botanist Augustin Pyramus de Candolle and elaborated on by Amerine and Winkler (1944). As such, an index of heat summation for California was developed that is now widely used as a guide for selecting appropriate grape varieties and for determining a given area's suitability to produce quality wine grapes.

The index is calculated by summing the GDDs for the period of April 1 through October 31 in a year. This time frame represents the growing season of grapevines in the Northern Hemisphere. Amerine and Winkler used this index to define five climatic zones, or Winkler regions, for California, with recommended grape varieties suitable for each region.

## Forecasting GDD

Daymet minimum and maximum daily temperature data, produced by the Oak Ridge National Laboratory, extends from January 1, 1980, to the current calendar year for North America. Once stored in a multidimensional raster, temperature data can be used in trend and predictive analysis tools in ArcGIS Pro. Linear trend analysis of the existing temperature data is used to forecast minimum and maximum daily temperatures for the year 2050. A heat summation index is calculated for the period of April 1 through October 31 using the GDD formula. With the ability to forecast future GDD distribution, the predictive analysis tools can help categorize the data into the five regions defined by Amerine and Winkler to generate a map of Winkler regions in 2050.

## Comparing the past to the future

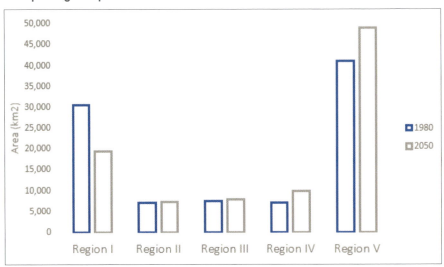

A comparison of Winkler regions from 1980 to the forecast regions in 2050 shows a significant loss of Region I, particularly in Northern California. Meanwhile, the warmest region is expected to grow northward and toward the coast. Regions II, III, and IV also show small increases. Therefore, grape varieties suitable to Region I are less likely to be healthy or profitable in California in 2050.

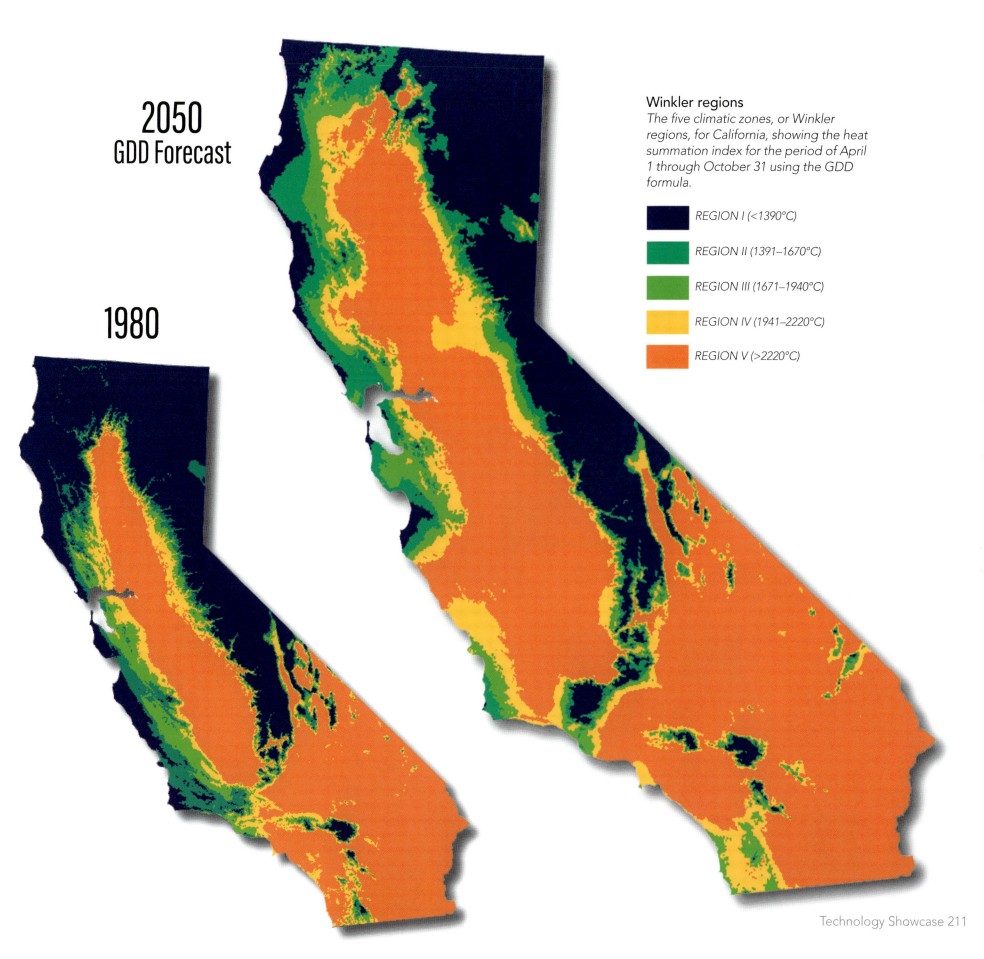

## 2050
### GDD Forecast

## 1980

**Winkler regions**
*The five climatic zones, or Winkler regions, for California, showing the heat summation index for the period of April 1 through October 31 using the GDD formula.*

- *REGION I (<1390°C)*
- *REGION II (1391–1670°C)*
- *REGION III (1671–1940°C)*
- *REGION IV (1941–2220°C)*
- *REGION V (>2220°C)*

# INTERACTIVE SUITABILITY MODELING

## A new approach using ArcGIS Pro Suitability Modeler

Kevin Johnston, Esri

Suitability modeling is a type of GIS analysis used to determine the best locations for something based on a set of spatial criteria and statistical parameters. With ArcGIS Pro Suitability Modeler, analysis becomes a flexible, interactive process in which the user adds and removes criteria and adjusts suitability parameters at any time to see how those changes affect the resulting model.

Criteria for suitability analysis typically include a set of spatial analysis maps, such as elevation, slope, land use types, and distance from physical or administrative areas, such as rivers and protected habitat. The criteria maps are added to Suitability Modeler, and the values behind the maps, such as elevation in meters and slope as a percentage, are quickly transformed into a common suitability scale, say 1–10. The common suitability scale allows the user to perform mathematical operations when overlaying different map layers.

In this example, Suitability Modeler is used to find the best locations for sustainable shrimp farms in western Costa Rica around the inlet waters of Golfo de Nicoya.

## Determine suitability criteria

For a shrimp farm to be operationally and financially viable, potential locations must include easy access to coastal waters, freshwater rivers, and serviceable roads but not infringe on protected mangrove forests. This example uses four raster maps as criteria, with each map classifying the distance from the input source data: shoreline, rivers, roads, and mangrove forests, respectively.

## Create a common suitability scale

Before modeling potential suitable locations, the data values in each criterion must be transformed to a common suitability scale. When the transformation function in Suitability Modeler is applied, a histogram of the input values, the transformation function, the transformed map, the histogram, and the final suitability map are displayed simultaneously, allowing the user to view and evaluate the quality and integrity of the result. Suitability Modeler users typically try different functions and change parameters for each criterion, creating dozens of comparative views before determining the best model.

## Weight the criteria

Another technique for adjusting the model is to weight each criterion. For example, nearness to rivers may be ranked more important than nearness to roads. As with the transformation stage, Suitability Modeler users can try different weighting schemes and receive immediate feedback in the form of histograms and maps. Once again, users may create many comparative views of the results.

## Define spatial requirements

Finally, Suitability Modeler users can define spatial requirements, such as the total area of a suitable location, the number of suitability areas within the study area, and the preferred distance between suitable areas. Like transformation and weighting, spatial requirements can be revisited and adjusted at any time.

Suitability Modeler offers a new approach to suitability analysis. At any point in the analysis, the user can adjust model parameters by adding more or different criteria, changing the common suitability scale, altering the weight of criteria, and redefining spatial requirements. With every adjustment, the user is presented with immediate visual and statistical feedback to evaluate and present to stakeholders.

*Study area for finding suitable shrimp farm locations near Golfo de Nicoya, Costa Rica.*

# CRITERIA

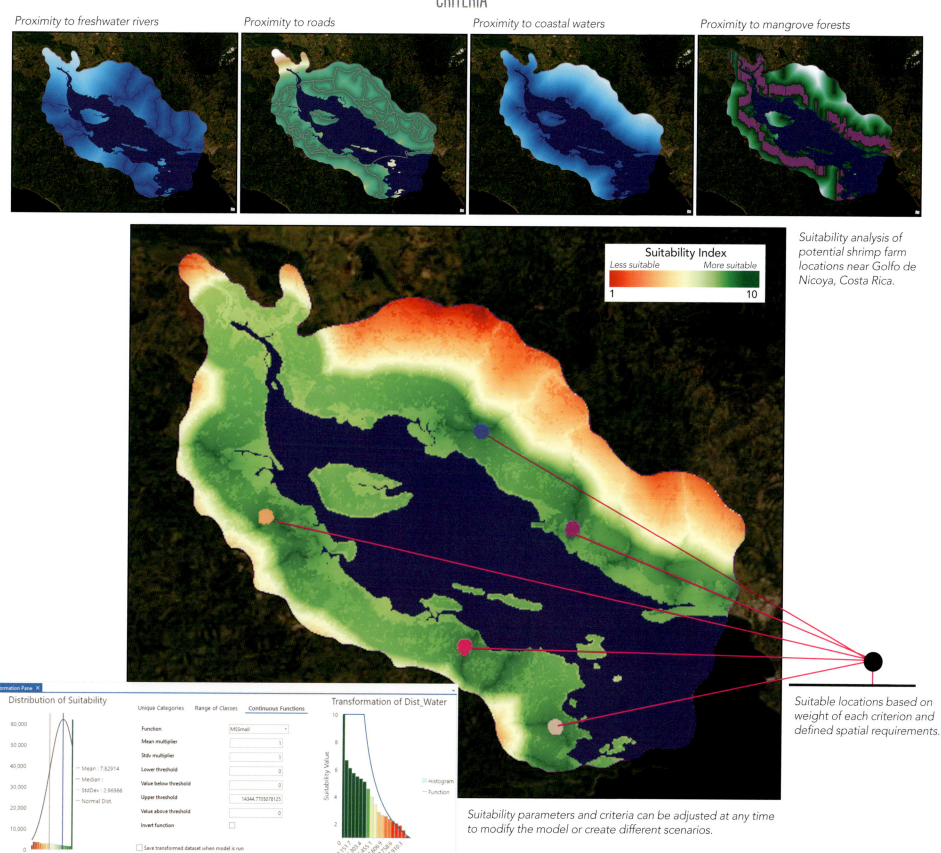

Proximity to freshwater rivers

Proximity to roads

Proximity to coastal waters

Proximity to mangrove forests

Suitability Index
Less suitable — More suitable
1 — 10

*Suitability analysis of potential shrimp farm locations near Golfo de Nicoya, Costa Rica.*

*Suitable locations based on weight of each criterion and defined spatial requirements.*

Transformation Pane ✕

Distribution of Suitability

Unique Categories   Range of Classes   **Continuous Functions**

Transformation of Dist_Water

| | |
|---|---|
| Function | MSSmall |
| Mean multiplier | 1 |
| Stdv multiplier | 1 |
| Lower threshold | 0 |
| Value below threshold | 0 |
| Upper threshold | 14344.7705078125 |
| Value above threshold | 0 |
| Invert function | ☐ |

Mean : 7.82914
Median :
StdDev : 2.96966
Normal Dist.

☐ Histogram
— Function

☐ Save transformed dataset when model is run

*Suitability parameters and criteria can be adjusted at any time to modify the model or create different scenarios.*

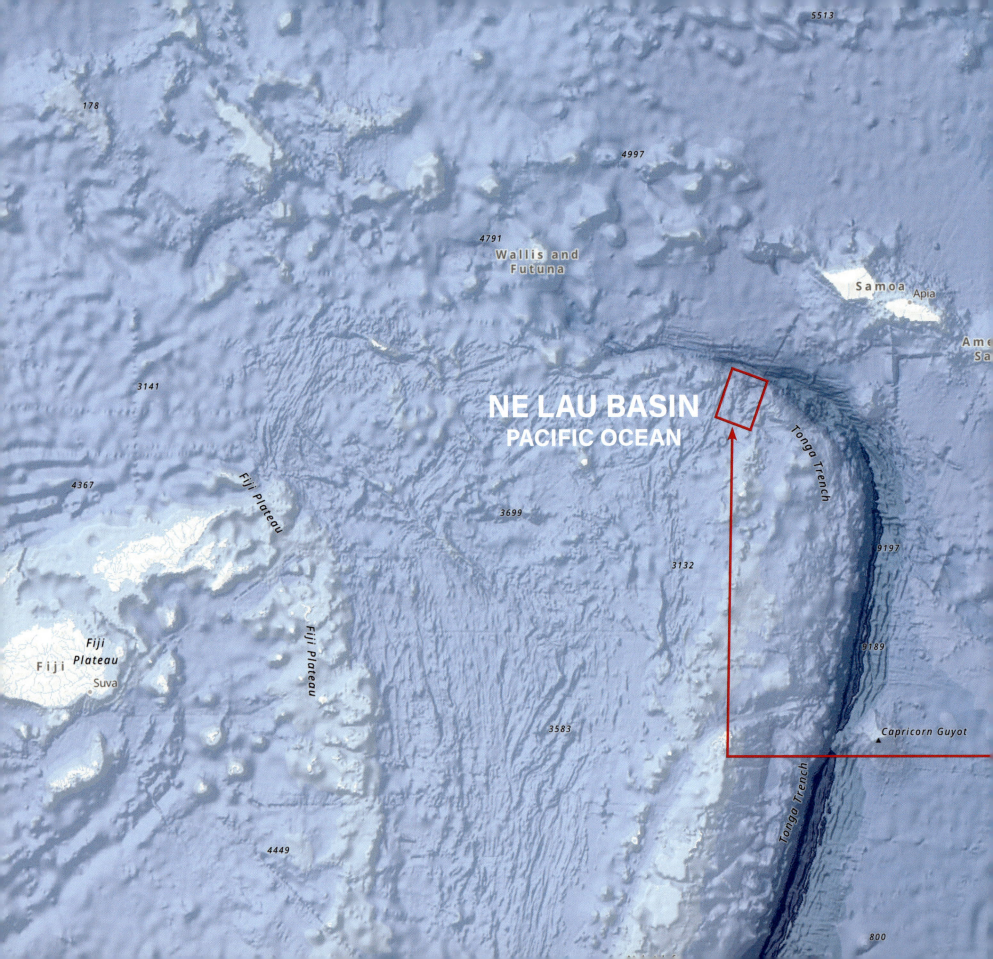

# INSIDE SUBMARINE VOLCANIC ERUPTIONS

## Using voxels to reveal anomalies

Neeti Nayak and Keith VanGraafeiland, Esri

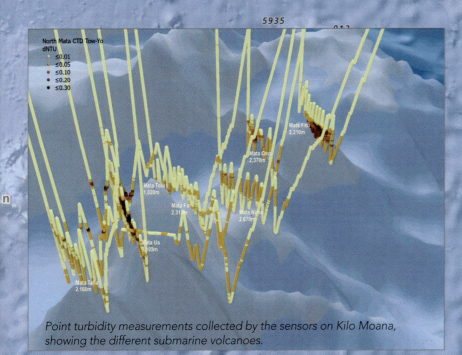

Point turbidity measurements collected by the sensors on Kilo Moana, showing the different submarine volcanoes.

Interpolated voxel layer from the point data, showing the different submarine volcanoes.

Data provided by the NE Lau Basin, R/V Kilo Moana expedition KM1008, April 28–May 10, 2010, cruise report.

The innate interdisciplinary nature of GIS demands that the results of a complex multidimensional data analysis should lend itself to digestible visualization with interactive tools for visual analytics. Voxel layers, which represent multidimensional spatial and temporal information in a 3D volumetric visualization, provide GIS users with a highly accurate volumetric analysis and an understanding of conditions and phenomena that can't be physically experienced.

Most of the volcanic activity on Earth occurs on the ocean floor. For example, the floor of the Lau Basin, located in the southwest Pacific Ocean, is one of the most volcanically active areas known to scientists. Exploration of Lau Basin by the National Oceanic and Atmospheric Administration (NOAA) Pacific Marine Environmental Laboratory (PMEL) Vent Program in 2008–2010 revealed numerous locations of hydrothermal activity and two active eruptions. Water quality characteristics were collected and mapped to better understand the range of the erupting undersea volcanoes, explore the spatial extents of volcanic ash plumes, and recognize the global impacts of submarine eruptions on ocean chemistry. The data included measurements collected by a conductivity, temperature, and depth (CTD) device, which was towed by the research vessel *Kilo Moana*.

The CTD device was raised and lowered through the water column while passing over active hydrothermal vents. The device collected temperature, conductivity, salinity, and turbidity measurements so that researchers could precisely determine the shape of hydrothermal volcanic plumes and source locations on the ocean floor.

NOAA/PMEL in Seattle, Washington, recorded tabular data representing locations in the north Mata project area (where turbidity anomalies were observed) and mapped the data as points. Using turbidity measurements taken above the erupting undersea volcanoes, the program performed a 3D geostatistical interpolation to predict turbidity levels throughout the study area. The results were exported to a multidimensional data file (in NetCDF format) for viewing as a voxel layer in GIS.

The voxel layer reveals anomalies, specifically high turbidity values corresponding to eruptions and the structure and symmetry of the different eruptions. Particle anomaly contour plots from CTD tows and the corresponding voxel visualization of the erupting peaks, along with a section diagram showing the voxel dissected into planes, show the distribution of turbidity values within the volume.

# SPATIOTEMPORAL MACHINE LEARNING
## Modeling transshipment patterns

Orhun Aydin, Esri

The term *transshipment* refers to the transfer of cargo, crew, or supplies from one vessel to another. A common type of transshipment occurs between fishing vessels and a refrigerated vessel, also known as a reefer. This practice allows shipping vessels to operate more effectively by reducing round trips to port. Despite its advantages for fishing, transshipment is also linked to human trafficking and the practice of forced labor, because it forces crews to stay onboard for extended hours. From a sustainability point of view, transferring a catch from one fishing vessel to another can obscure the actual location of the catch, making it relatively easy to evade quota requirements and regulations. Transshipment also poses a growing challenge for managing fisheries due to undocumented transactions of catch in international waters, which in turn facilitates illegal catch entering the seafood market. Transshipment activities often occur in international waters where the policy to counter this practice or regulate it is cumbersome.

## Exploring and wrangling transshipment data

Understanding the spatial and temporal patterns of transshipment data requires exploring two distinct types of movement patterns of vessels: encountering and loitering. An encountering event is characterized as two vessels remaining close and moving together slowly for minutes or hours. A loitering event occurs when a vessel capable of transporting goods to the port travels at low speed, waiting for other vessels to approach.

## Spatial clusters of transshipment

Defining data-driven spatial clusters of transshipment events reveals that some areas of international waters need monitoring. These clusters can be used to derive actionable maps for policy. Understanding "transshipment hotbeds" raises a pertinent question: In which areas are transshipments concentrated? Extensive transshipment occurs when a reefer can meet multiple vessels with ease. So it is vital to understand areas where encountering and loitering events are concentrated. In GIS, visualizing concentrations of this type can be accomplished using density-based clustering and HDBScan tools in ArcGIS Pro.

Using ArcGIS Pro to analyze data provided by *Global Fishing Watch*, encountering and loitering events are displayed as a data clock or a radial histogram to visualize the events temporally.

## Spatiotemporal colocation of encountering and loitering vessels

Two ships must be at the same location at the same time to transfer goods. Data science methods that can mine proximity from two space-time data sources are required. Spatiotemporal colocation of encountering and loitering events summarizes space-time neighborhoods where two events occur significantly and frequently. If the spatiotemporal configuration indicates significant clustering in space and time, significant colocation is indicated. In contrast, events that occur significantly far from each other indicate significant isolation. In transshipment, isolation areas correspond to movement corridors where vessels and reefers are progressively separated in space. Colocation corresponds to areas where significant time has been spent when vessels were nearby, pointing to catch transfer locations.

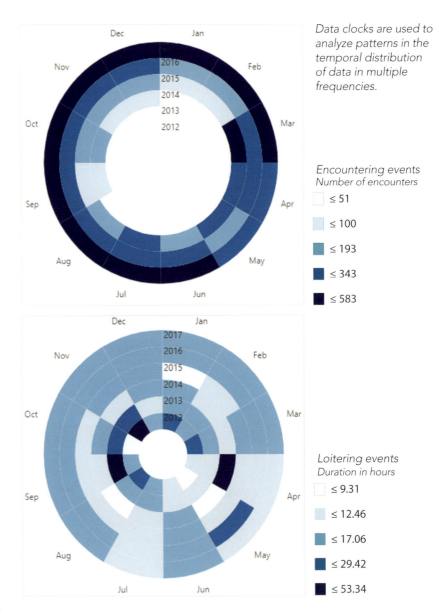

*Data clocks are used to analyze patterns in the temporal distribution of data in multiple frequencies.*

*Encountering events*
Number of encounters

- ≤ 51
- ≤ 100
- ≤ 193
- ≤ 343
- ≤ 583

*Loitering events*
Duration in hours

- ≤ 9.31
- ≤ 12.46
- ≤ 17.06
- ≤ 29.42
- ≤ 53.34

## Spatiotemporal locations of encountering events

## Spatiotemporal locations of loitering events

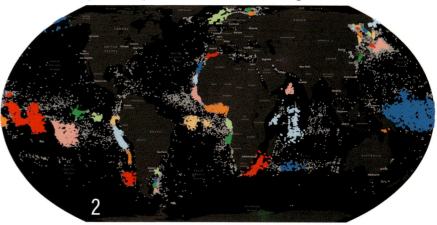

This map pair represents the spatial density clusters of 1) transshipment encountering events and 2) loitering events. The spatial extent of loitering clusters is more extensive because these events are analogous to service areas for reefers. Spatial density clusters show that reefers loiter within large areas while waiting for fishing vessels to approach them to transfer goods. Spatial clusters of loitering and encounters correspond at a small distance to the shore, where licensed reefers can haul goods to the port, and at large distances where transfers are made in international waters.

## Spatiotemporal colocation of encountering and loitering events

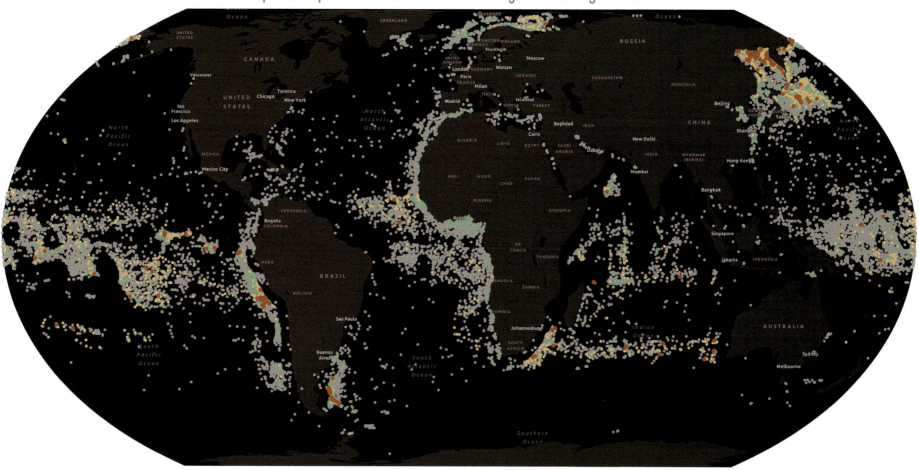

In this map, spatiotemporal colocation analysis indicates that international waters are hotbeds of catch transfer, with some exceptions near exclusive economic zones. Significant colocation of encounters and loitering is shown in brown. Significant isolation, where encounters and loitering events occur far from each other, is shown in green.

Check out

# GIS for SCIENCE

## Volumes 1 and 2

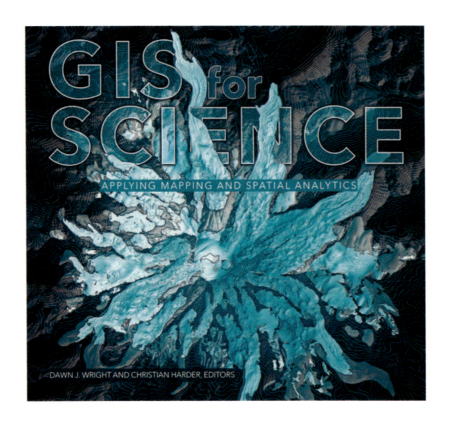

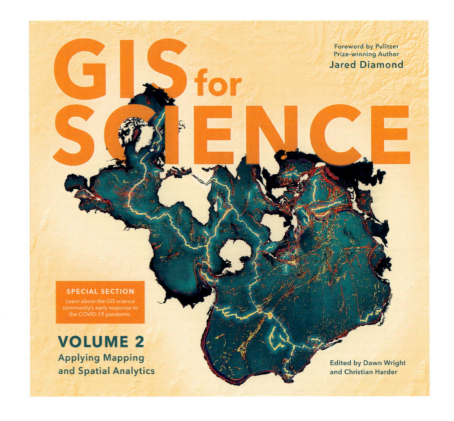

Visit esri.com/en-us/esri-press